ABANDONED COLD WAR PLACES

ABANDONED COLD WAR PLACES

Nuclear bunkers, submarine bases, missile silos, airfields and listening posts from the world's most secretive conflict

ROBERT GRENVILLE

amber BOOKS

This Amber Book published in 2019

Published by
Amber Books Ltd
United House
North Road
London N7 9DP
United Kingdom
www.amberbooks.co.uk
Instagram: amberbooksltd
Facebook: www.facebook.com/amberbooks
Twitter: @amberbooks
Pinterest: Amberbooksltd

978-1-78274-917-2

Project Editor: Michael Spilling
Designer: Mark Batley
Picture Research: Justin Willsdon

Contents

Introduction

The relationship between the Soviet Union and her Western Allies during World War II had always been more of a marriage of convenience than a love match. Once they had bested their Axis enemies, what can be seen (with hindsight) as the beginnings of the Cold War set in as the victors' differing visions of Europe's future led to clashes of increasing animosity. By the time Winston Churchill spoke in 1946 of an 'Iron Curtain' coming down across Europe, the battle lines for this 'war that wasn't' were drawn.

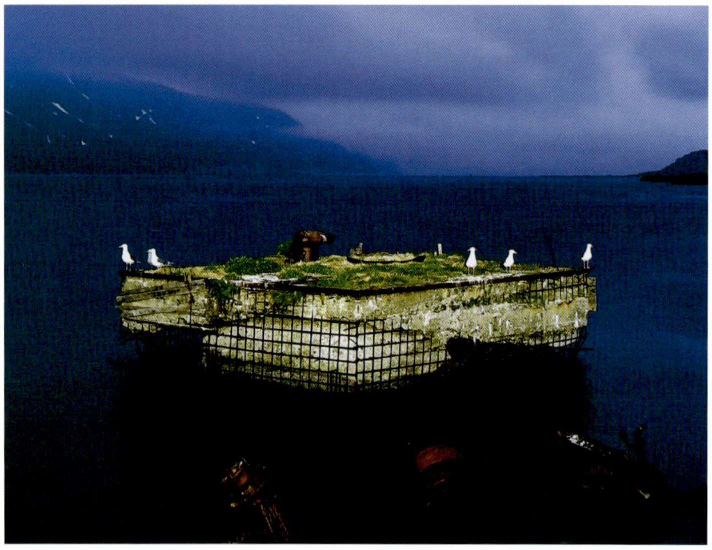

For the next four decades or so, the two sides would prepare for a conflict that thankfully would never happen. When the Berlin Wall came down in 1989, and the Soviet Union collapsed two years later, the Cold War came to an end, but it has left its mark around the globe. Decades later, traces of these mighty war machines still remain.

ABOVE:
Disused Quay, Bechevinskaya Bay, Kamchatka, Russia
A disused quay at the former Soviet naval base in Bechevinskaya Bay on the Kamchatka Peninsula in eastern Russia, now abandoned to wildlife and tourists.

OPPOSITE:
Abandoned Buildings, RAF Upper Heyford, Oxfordshire
Upper Heyford was home to U-2 aircraft used to overfly the Soviet Union on reconnaissance missions. The U-2 pilots only took off at night and their hangars were kept as far away from the public's gaze as possible. The shooting down of Gary Powers' U-2 and increased satellite use led to more tactical reconnaissance aircraft on the base.

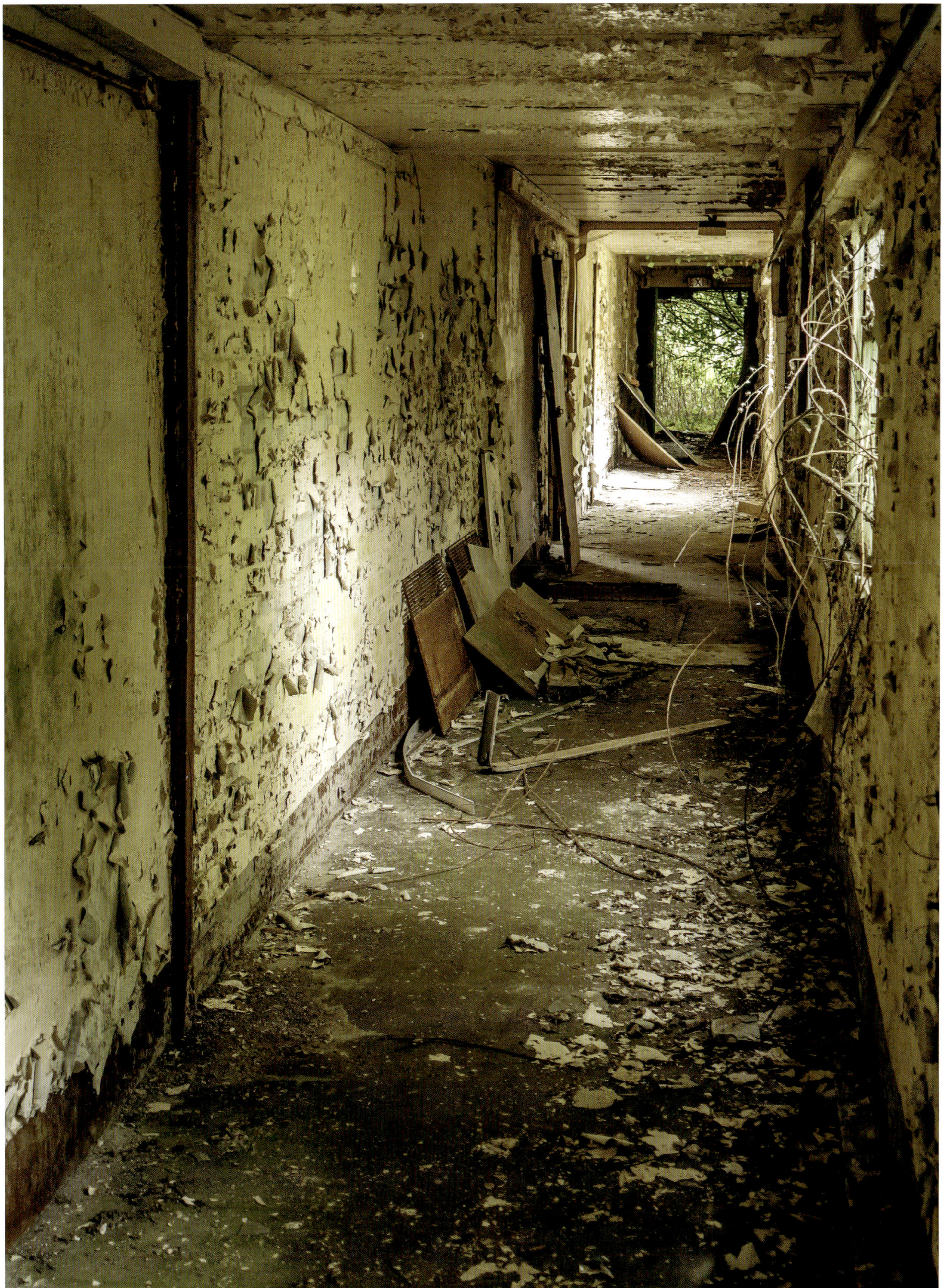

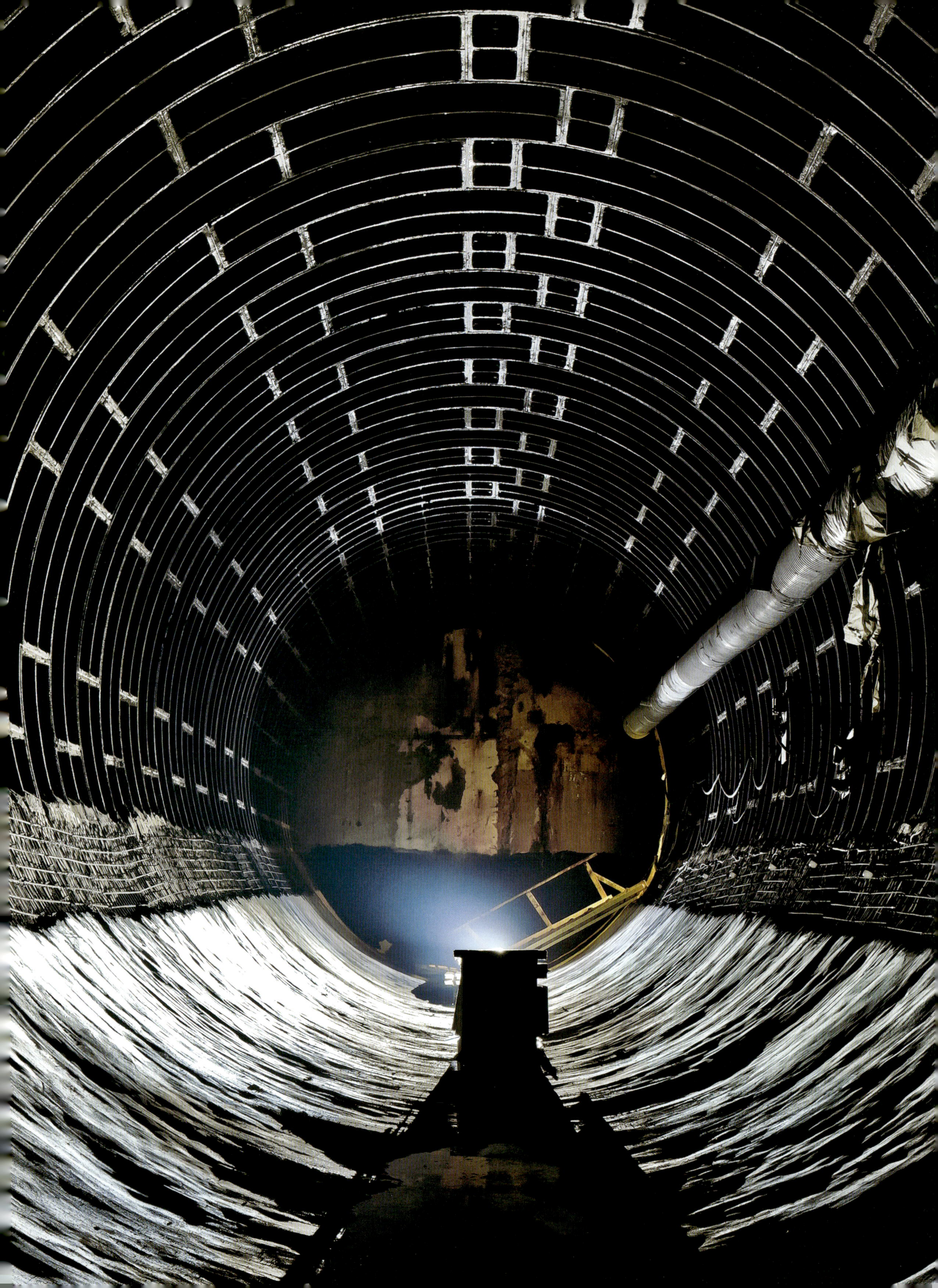

Russia

After twice being attacked by Germany and her European allies in two generations, the Soviet Union was not prepared to risk her security again. Although World War II's requirements for arms to fight the Third Reich had helped to industralize the country, in most respects, Soviet technology lagged behind the West. Initial Cold War weapons systems were either directly copied from American, British or German designs, or inspired by them, with former Nazi scientists assisting their development. A prime example was the first Soviet atomic bomb, successfully tested in 1949 and based on the 'Fat Man' bomb design that was used on Nagasaki. By 1957 the Soviet Union tested the R-7, the first Soviet ICBM. The emphasis quickly switched from bombers dropping nuclear weapons to missiles launched from prepared sites to those launched from silos, mobile launchers and submarines, and tactical nuclear weapons.

Equally, priority was given to new types of radar capable of detecting ICBMs as soon as possible and anti-ballistic missiles that could destroy warheads before they reached their targets. It has been estimated that at its peak in 1986 the Soviet nuclear arsenal had approximately 45,000 warheads. The enormous cost of the Soviet military, around a quarter of the Soviet Union's gross domestic product, was unsustainable, and the economy collapsed, bringing the Cold War and the Soviet Empire to an end.

OPPOSITE:
Underground shelter, Moscow
As the former Soviet capital, Moscow has hundreds – if not thousands – of Cold War-era government shelters lying deep beneath its streets. Some remain in active use by government agencies, but others left abandoned have been taken over by squatters or clandestine groups as meeting places.

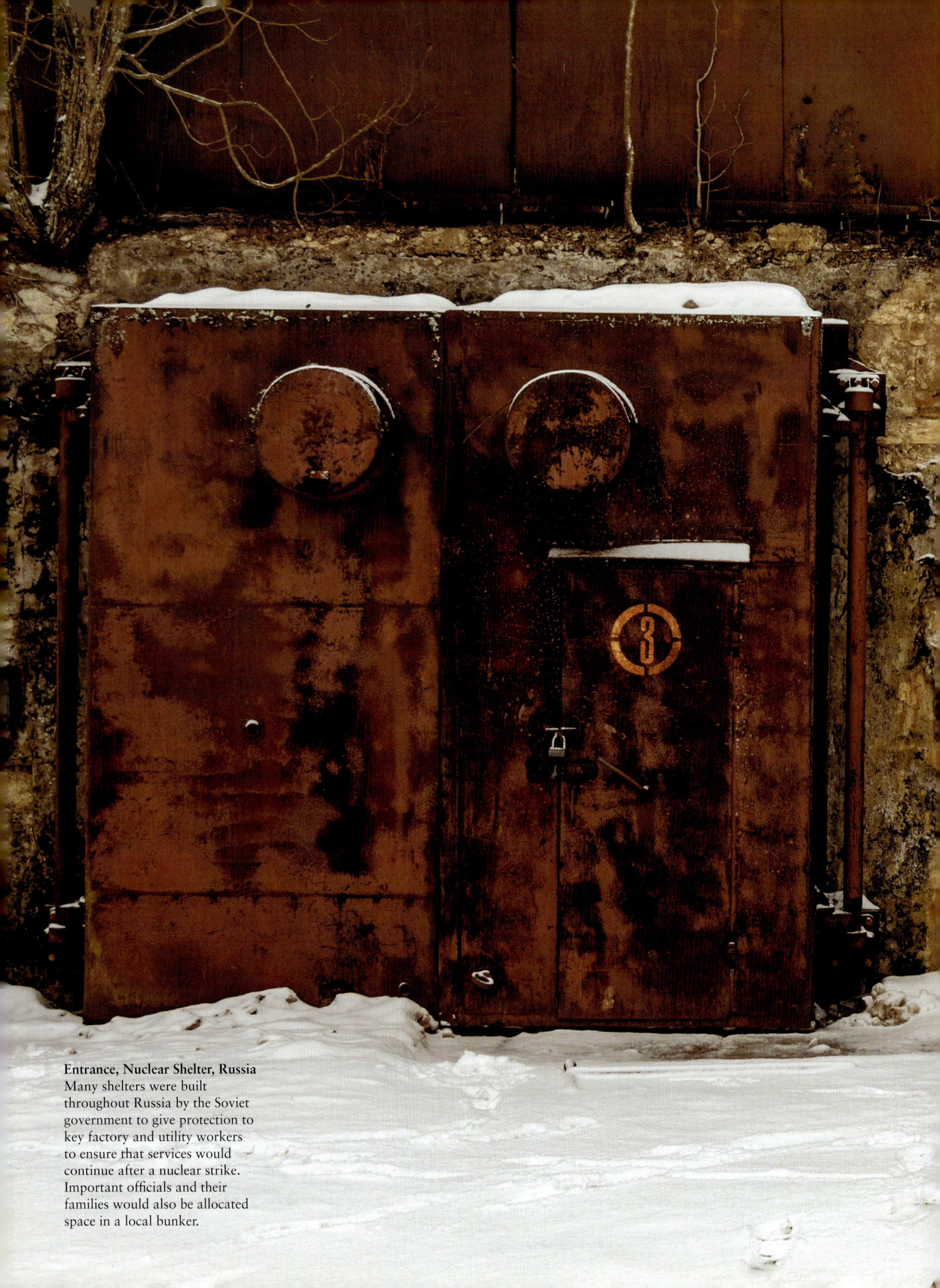

Entrance, Nuclear Shelter, Russia
Many shelters were built throughout Russia by the Soviet government to give protection to key factory and utility workers to ensure that services would continue after a nuclear strike. Important officials and their families would also be allocated space in a local bunker.

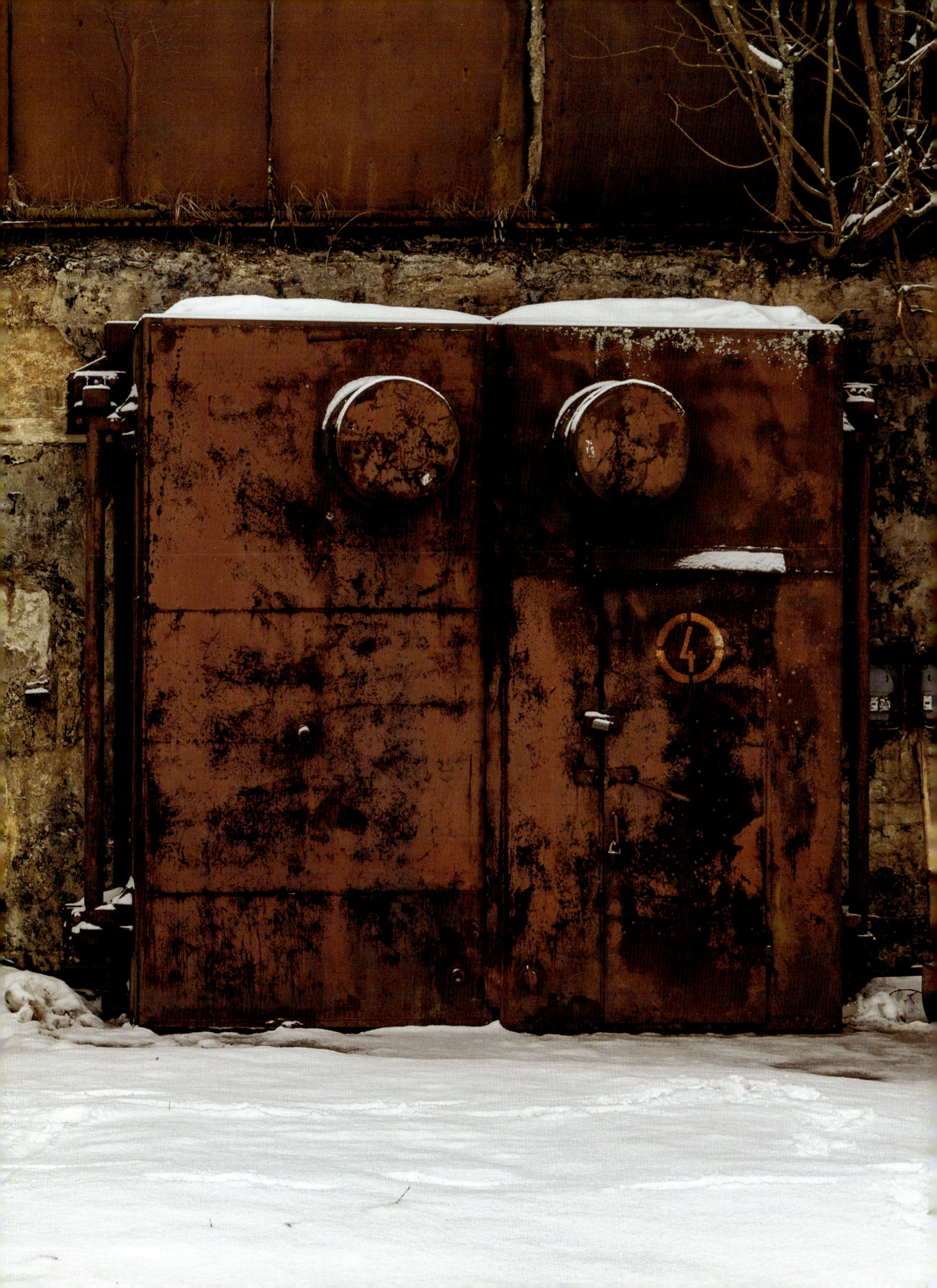

Bunk Beds, Interior, Shelter, Russia

Conditions for those housed in a shelter after a nuclear attack would have been far from luxurious for all but the most privileged members of the Soviet elite. Many apartment blocks constructed after the end of World War II were built with a shelter underneath them for their occupants to use should a bomb be detonated nearby.

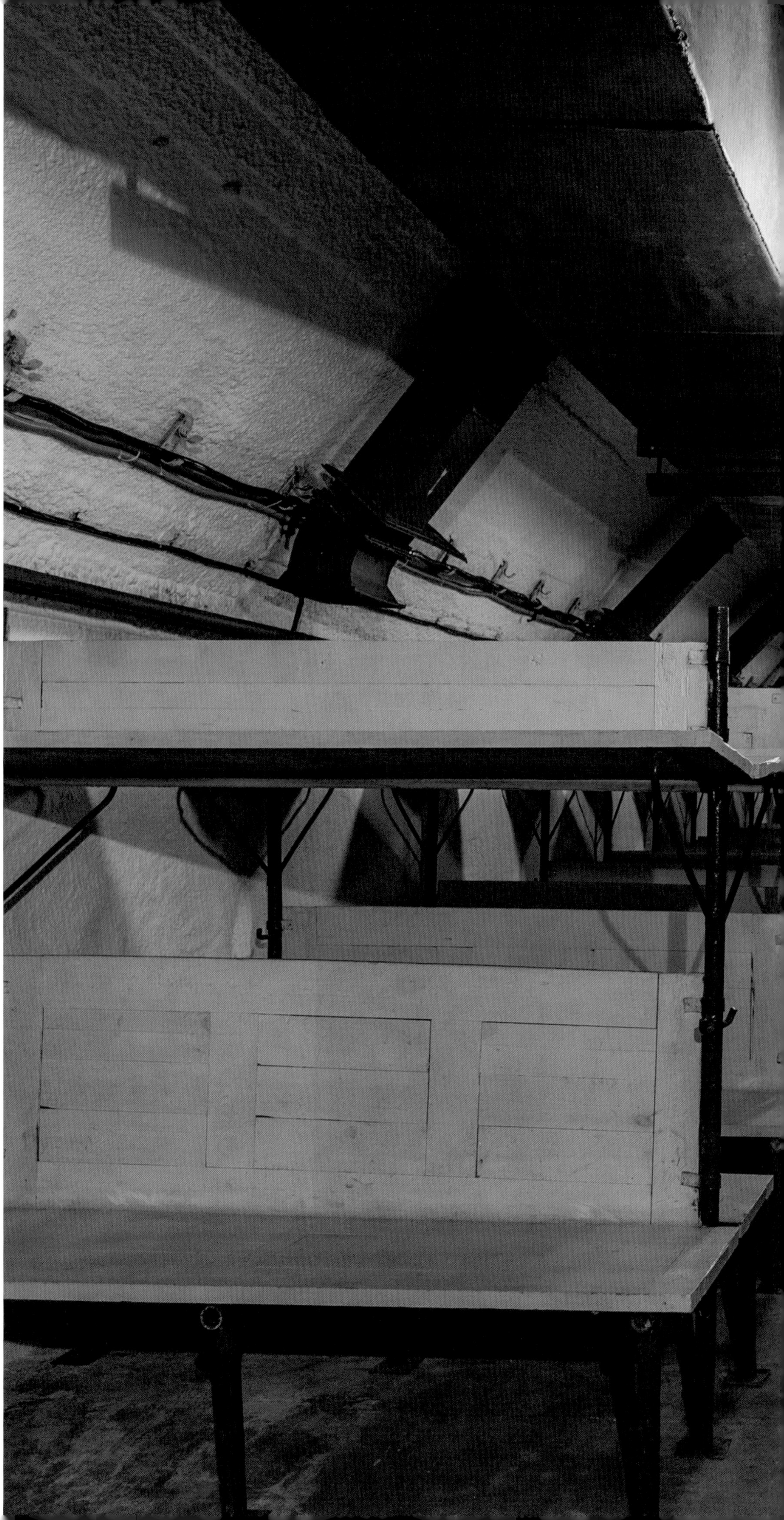

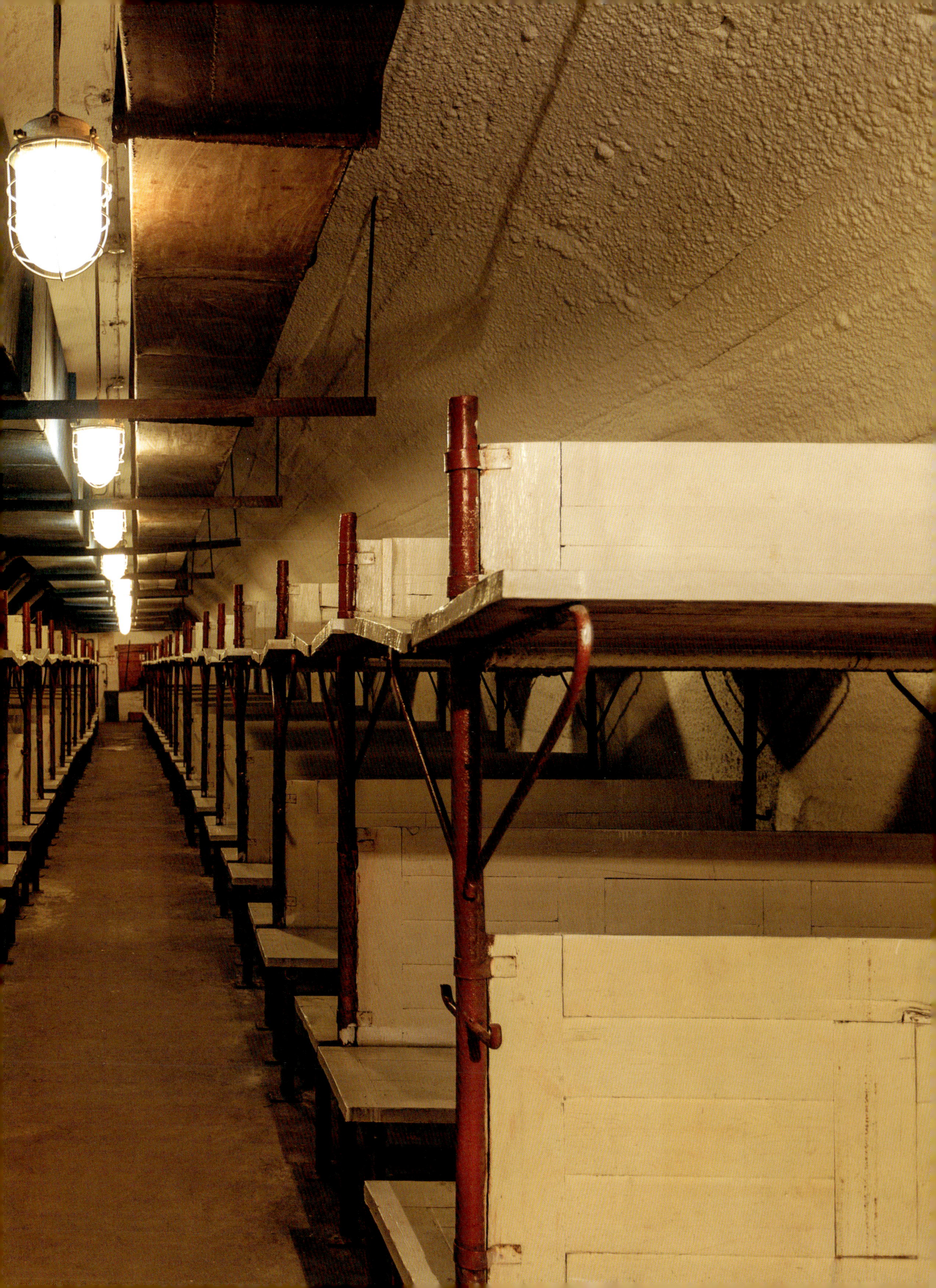

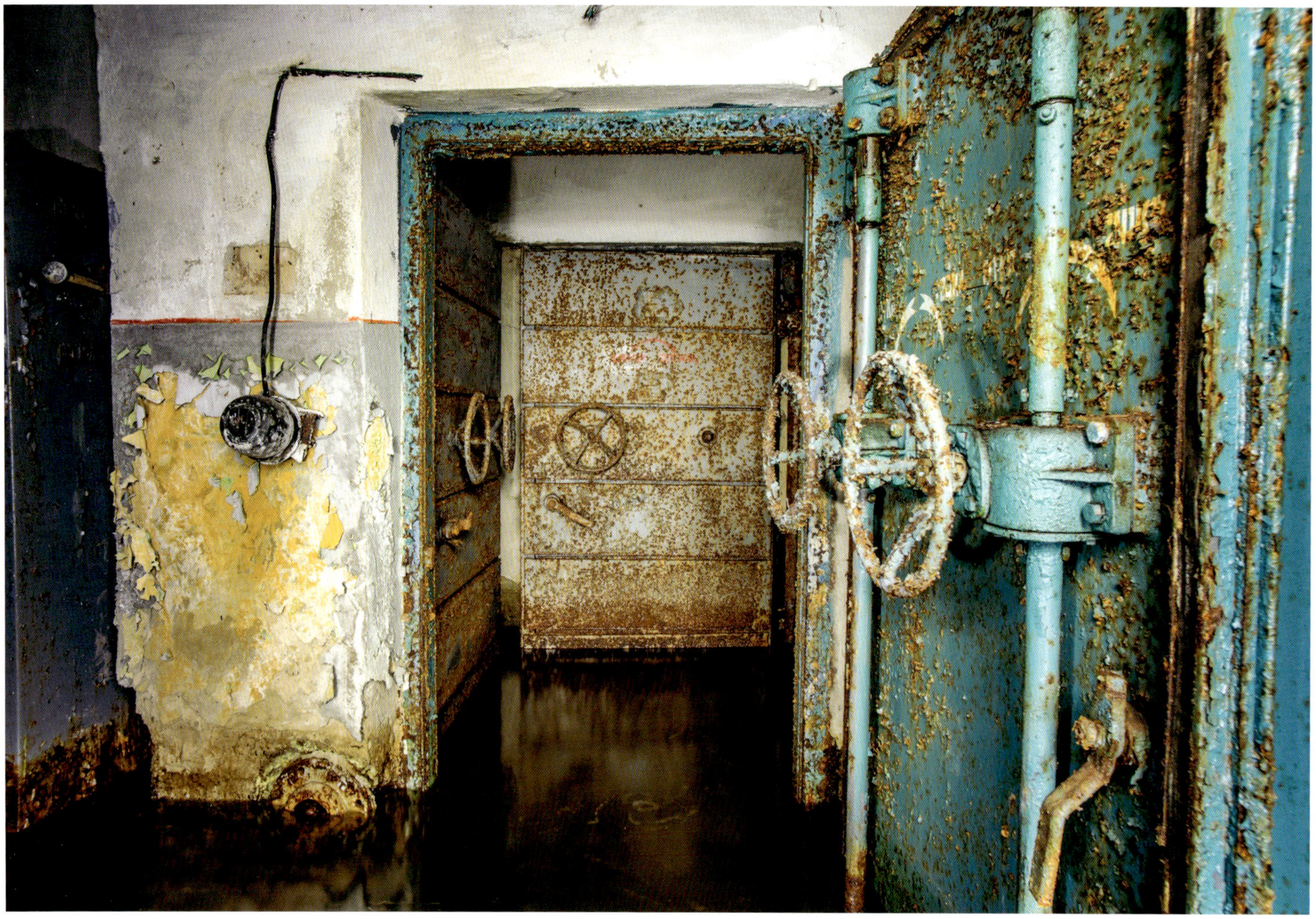

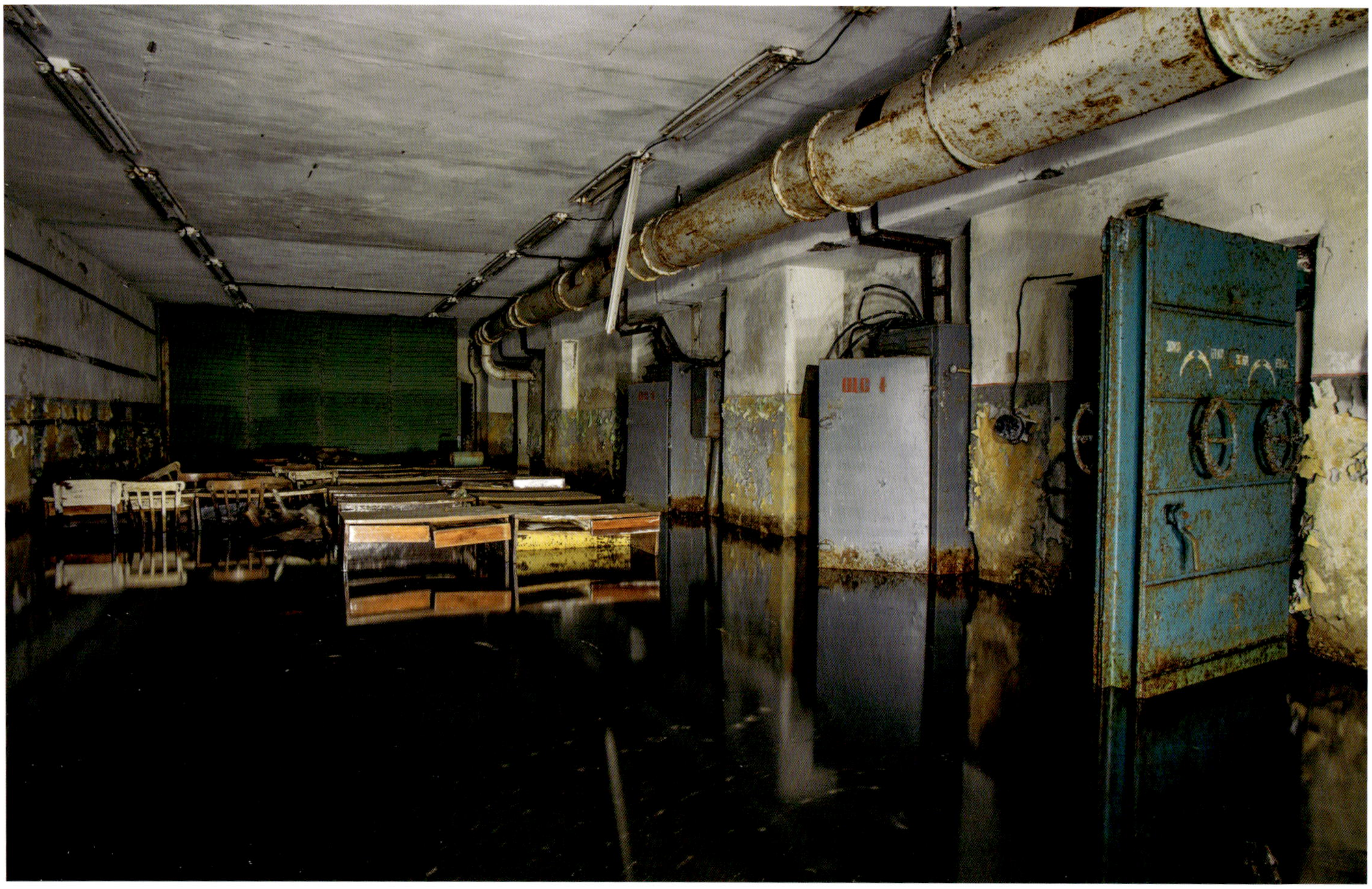

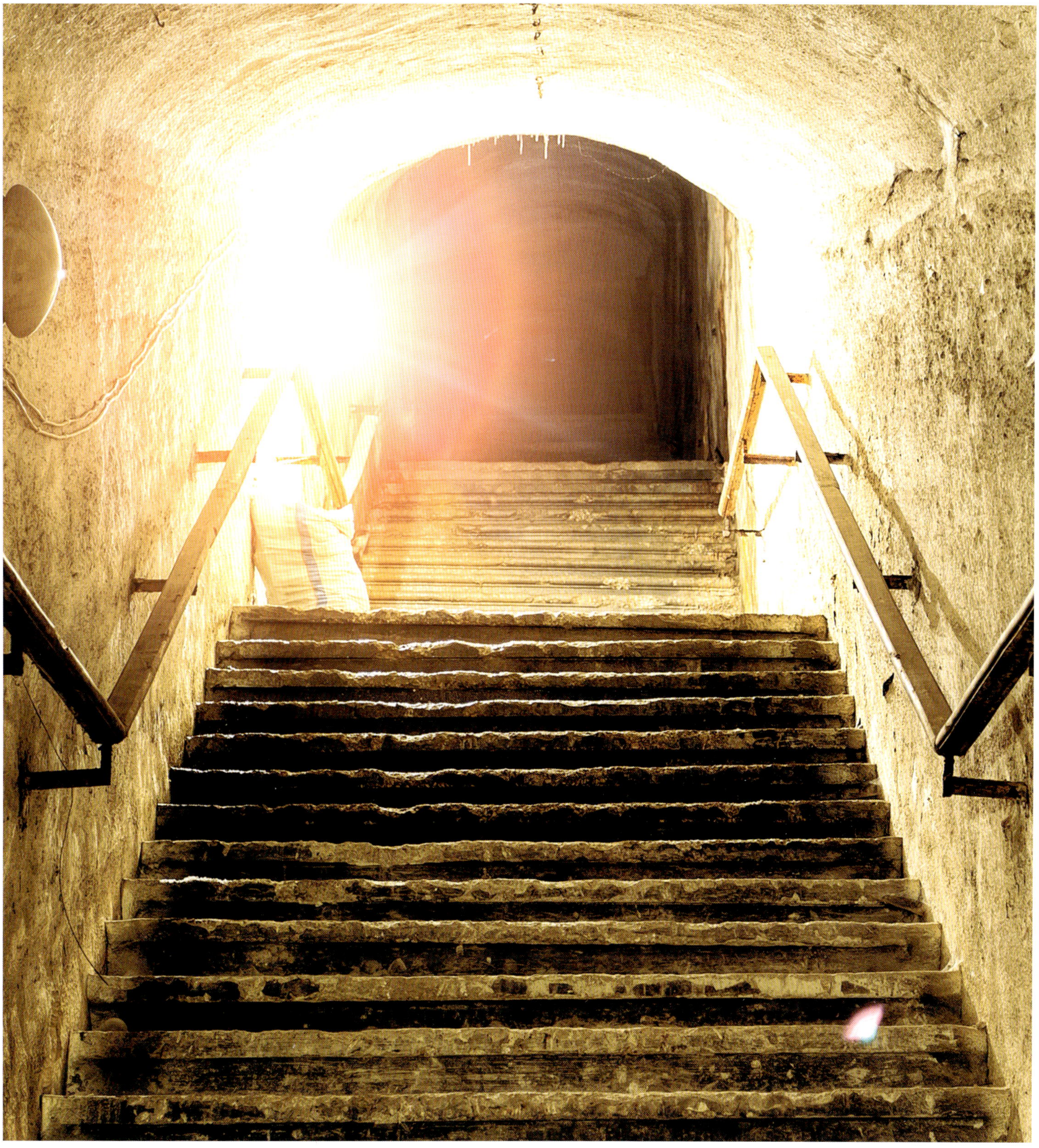

OPPOSITE TOP AND BOTTOM:
Flooded Shelter, Russia
Shelters were fitted with reinforced blast doors secured with thick bolts designed to keep out the effects of a nuclear blast and its aftermath. This abandoned shelter has flooded, either from a pipe leak or the groundwater pumps being switched off, but the half-submerged desks suggest this room would have been used for clerical work.

ABOVE:
Staircase, Former Soviet Military Bunker, Russia
An uneven concrete staircase provides a route to the surface. Bunkers were built with at least one alternative exit in case the main exit became blocked by a blast, and they were deep enough underground to escape destruction from all but the largest nuclear warheads. Tunnels were also commonly used to connect bunkers together.

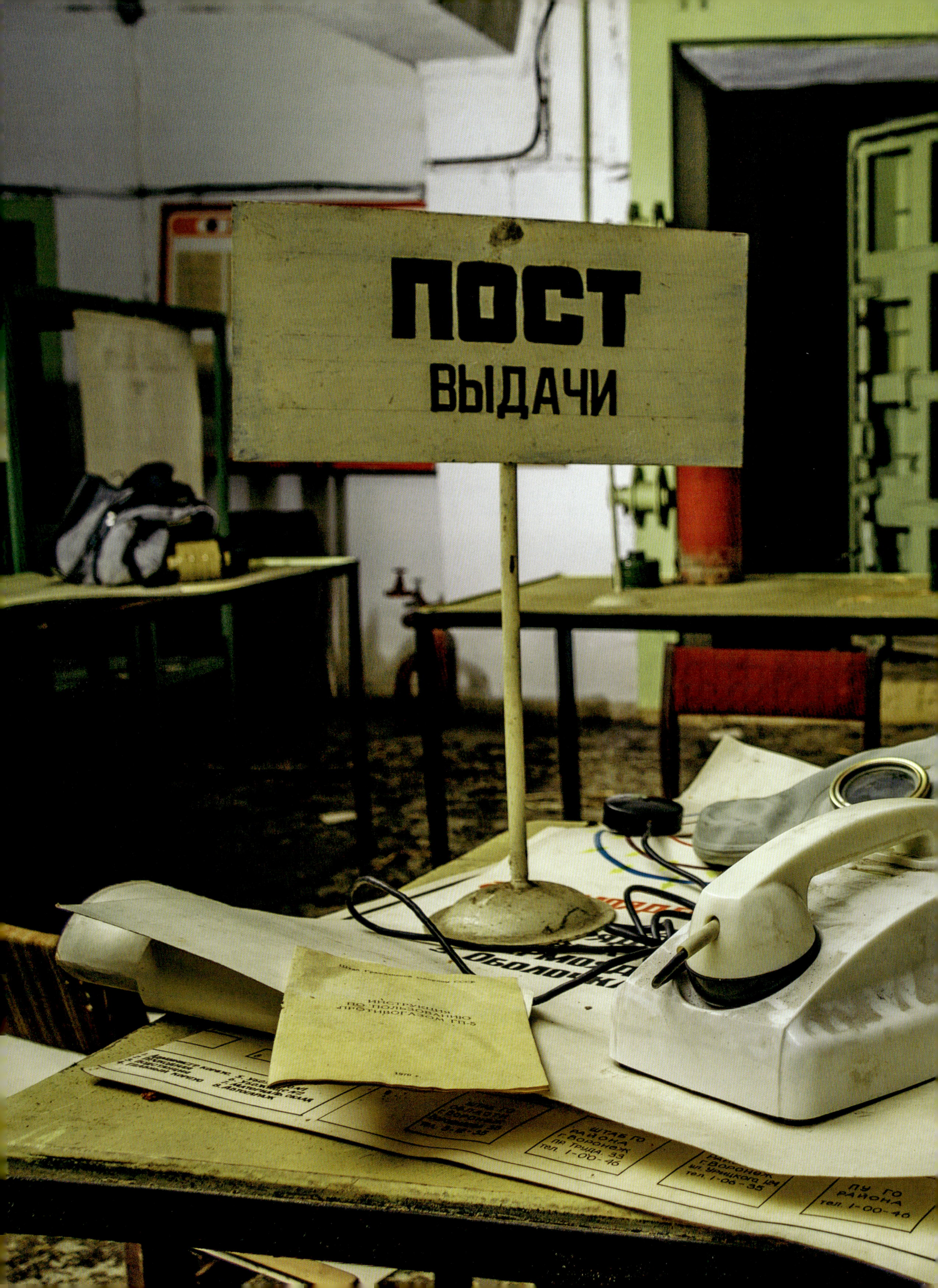
ПОСТ
ВЫДАЧИ

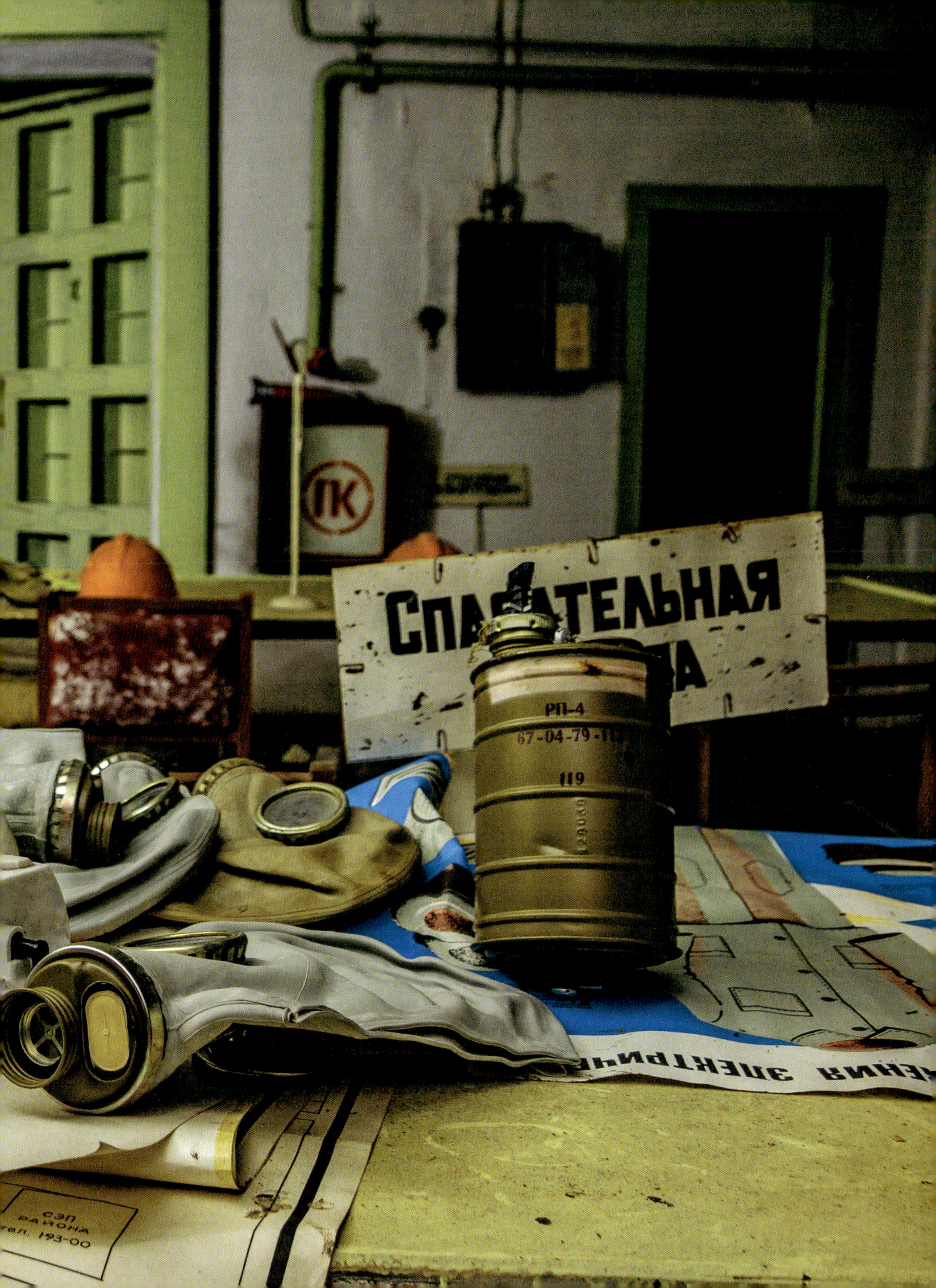
ПК
РП-4
119
СЭП
РАЙОНА
тел. 193-00

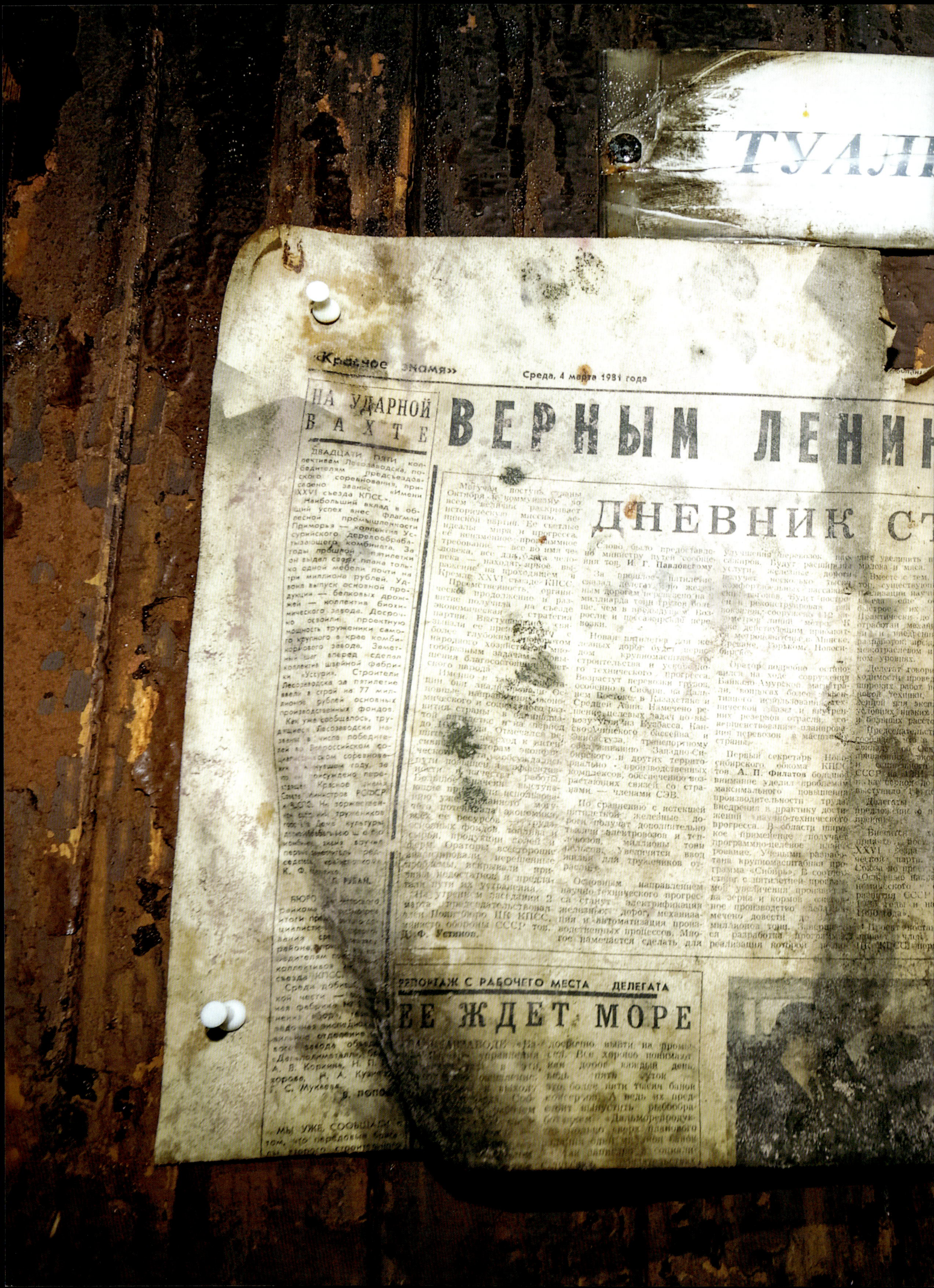
«Красное знамя»
Среда, 4 марта 1981 года
НА УДАРНОЙ ВАХТЕ
РЕПОРТАЖ С РАБОЧЕГО МЕСТА ДЕЛЕГАТА

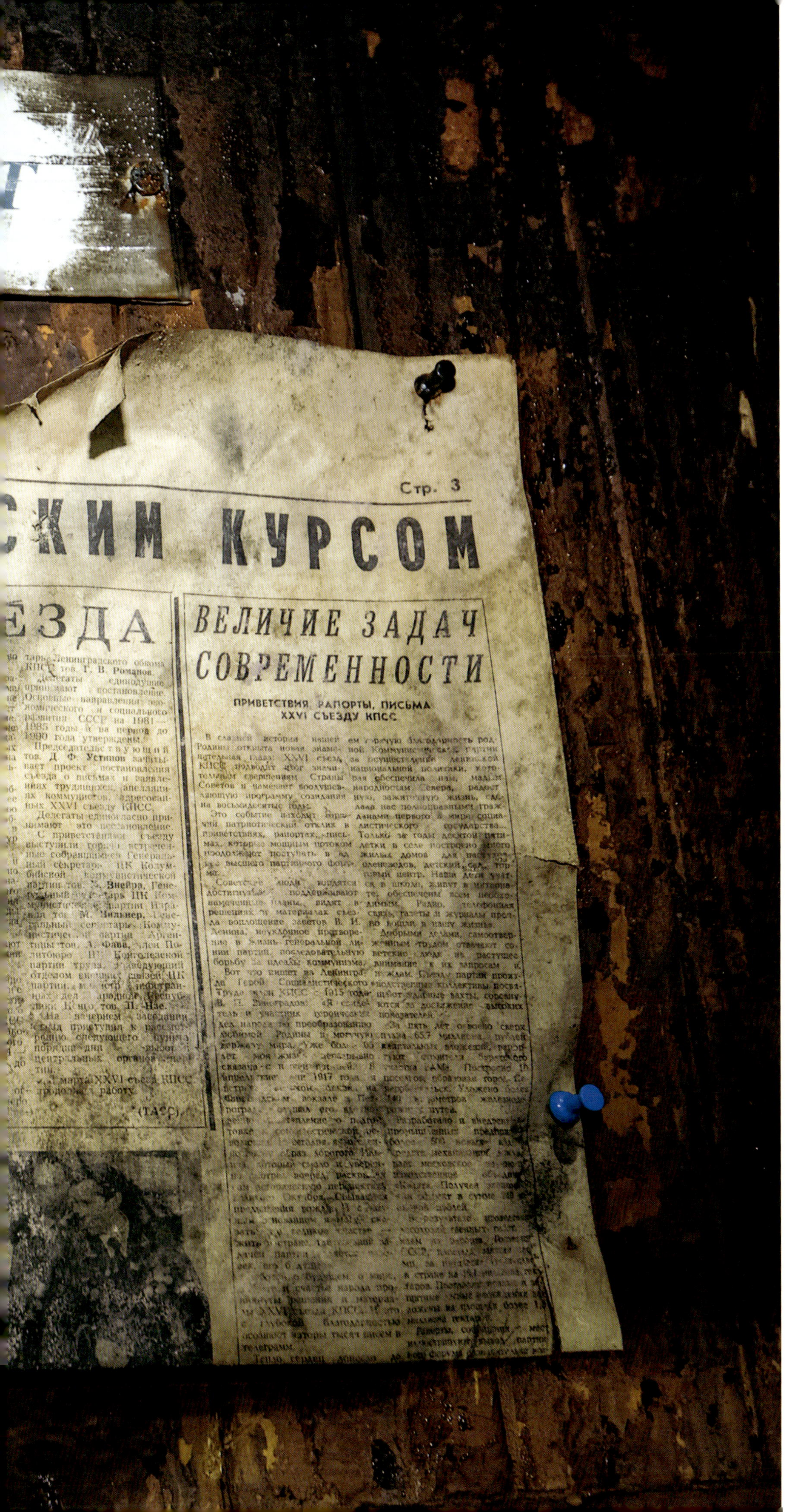

PREVIOUS PAGES:
Equipment Issuing Office, Shelter, Voronezh
The official at this desk would have distributed protective equipment such as the masks and their replaceable air filters. Strict measures would have been used to keep the shelter free from radioactive dust, and many shelters kept their internal air pressure higher than that outside to ensure no contaminated air could enter. Large chemical filters similar to those used in the space programme would have been used to clean the shelter's air of carbon dioxide.

LEFT:
Restroom, Communist Party Shelter, Russia
Pinned to this peeling toilet door is a copy of Red Banner from 4 March 1981 reporting on the latest events in the 26th Soviet Congress. The Politburo, led by Leonid Brezhnev, had just been re-elected for another five-year term. One of those re-elected was Mikhail Gorbachev, the future Soviet leader who helped end the Cold War.

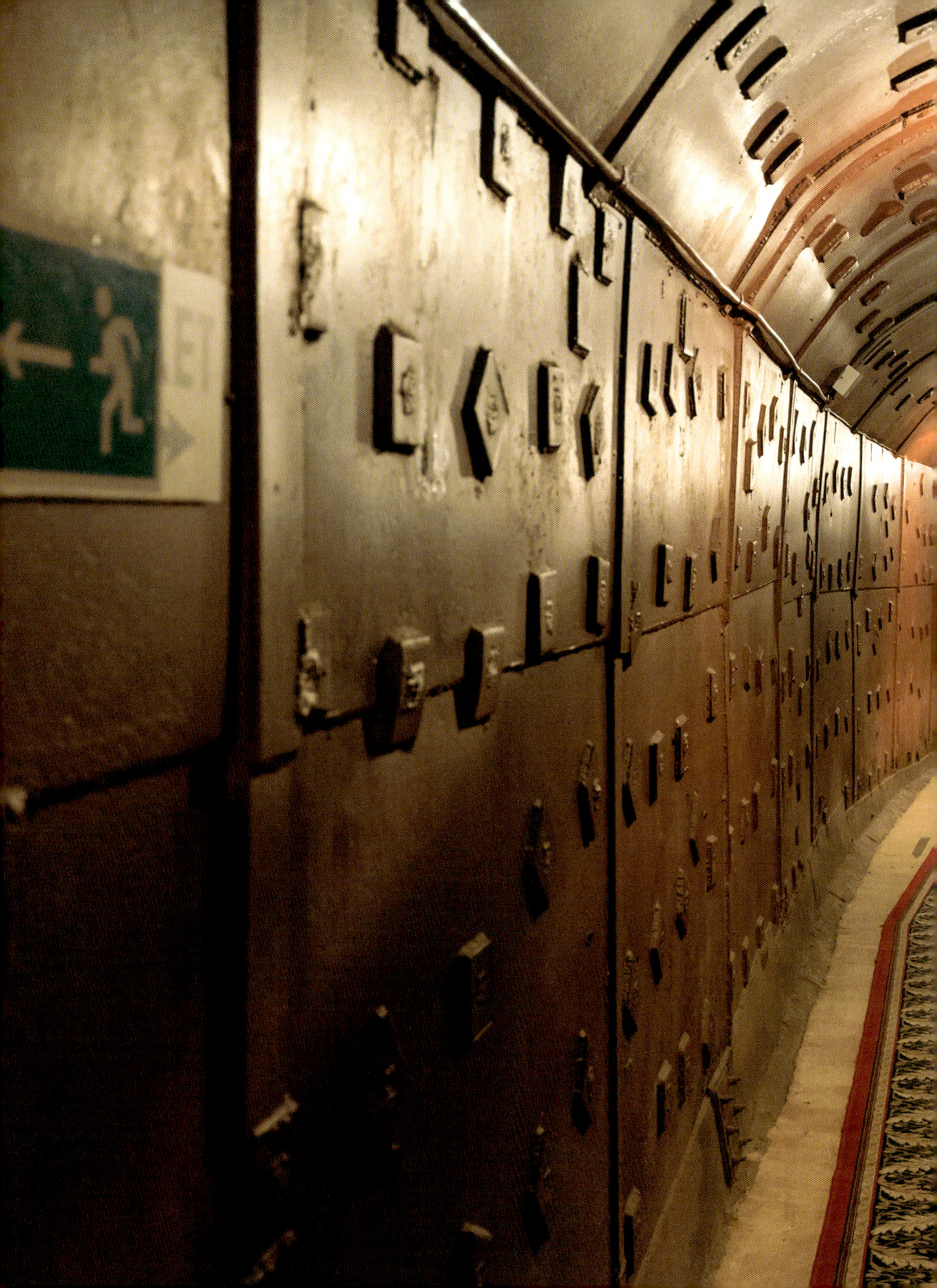

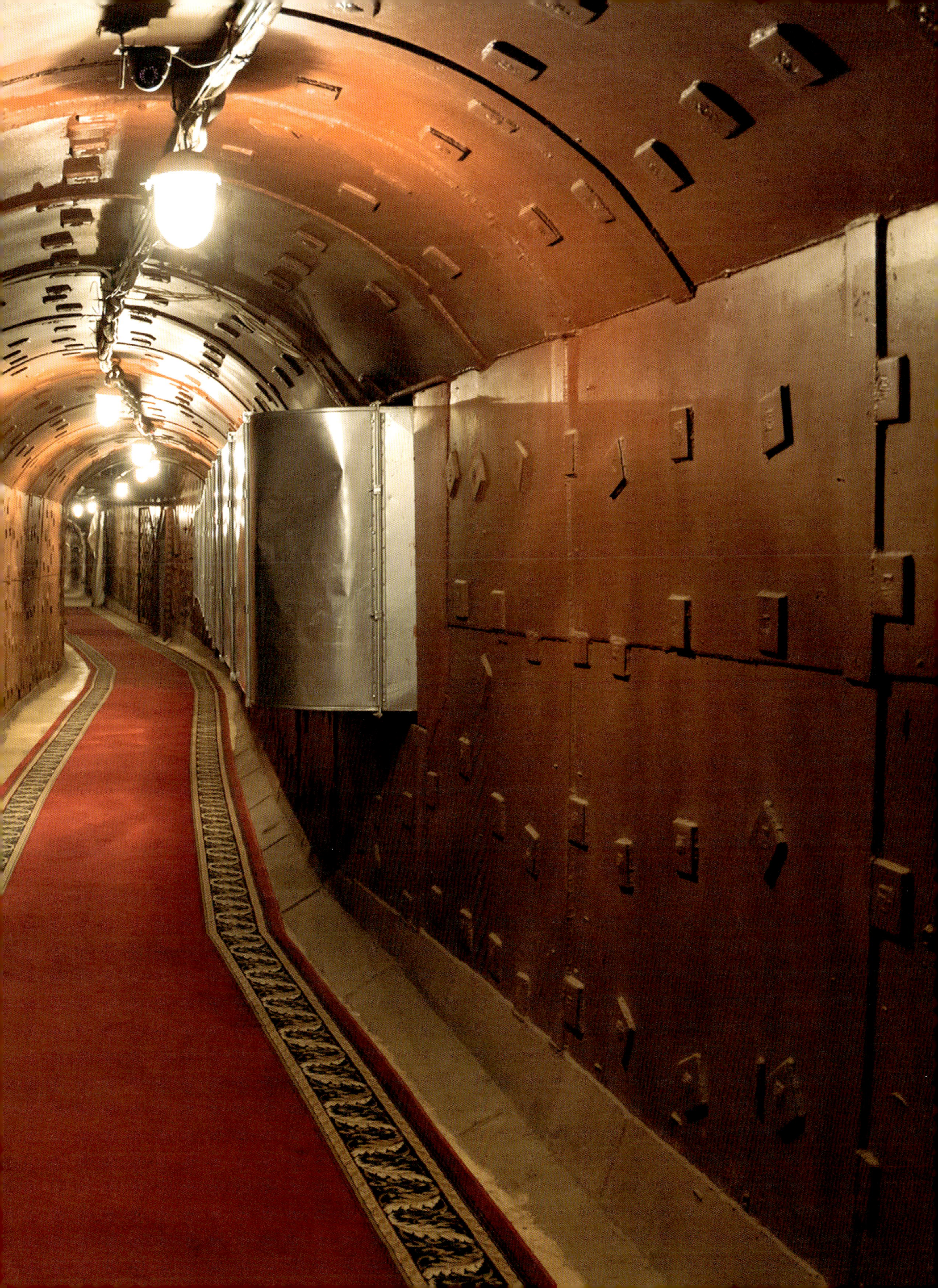

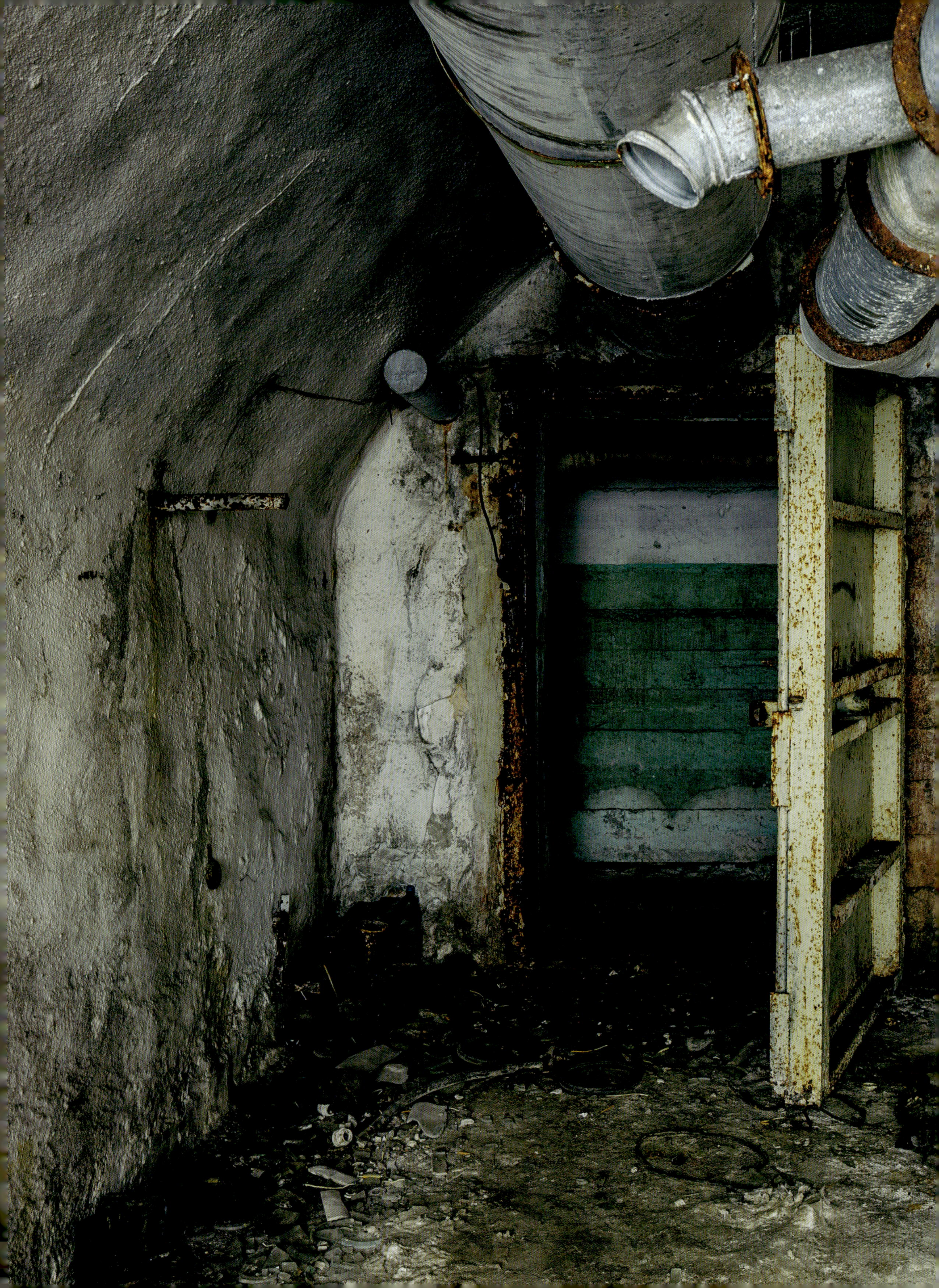

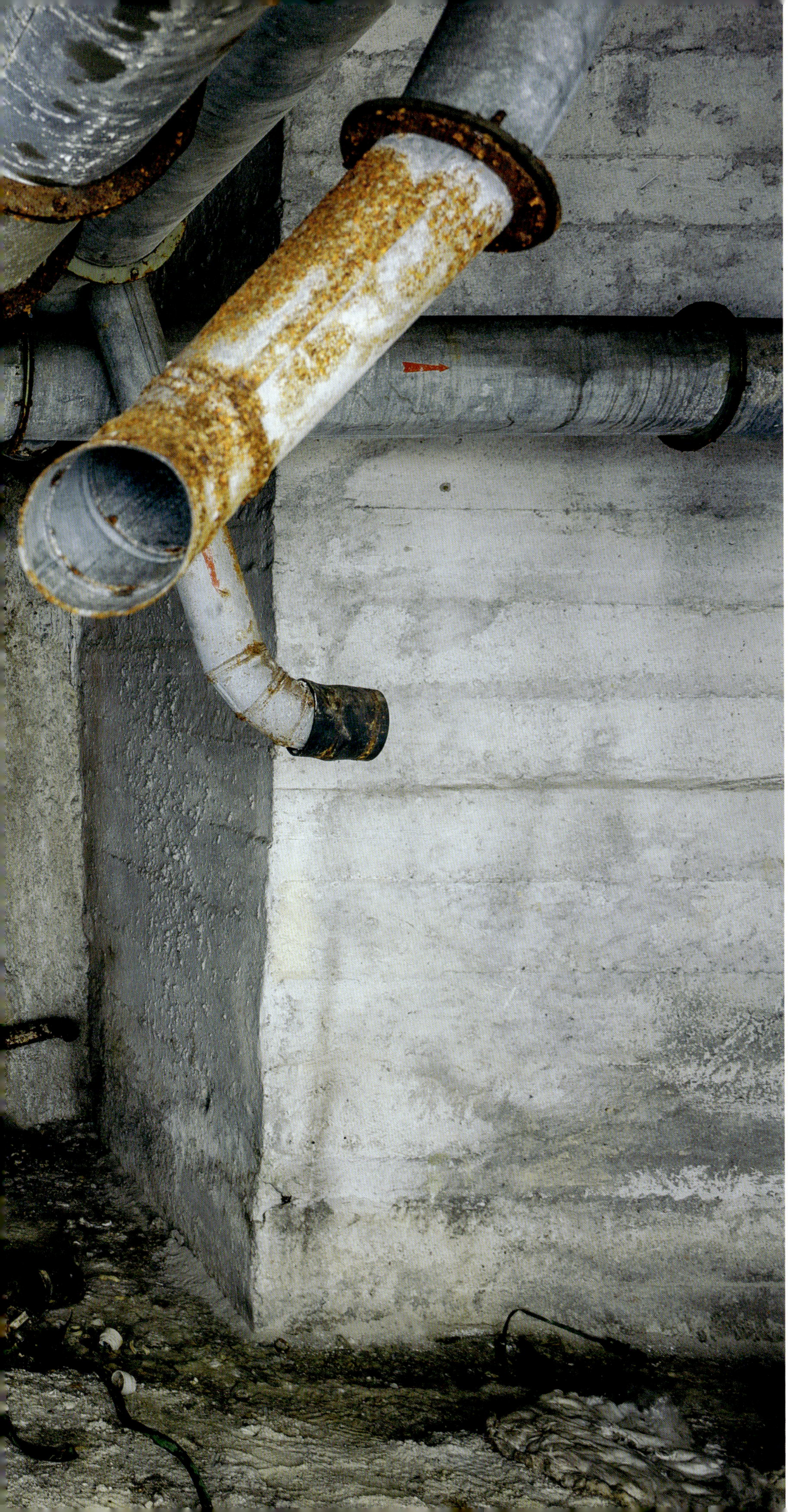

PREVIOUS PAGES:

Alternate Command Post, Long-Range Aviation, Moscow

Long-Range Aviation was the arm of the Soviet Air Force responsible for long-range nuclear bomber strikes, equivalent to the USAF's Strategic Air Command. As Soviet military command structures were likely to be targeted by a NATO nuclear strike, alternative locations like the one shown were constructed to ensure that military operations such as a Soviet counterstrike could continue unhindered.

LEFT:

Bunker, Sevastopol, Crimea

Services pipes – water, waste and air conditioning – crowd the concrete walls of this former bunker. Sevastopol, whose shipyard was first established by the Imperial Russian Navy in 1783, was a key port for the Soviet Navy during the Cold War, and home of the Black Sea Fleet.

TOP RIGHT:
Mil Mi-6 Helicopter, Former Airfield, Russia
First flown in 1957, the Mi-6 was a heavy transport helicopter capable of carrying 65 troops. Despite its size, it was the world's fastest helicopter for a time, capable of 190mph (300km/h).

BOTTOM:
Mil Mi-2 Helicopters, Former Airfield, Russia
Over 5000 Mi-2 helicopters were built in Poland after 1961. These belonged to a Soviet paramilitary sports society that helped prepare reserves for active service.

OPPOSITE TOP:
Aero L-29 Delfin Trainers, Former Airfield, Russia
Over 2000 Czech-designed and built Delfins were used as the Soviet Air Force's standard jet trainer aircraft for new pilots.

OPPOSITE BOTTOM:
Antonov An-12 Transport, Former Airfield, Russia
First flown in 1957, the An-12 was widely used for transport and liaison purposes.

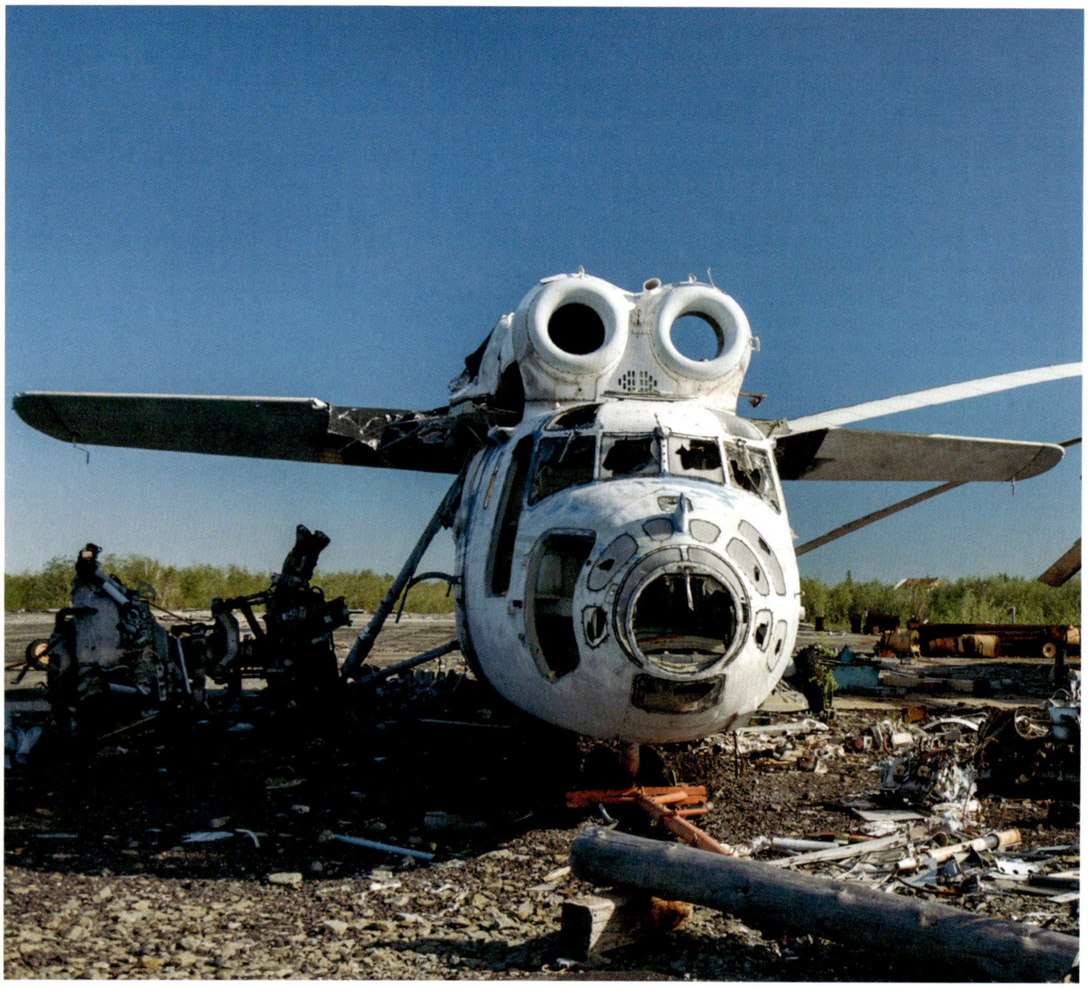

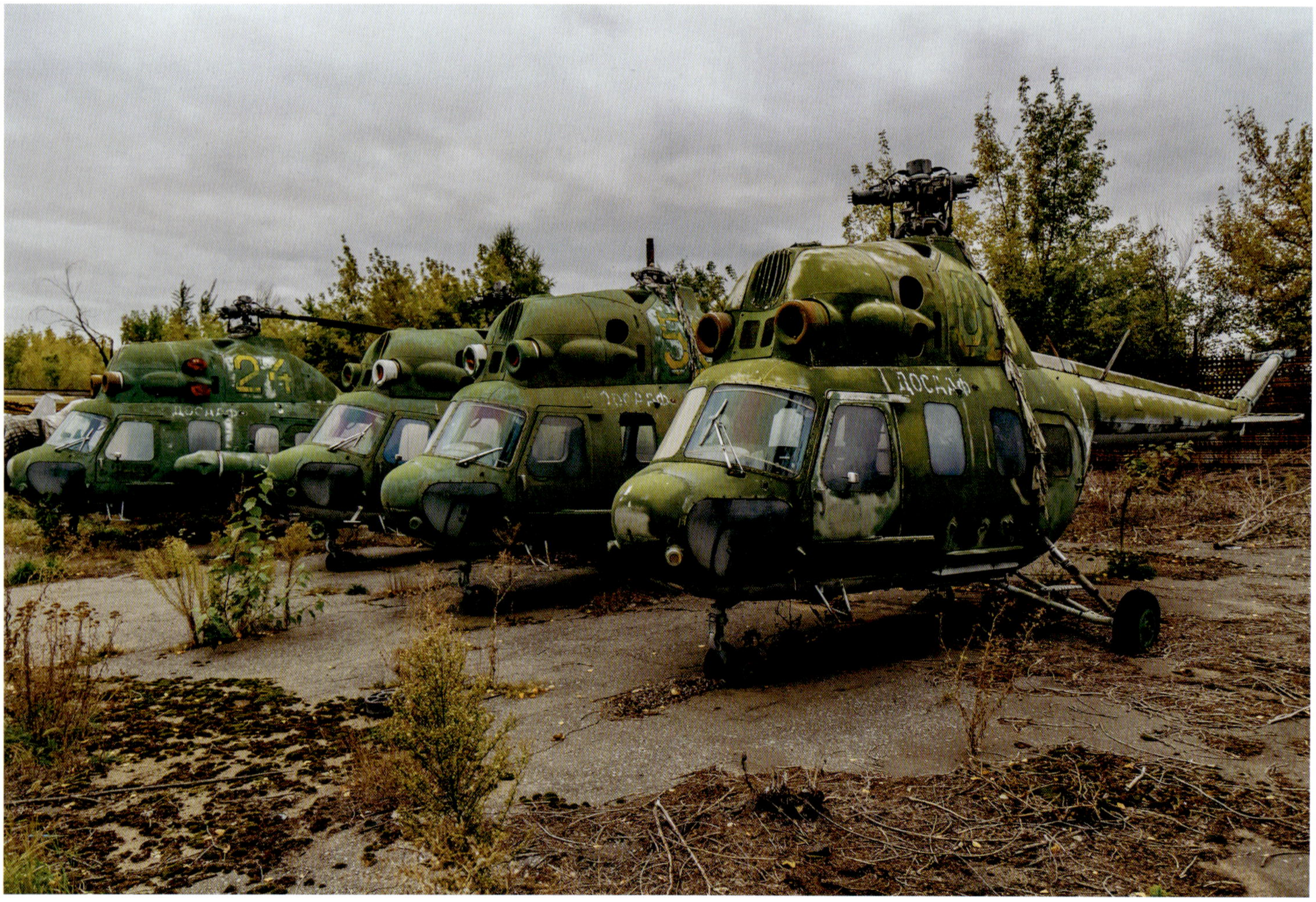

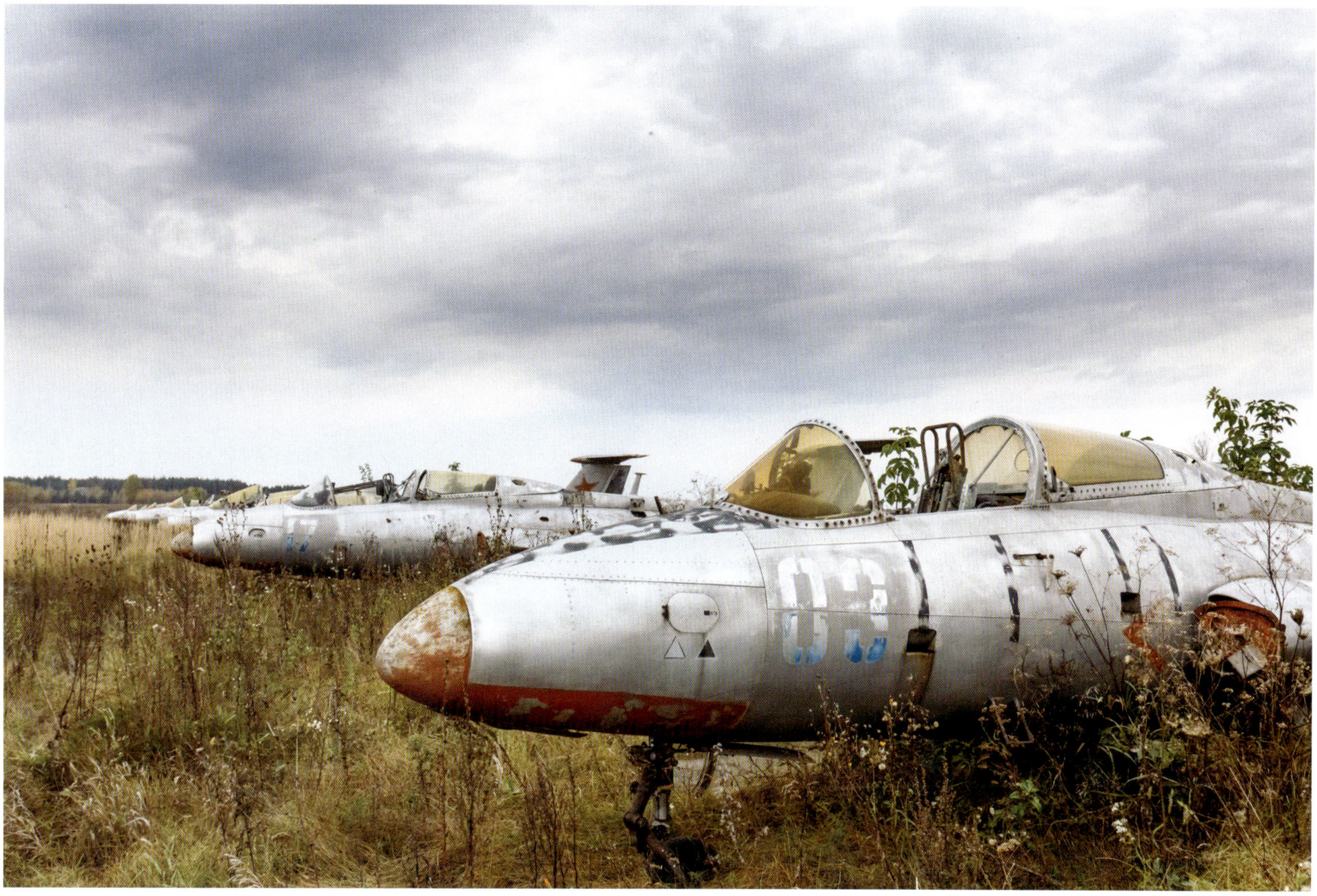
03

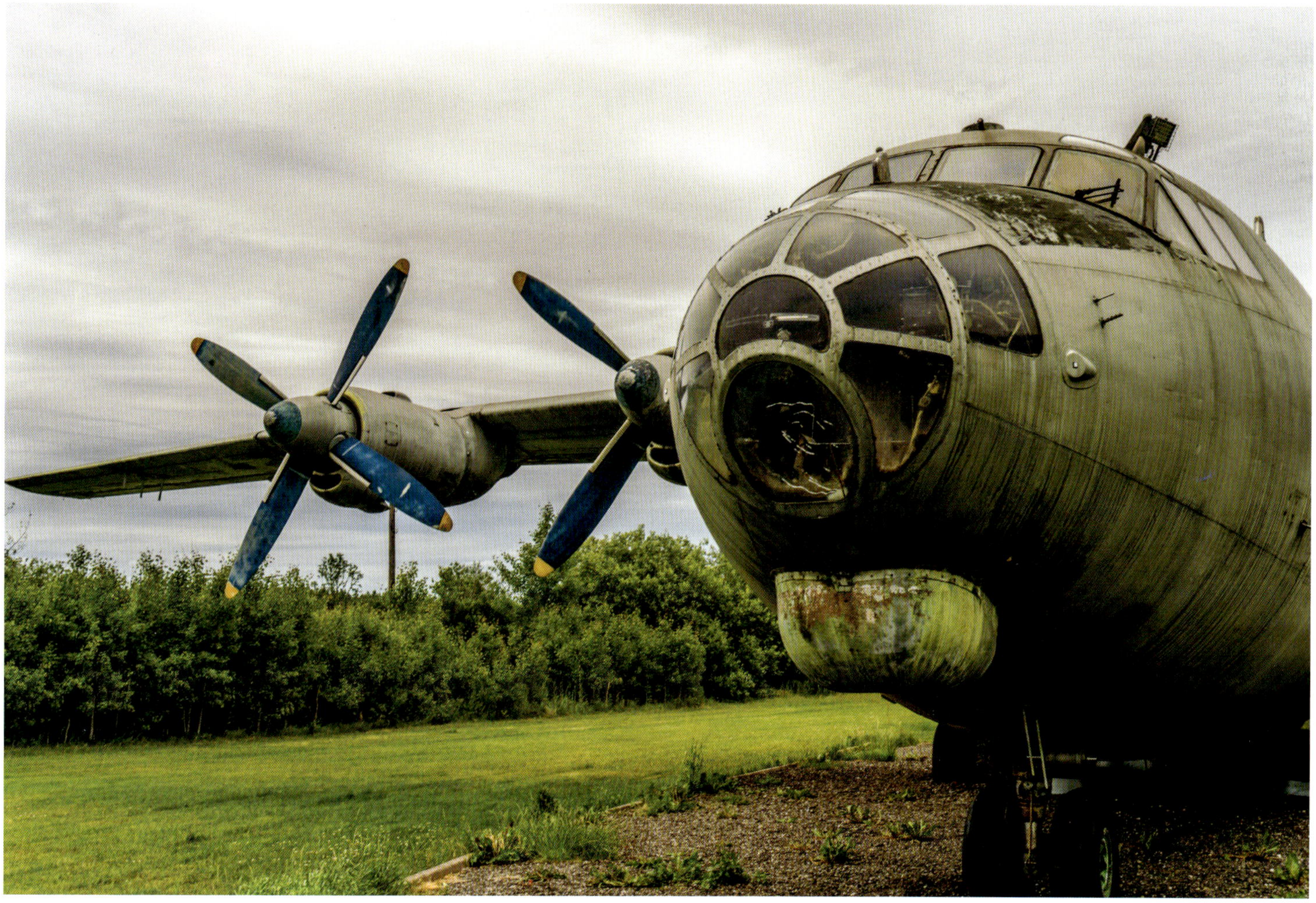

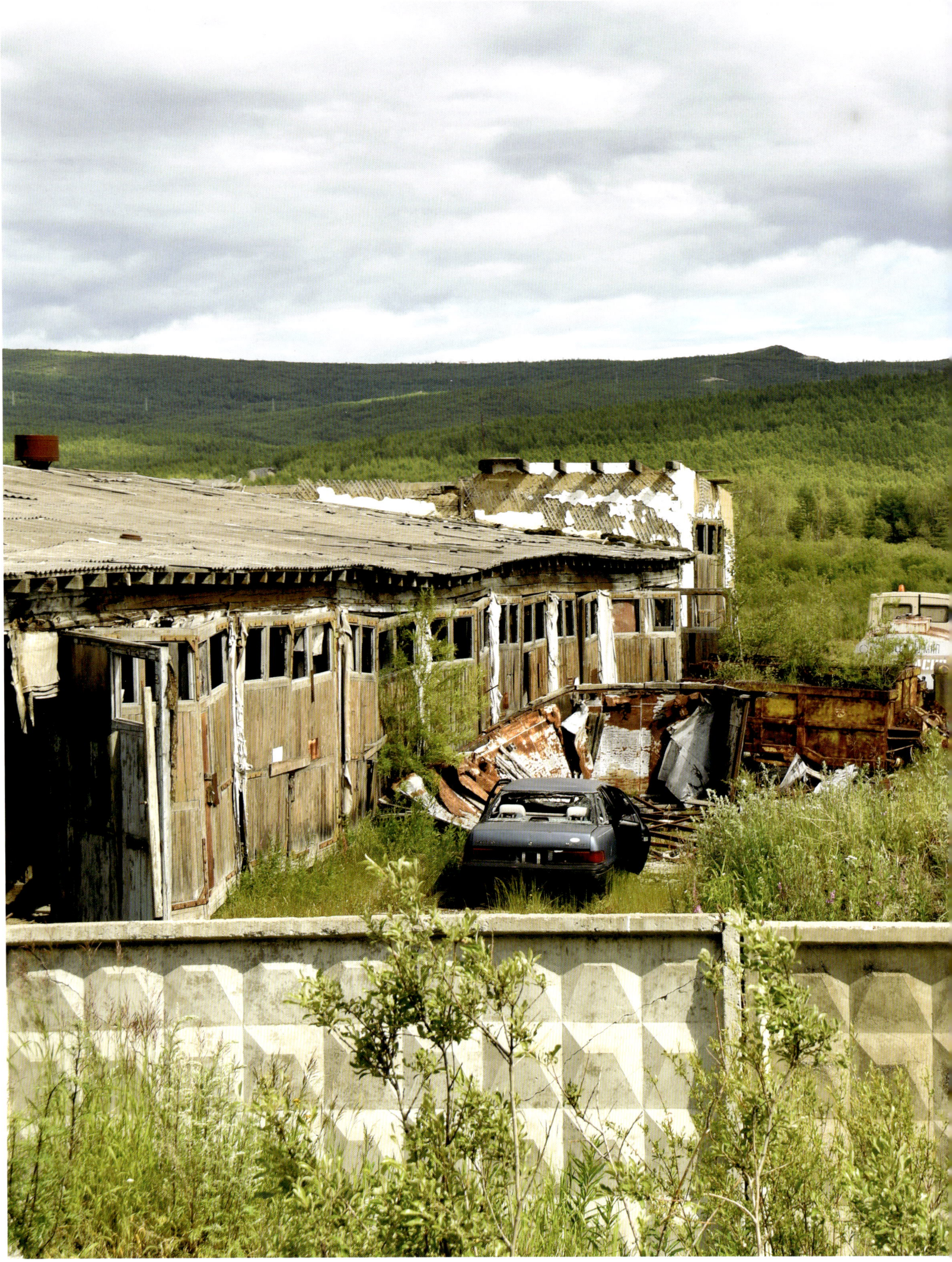

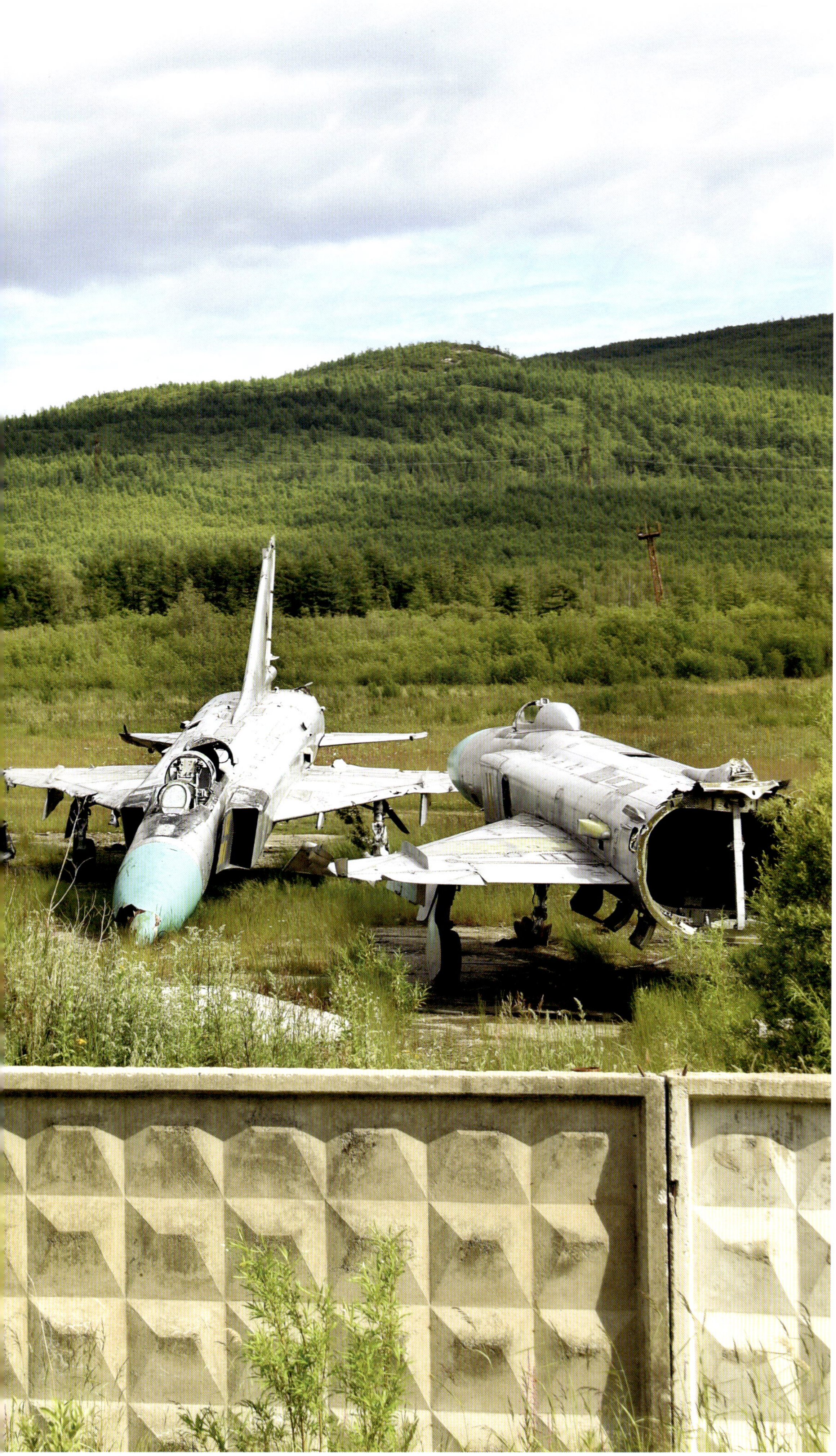

Magadan, Kolyma Region, Siberia
Their engines removed, the airframes of two Sukhoi Su-15 Flagon interceptors await scrapping. Magadan was in the front line of the Cold War, close to the Pacific Ocean. The Su-15 was developed as a counter to the US B-52 strategic bomber, and could travel over twice the speed of sound. It was a Soviet Su-15 aircraft that shot down Korean Air Flight 007 in 1983 after it flew into restricted airspace, killing all on board.

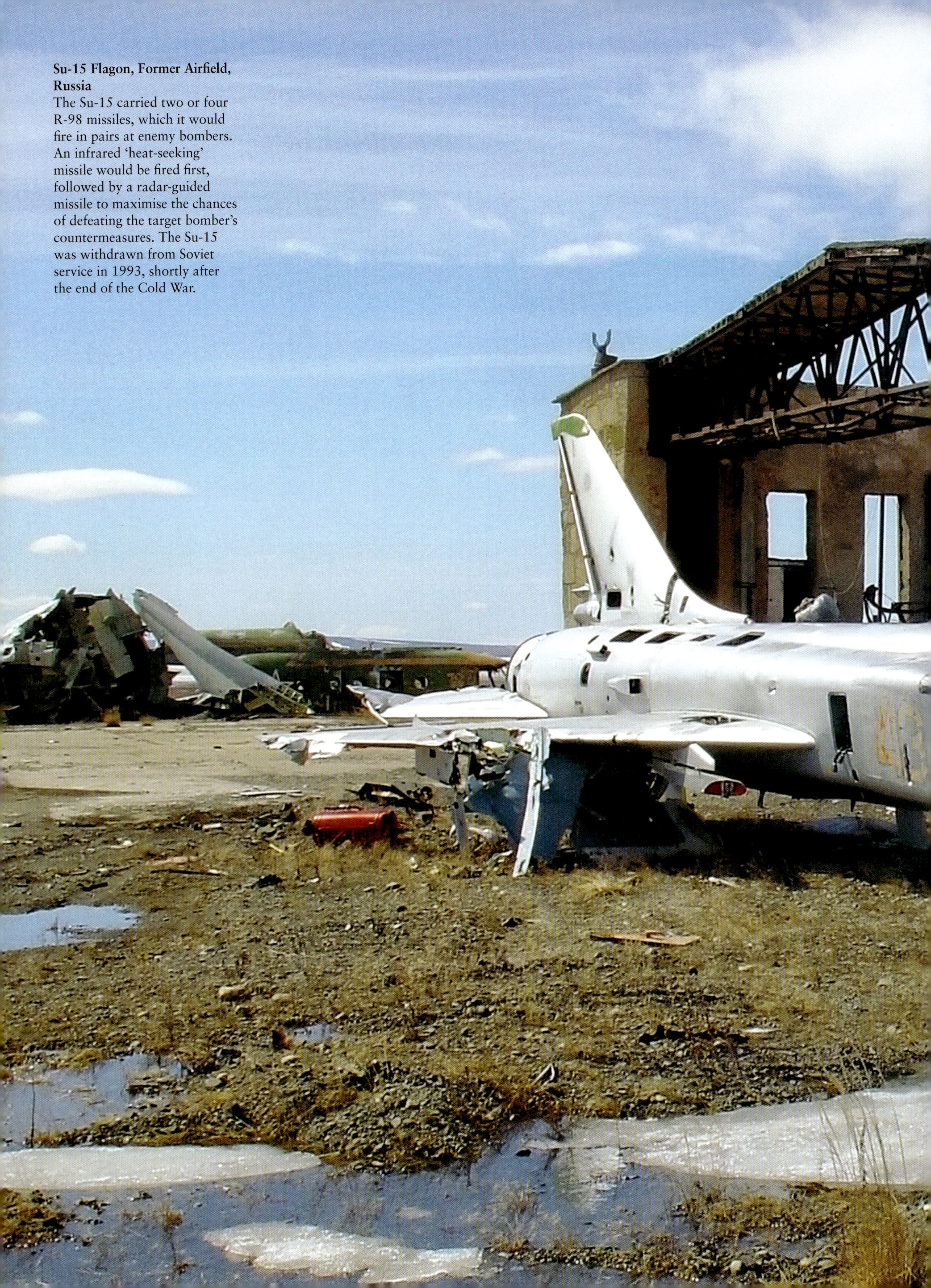

Su-15 Flagon, Former Airfield, Russia
The Su-15 carried two or four R-98 missiles, which it would fire in pairs at enemy bombers. An infrared 'heat-seeking' missile would be fired first, followed by a radar-guided missile to maximise the chances of defeating the target bomber's countermeasures. The Su-15 was withdrawn from Soviet service in 1993, shortly after the end of the Cold War.

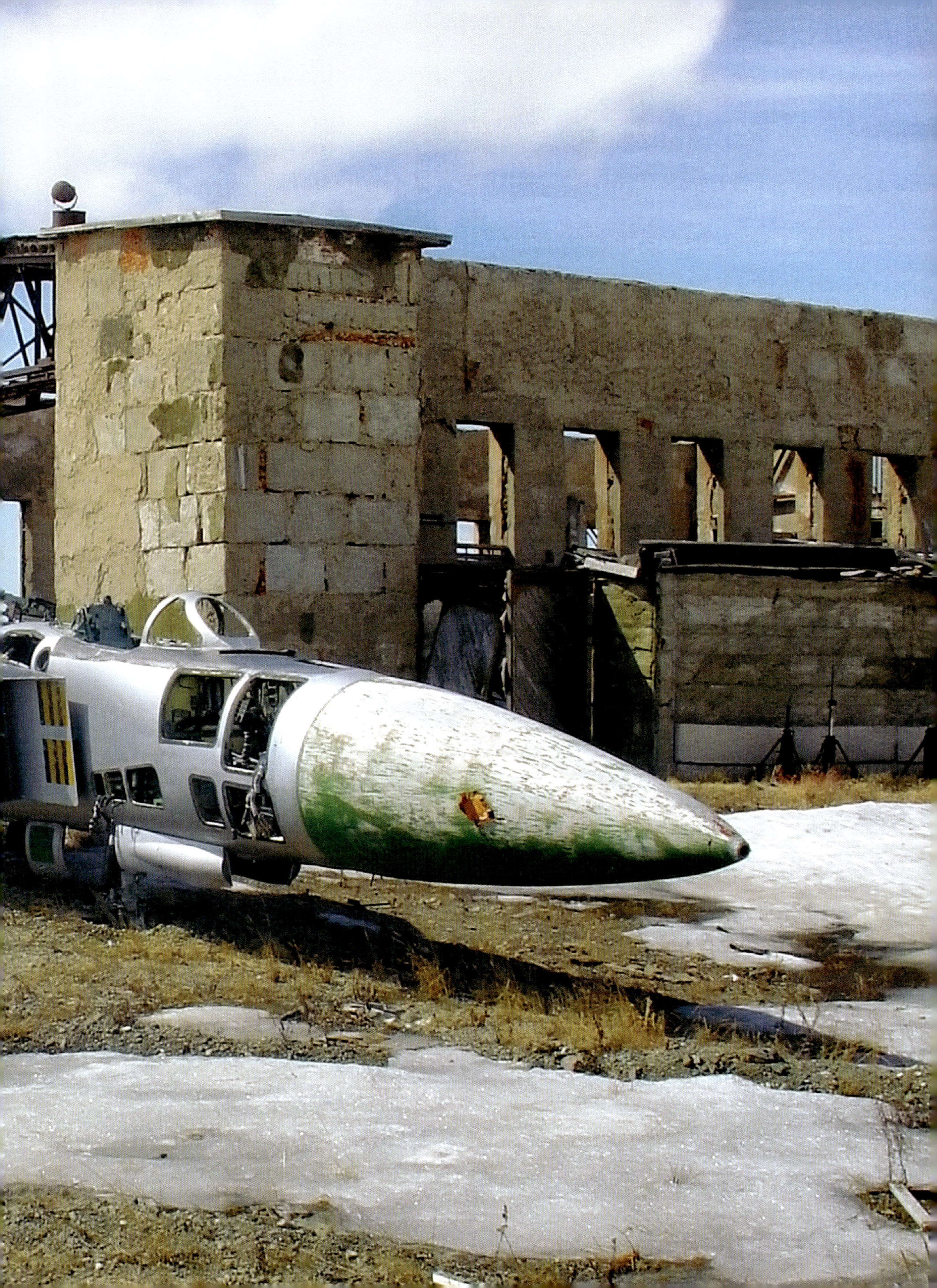

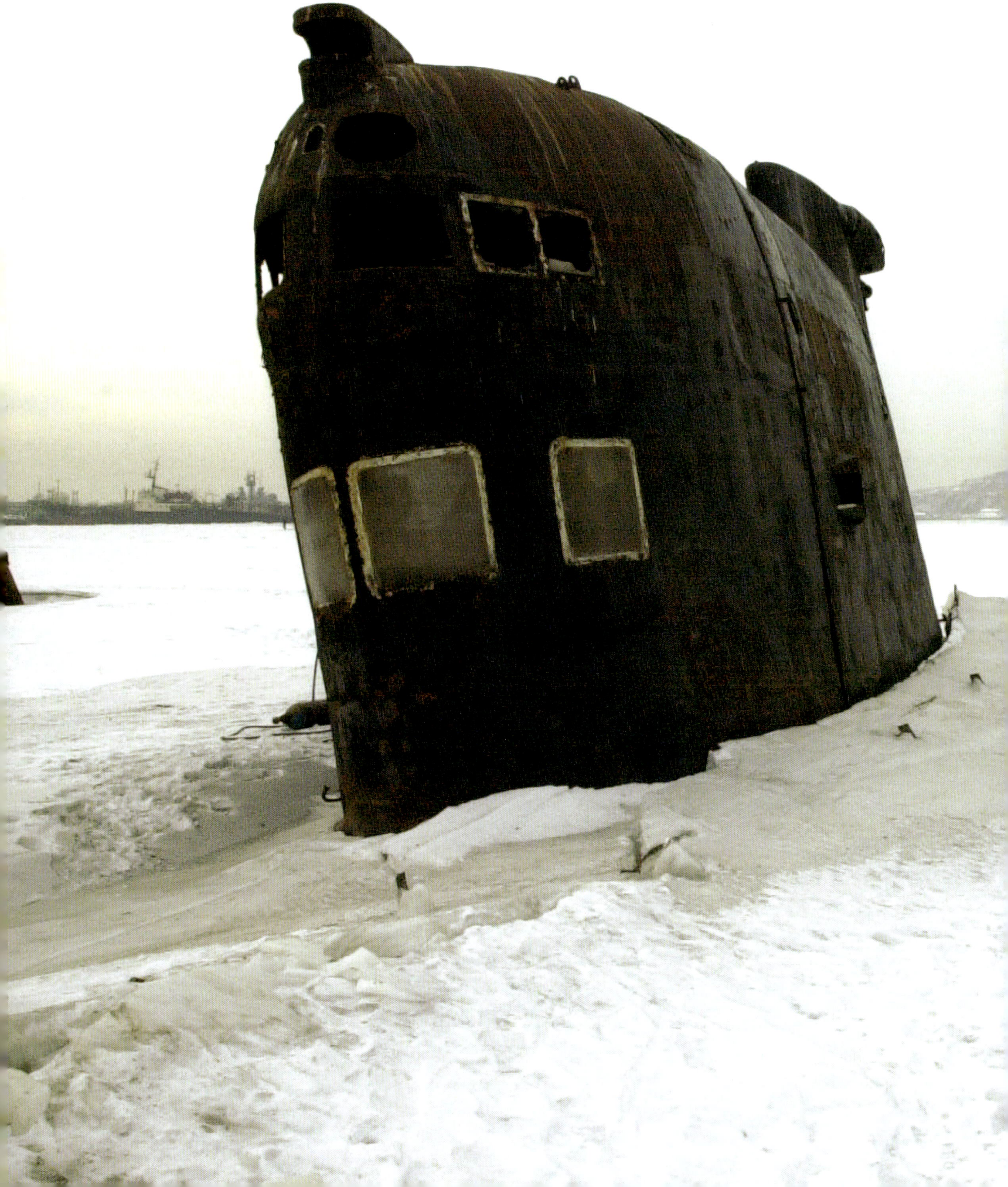

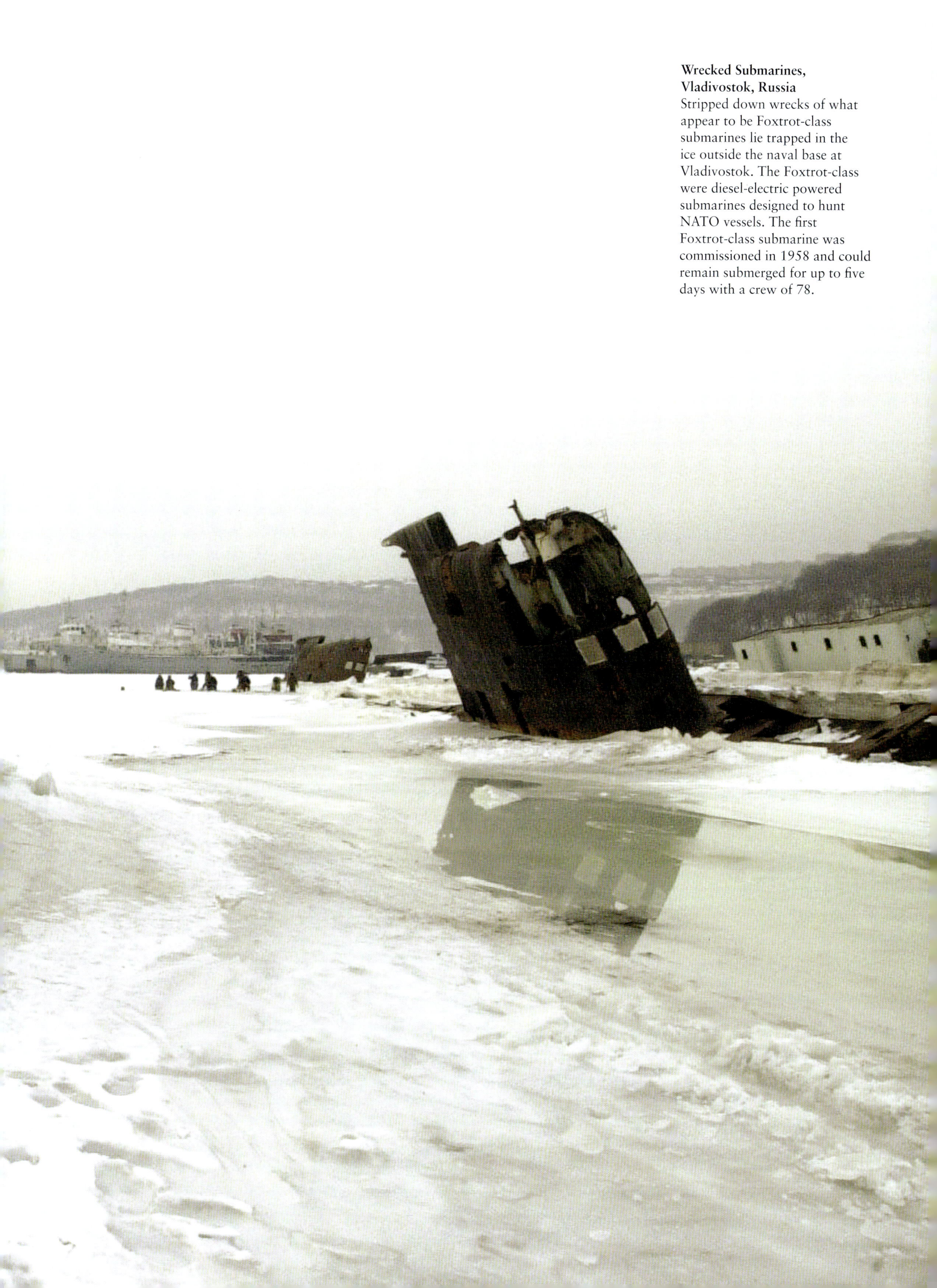

Wrecked Submarines, Vladivostok, Russia
Stripped down wrecks of what appear to be Foxtrot-class submarines lie trapped in the ice outside the naval base at Vladivostok. The Foxtrot-class were diesel-electric powered submarines designed to hunt NATO vessels. The first Foxtrot-class submarine was commissioned in 1958 and could remain submerged for up to five days with a crew of 78.

Abandoned Anti-Aircraft Gun, Former Submarine Base, Simushir Island, Kuril Islands
The Kuril Islands in the Pacific Ocean belonged to Japan for much of their history, but were surrendered to the Soviet Union at the end of World War II. Between 1987 and 1994, the Soviet Navy used Broutona Bay as a submarine base, but the island is now no longer inhabited.

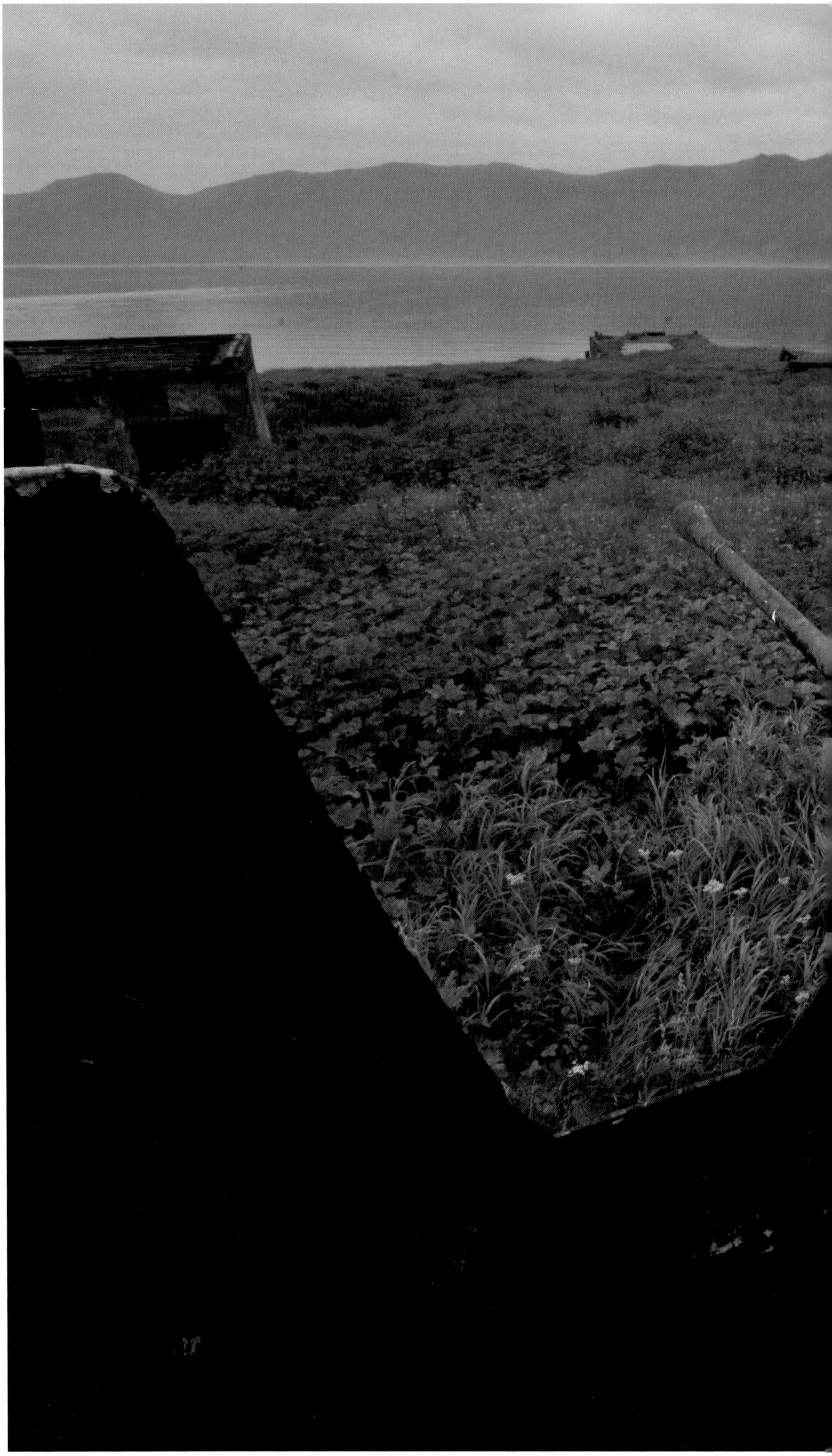

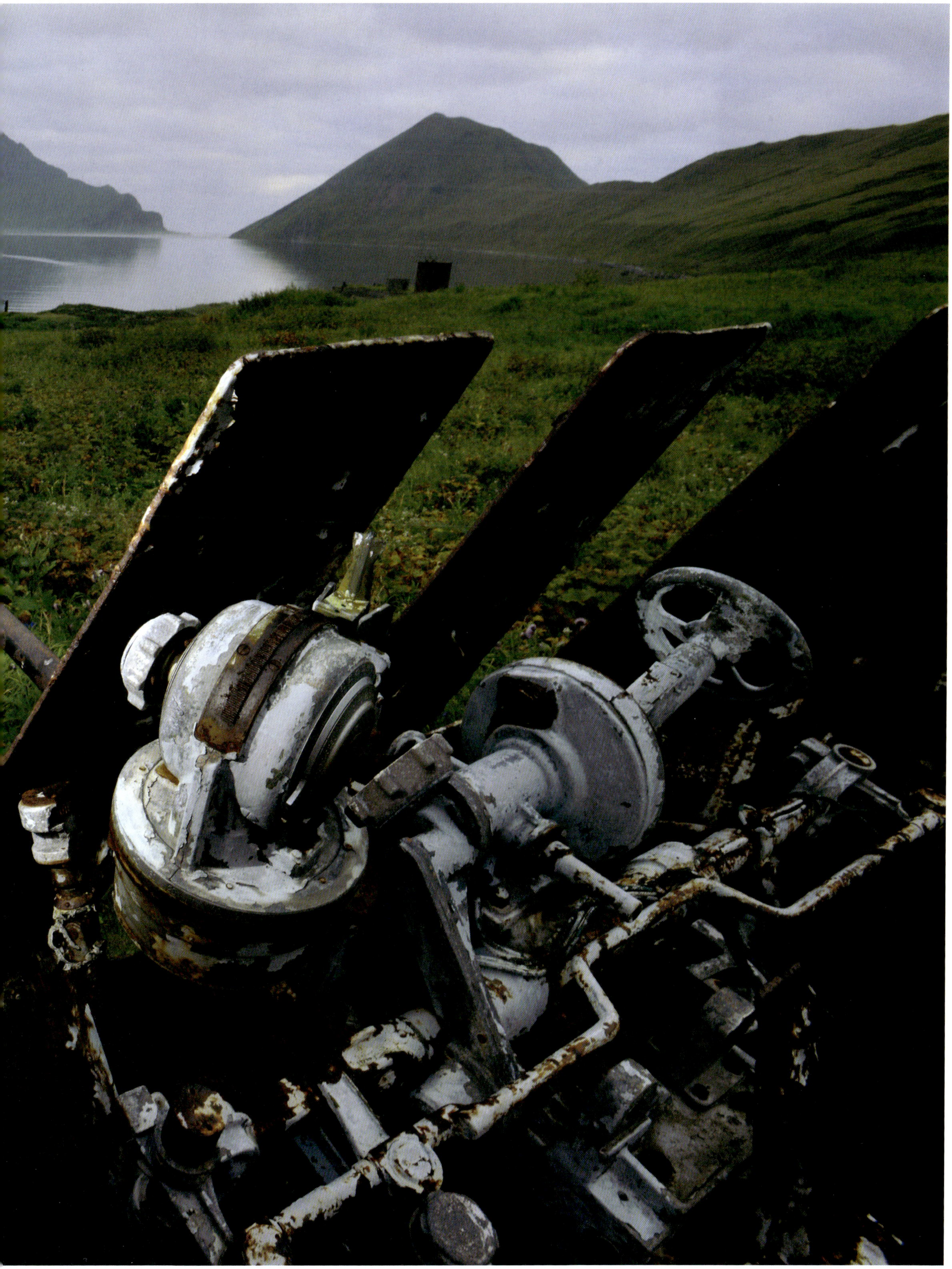

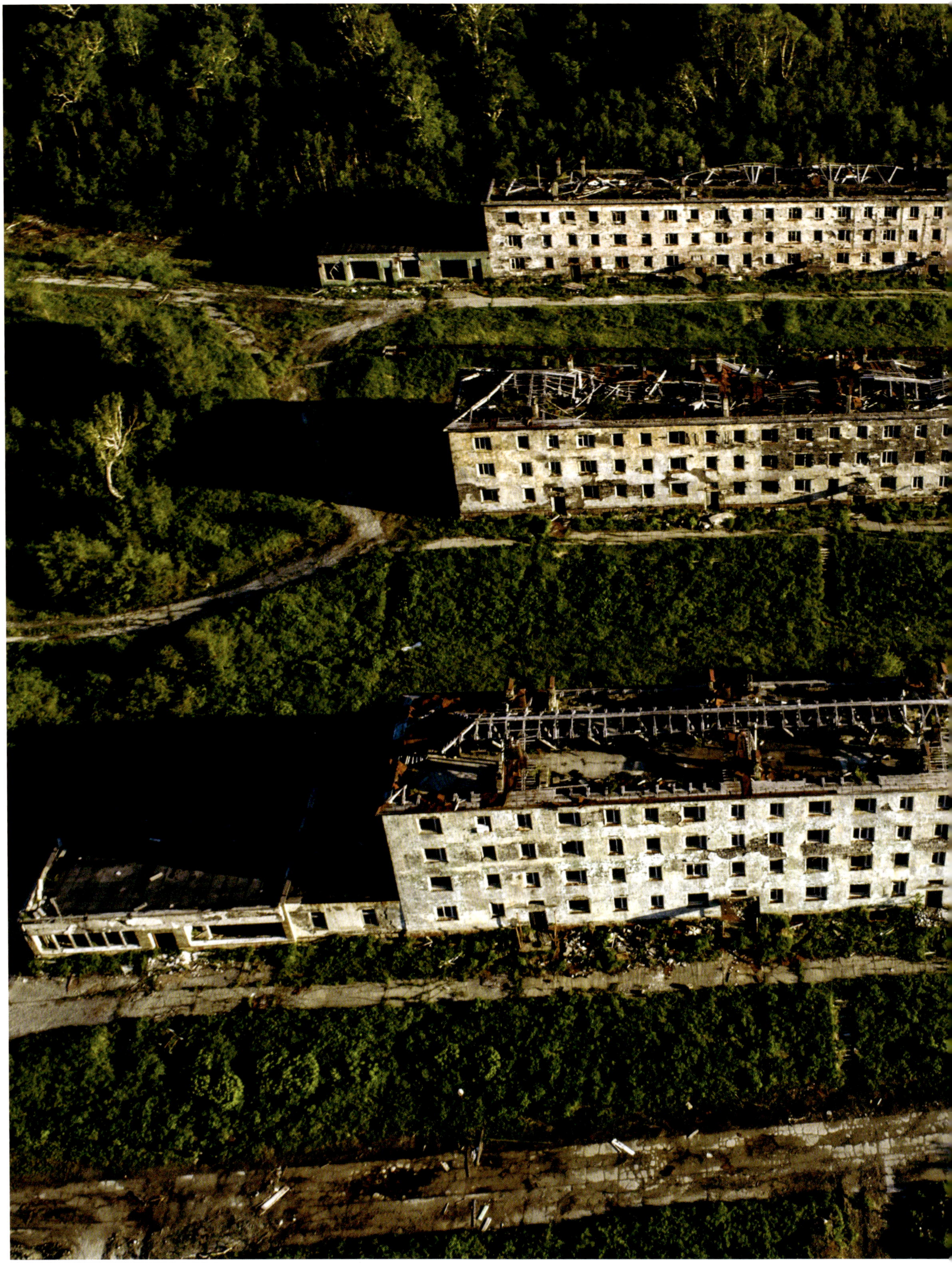

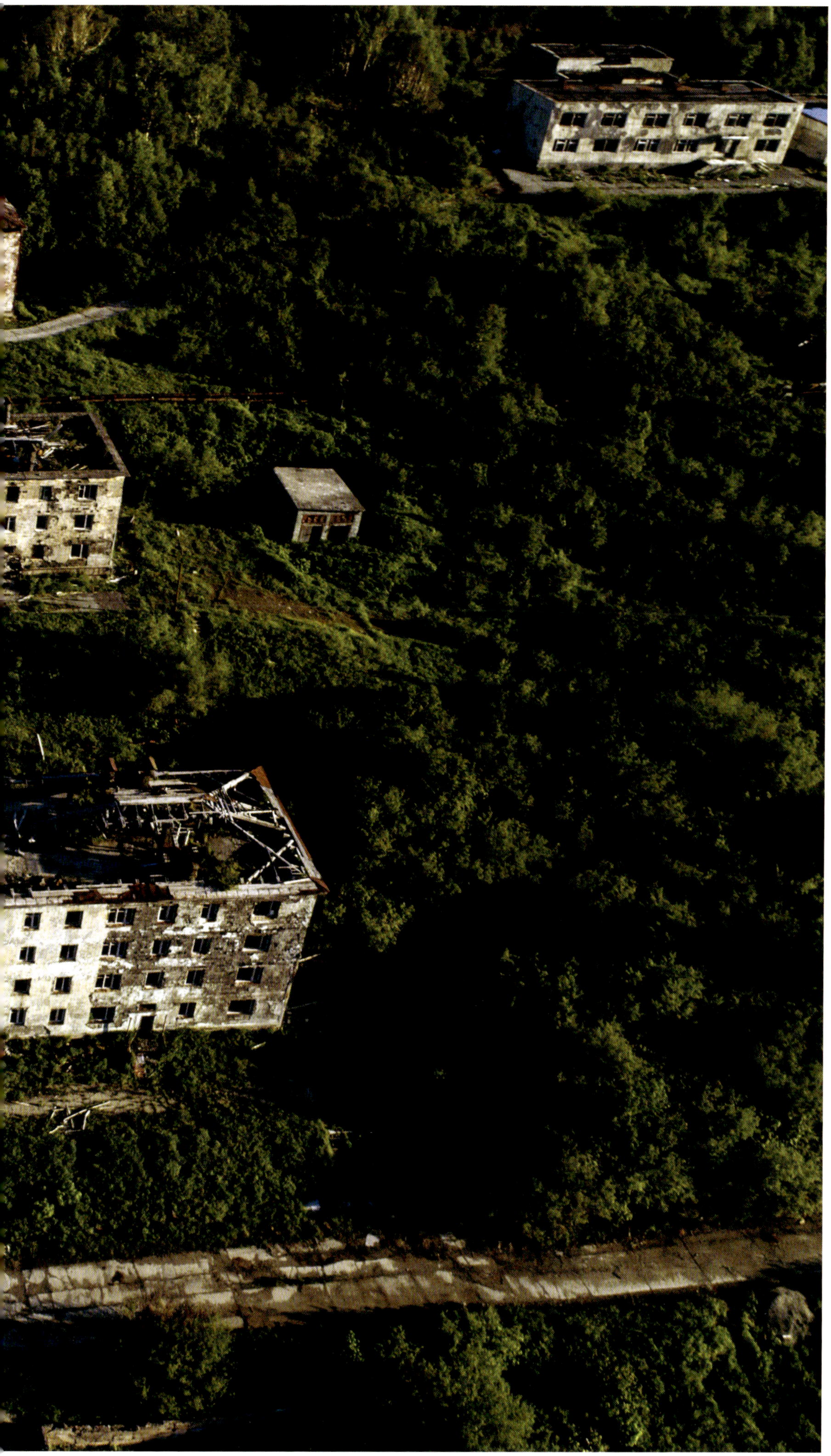

Bechevinka, Kamchatka Peninsula, Russia
An aerial view of the housing built for the staff of this former base known as Petropavlovsk-Kamchatsky-54. Established in 1960, it was an important naval base for much of the Cold War, but was closed in 1996 and is now a remote tourist destination on Russia's Pacific coast.

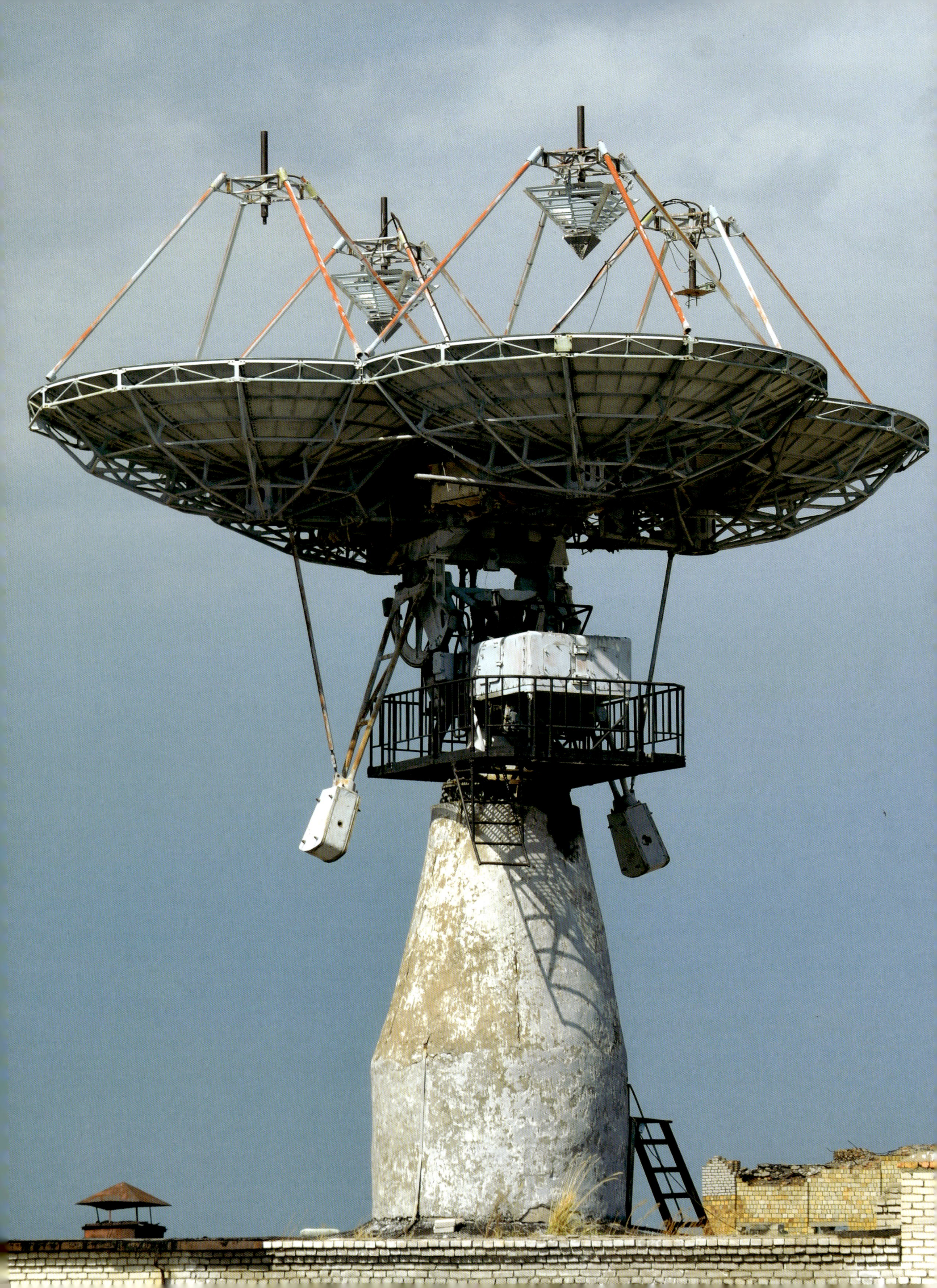

Eastern Europe and the Soviet Bloc

The three leaders of the major Allied powers, Stalin, Churchill and Roosevelt, had drawn up zones of control for post-war Europe at the Yalta conference in February 1945, and these were refined at the later Potsdam conference in July that year, with Churchill replaced by the new British Prime Minister, Clement Attlee, and President Harry S. Truman replacing Roosevelt, who had died in April. In the American, British and French zones of control, investment and reconstruction schemes were quickly put in place, but the Soviets used eastern Germany to extract reparations for themselves and the Polish government for the actions of the Third Reich.

In Poland, Hungary, Romania, Bulgaria, Czechoslovakia, Albania and, later, East Germany itself, communist governments were elected under Soviet influence. (Yugoslavia elected its own communist government led by Tito, the former Partisan leader.) The Soviet sphere of control was formalised by the creation of the Warsaw Pact in May 1955, but it was not universally popular: there were significant uprisings against Soviet control in East Germany in 1953, Hungary in 1956, and Czechoslovakia in 1968. After the collapse of the Soviet Union in 1991, the new democratic governments of Eastern European states inherited hundreds of former Soviet bases stripped of anything valuable by their garrisons as they left.

OPPOSITE:
Former Soviet Anti-Ballistic Missile Testing Range, Kazakhstan
Built on the shore of Lake Balkhash in the late 1950s, Sary Shagan was the anti-ballistic missile testing range for the Soviet Union. The world's first successful intercept of a warhead by a missile took place in March 1961. The facility tested missiles, radars and control systems; and, in the 1980s, early laser weapons.

Former Anti-Ballistic Missile Testing Range, Kazakhstan
Many of the original test antennae and radomes remain on the site. The A-35 anti-ballistic missile system that protected Moscow for two decades from 1971 was tested and developed here. The interceptor missile itself carried a nuclear warhead.

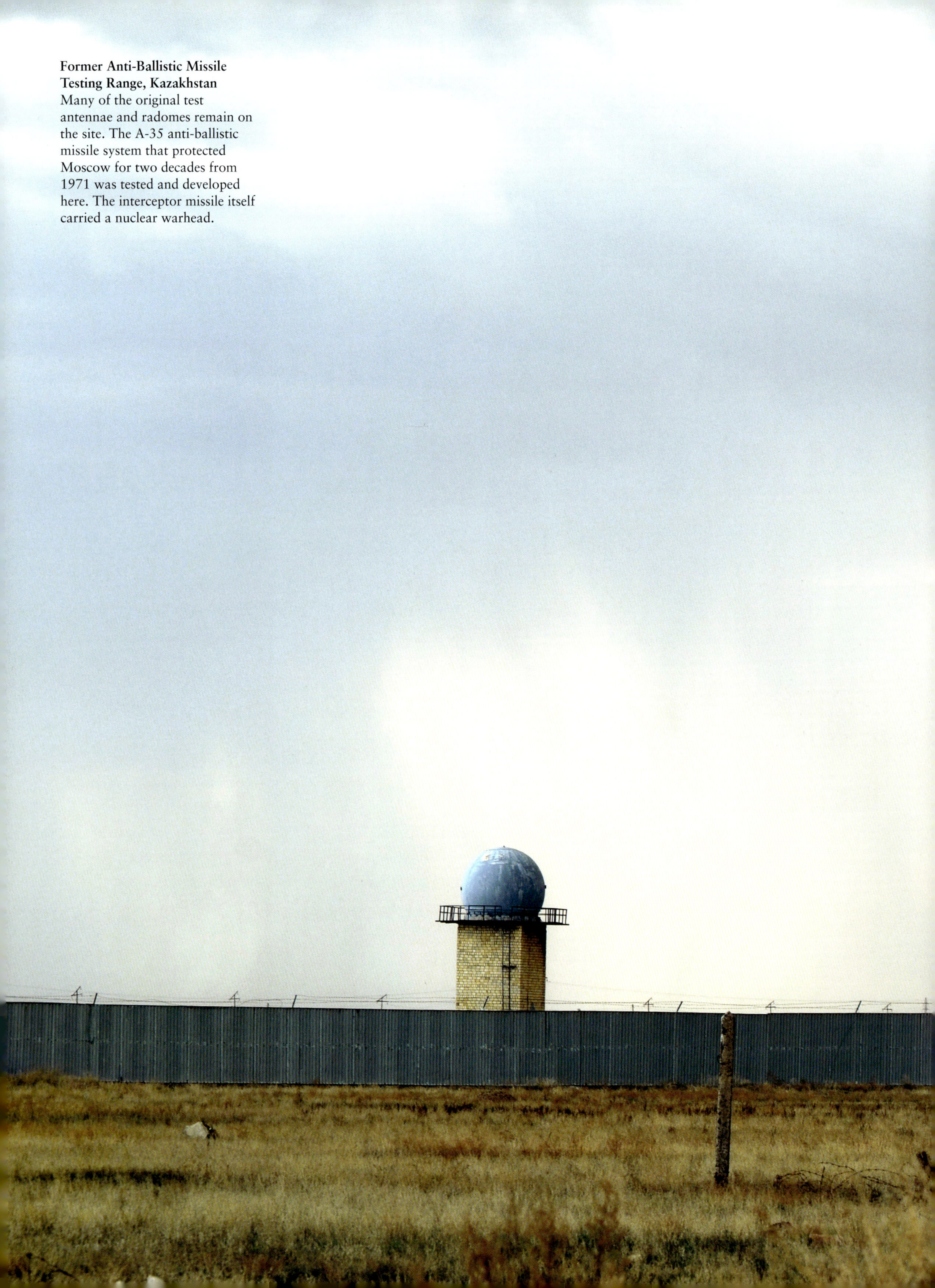

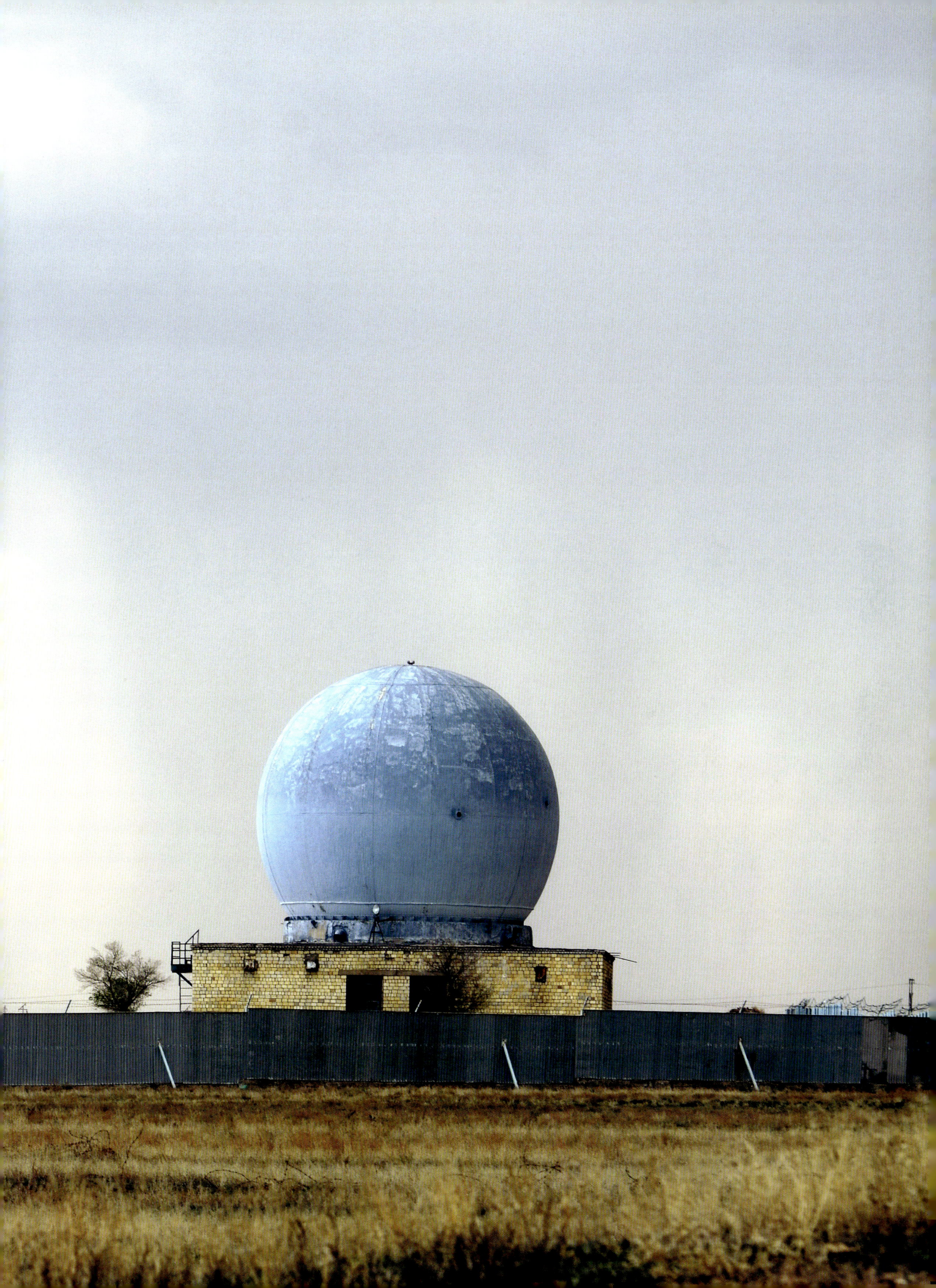

Abandoned Buran Transport, Baikonur, Kazakhstan
In response to the US Space Shuttle programme, the Soviet Union developed the Buran reusable spacecraft. This giant transporter carried the Buran and its Energia launch rocket to the launchpad. Only one unmanned orbital flight took place in 1988 before the programme was cancelled in 1993.

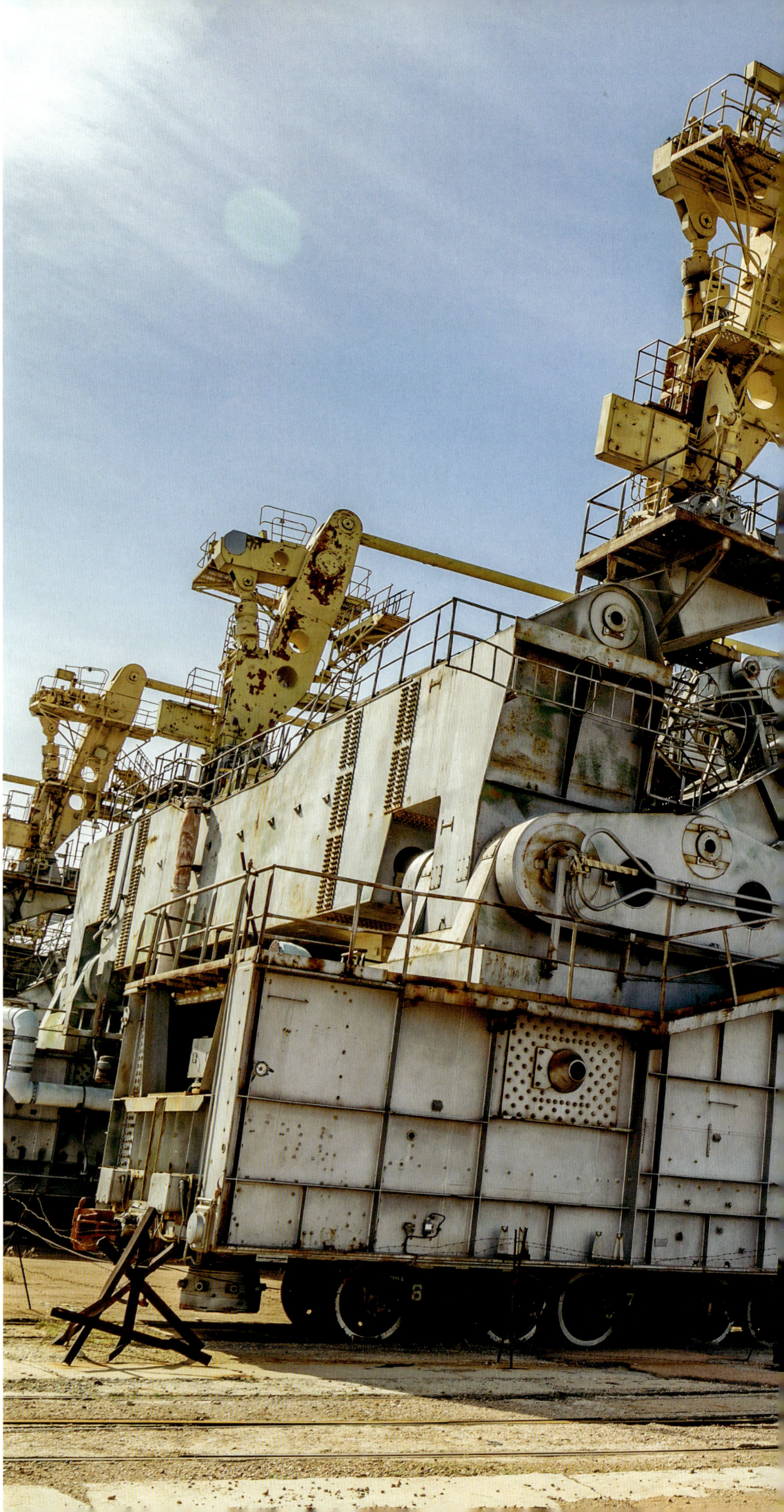

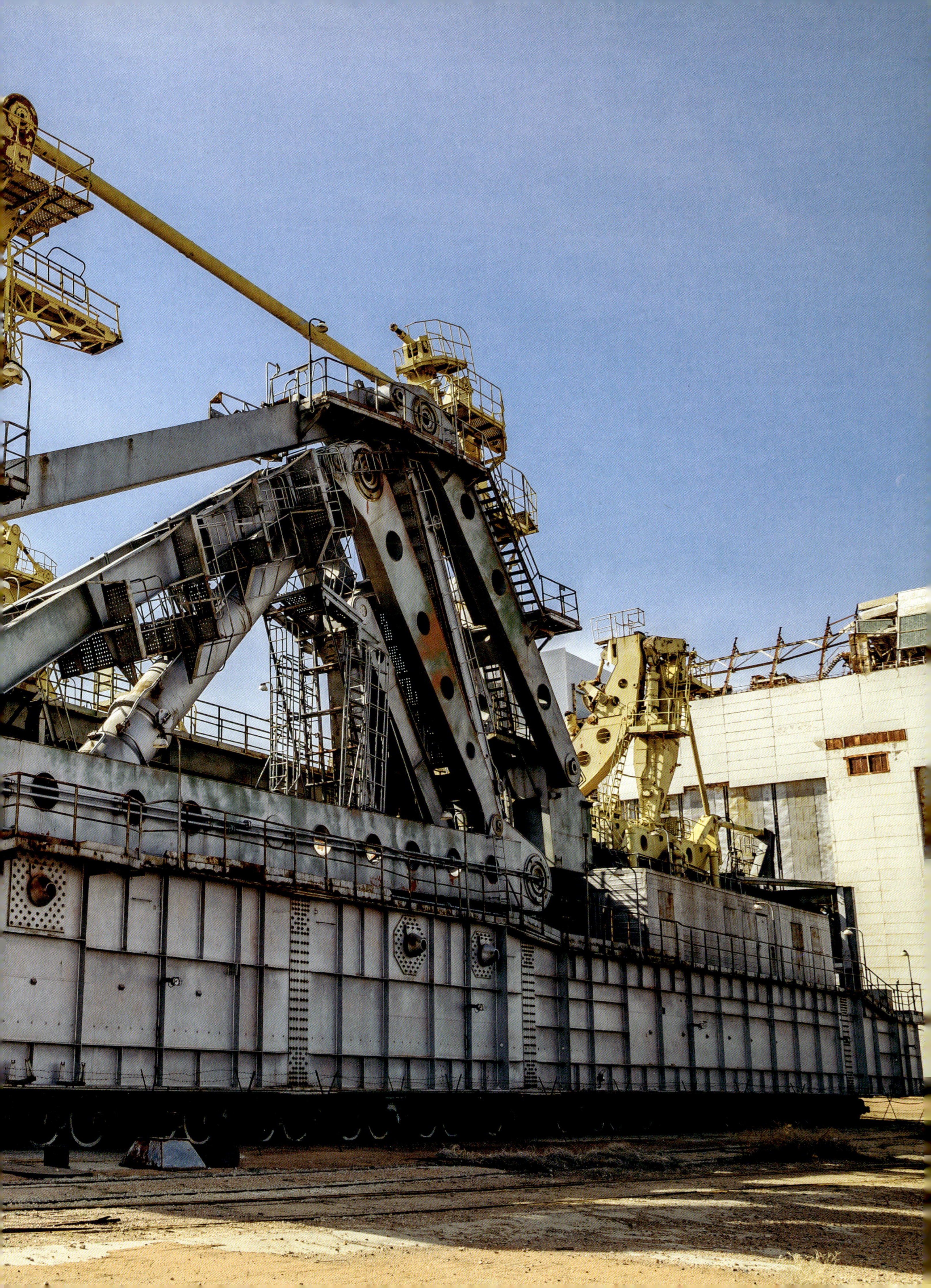

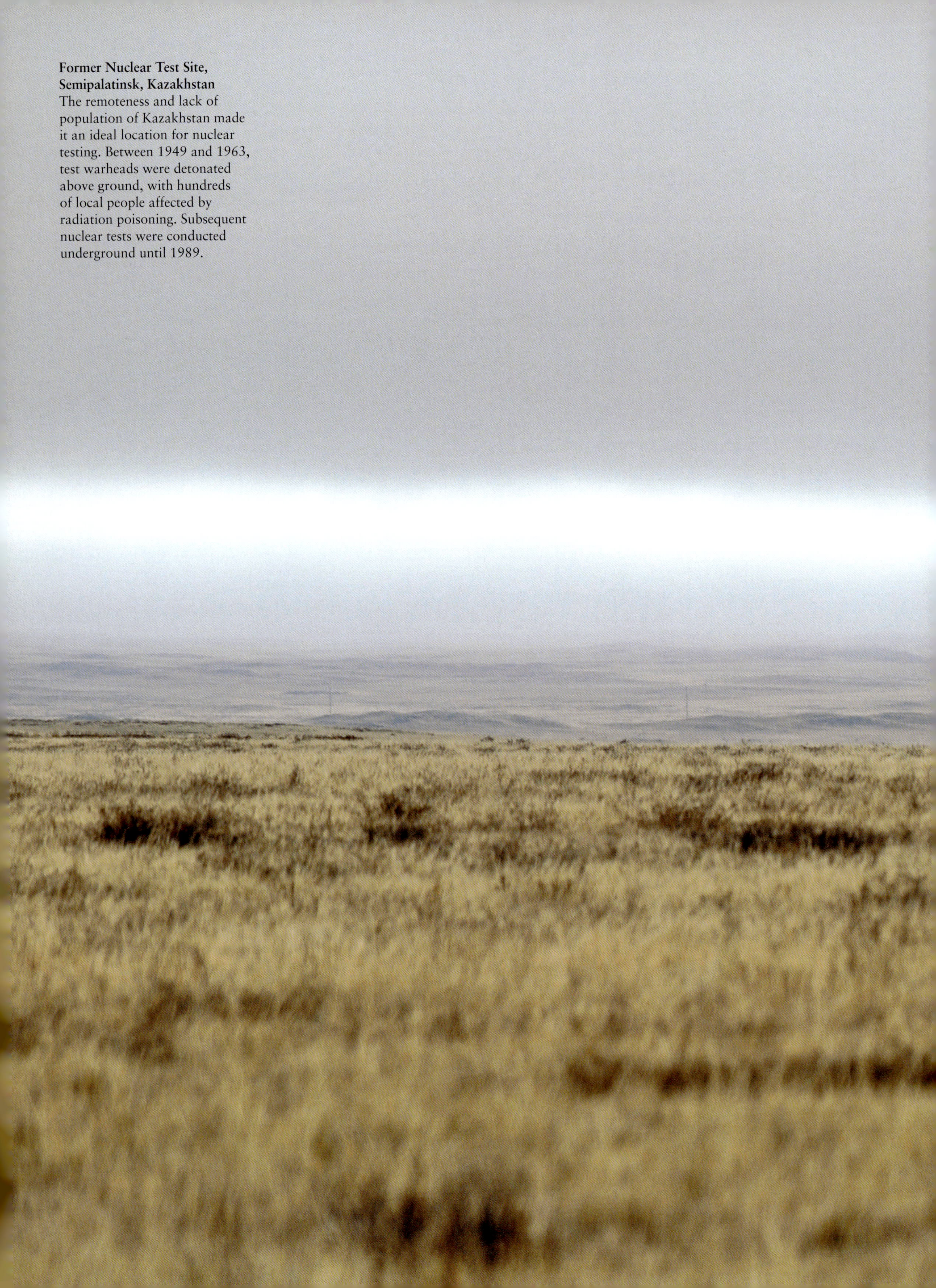

Former Nuclear Test Site, Semipalatinsk, Kazakhstan
The remoteness and lack of population of Kazakhstan made it an ideal location for nuclear testing. Between 1949 and 1963, test warheads were detonated above ground, with hundreds of local people affected by radiation poisoning. Subsequent nuclear tests were conducted underground until 1989.

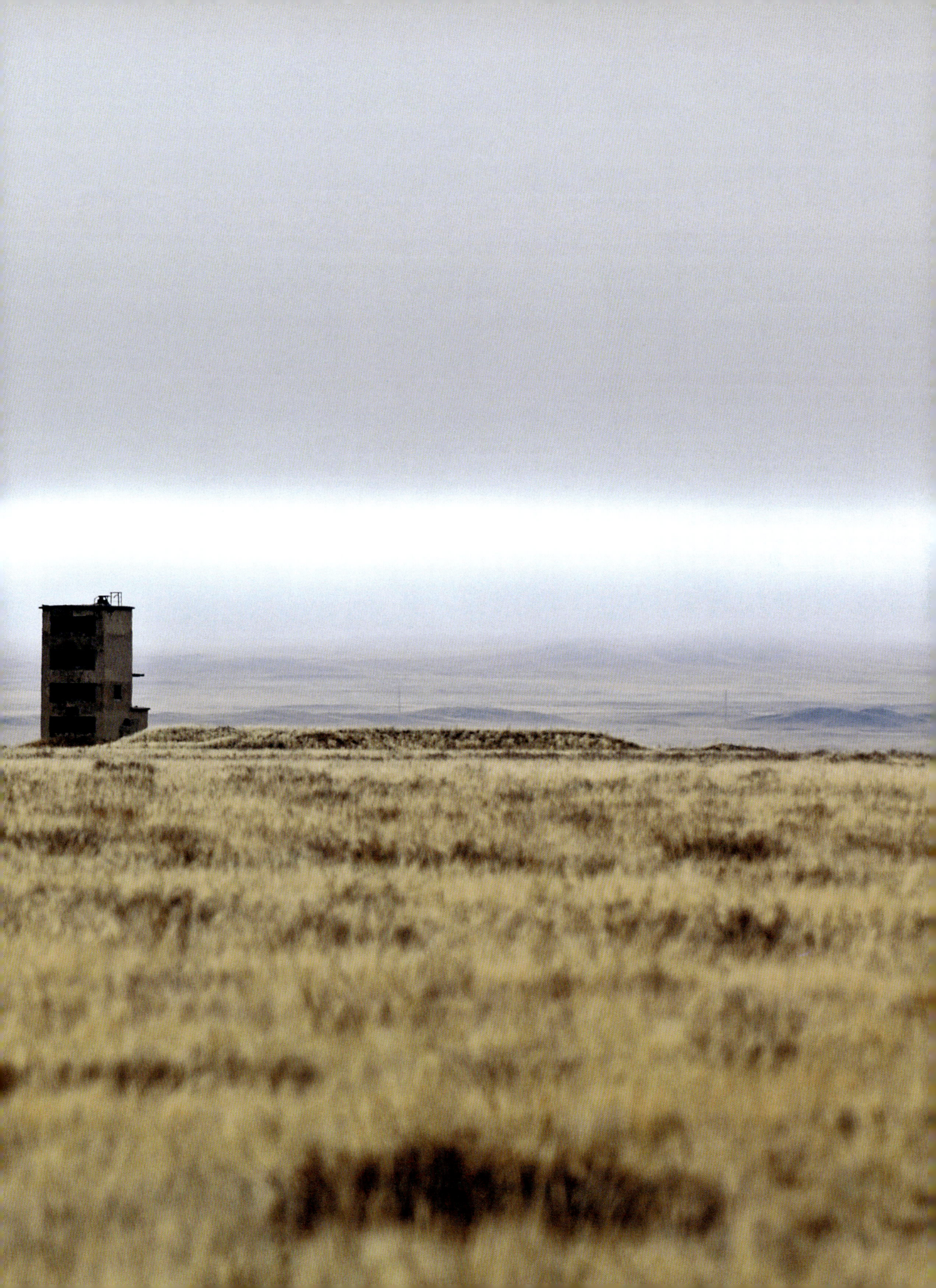

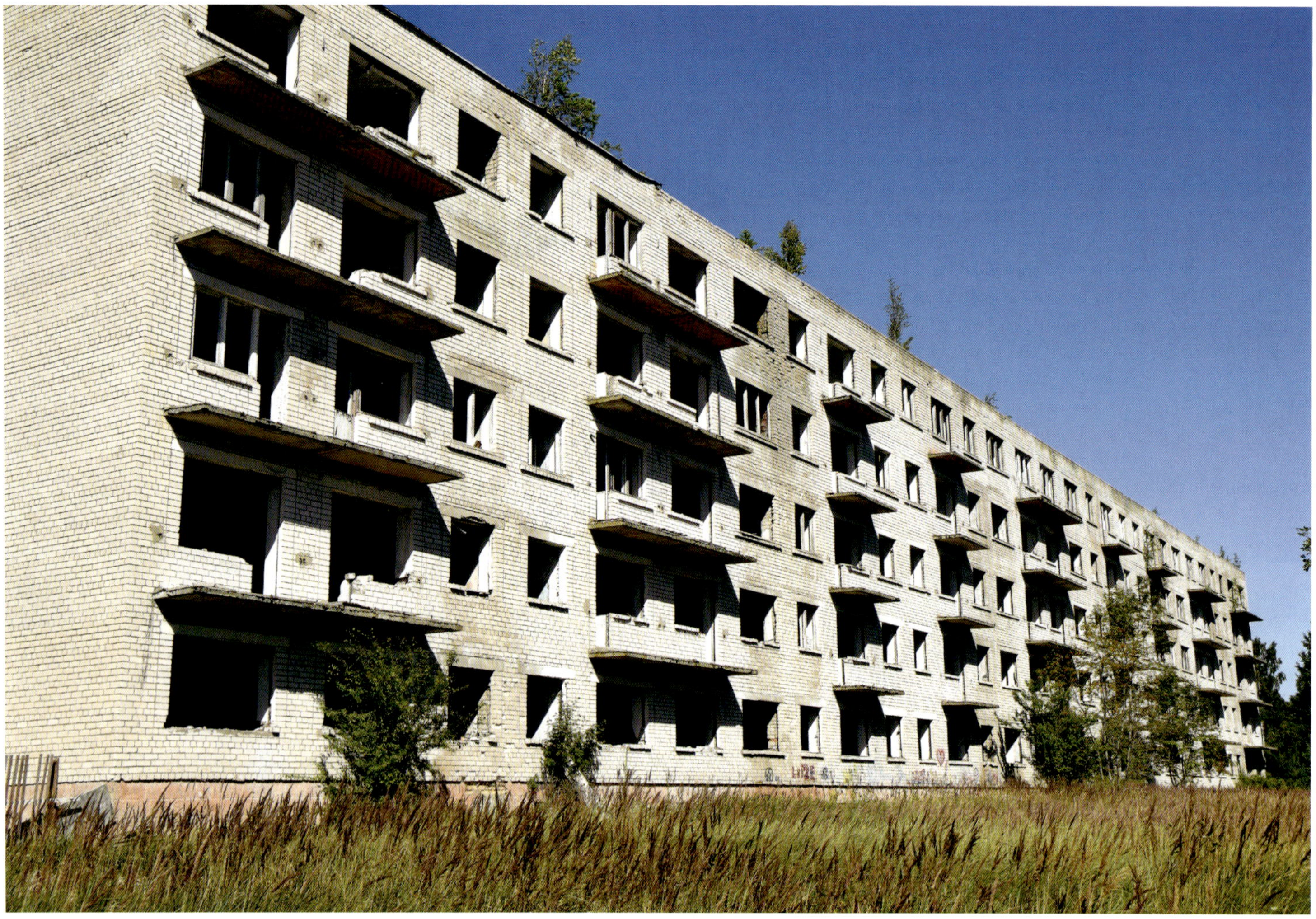

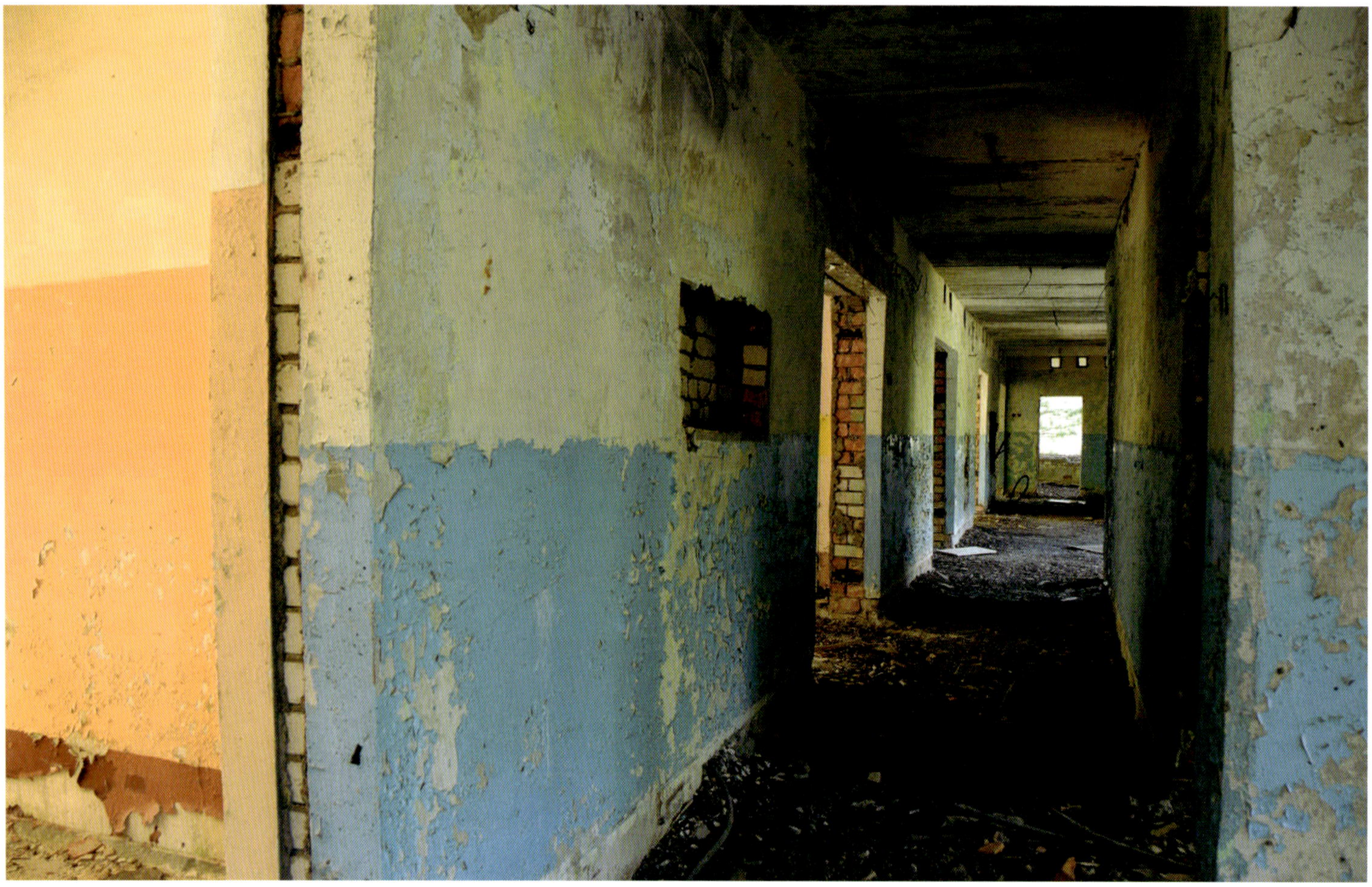

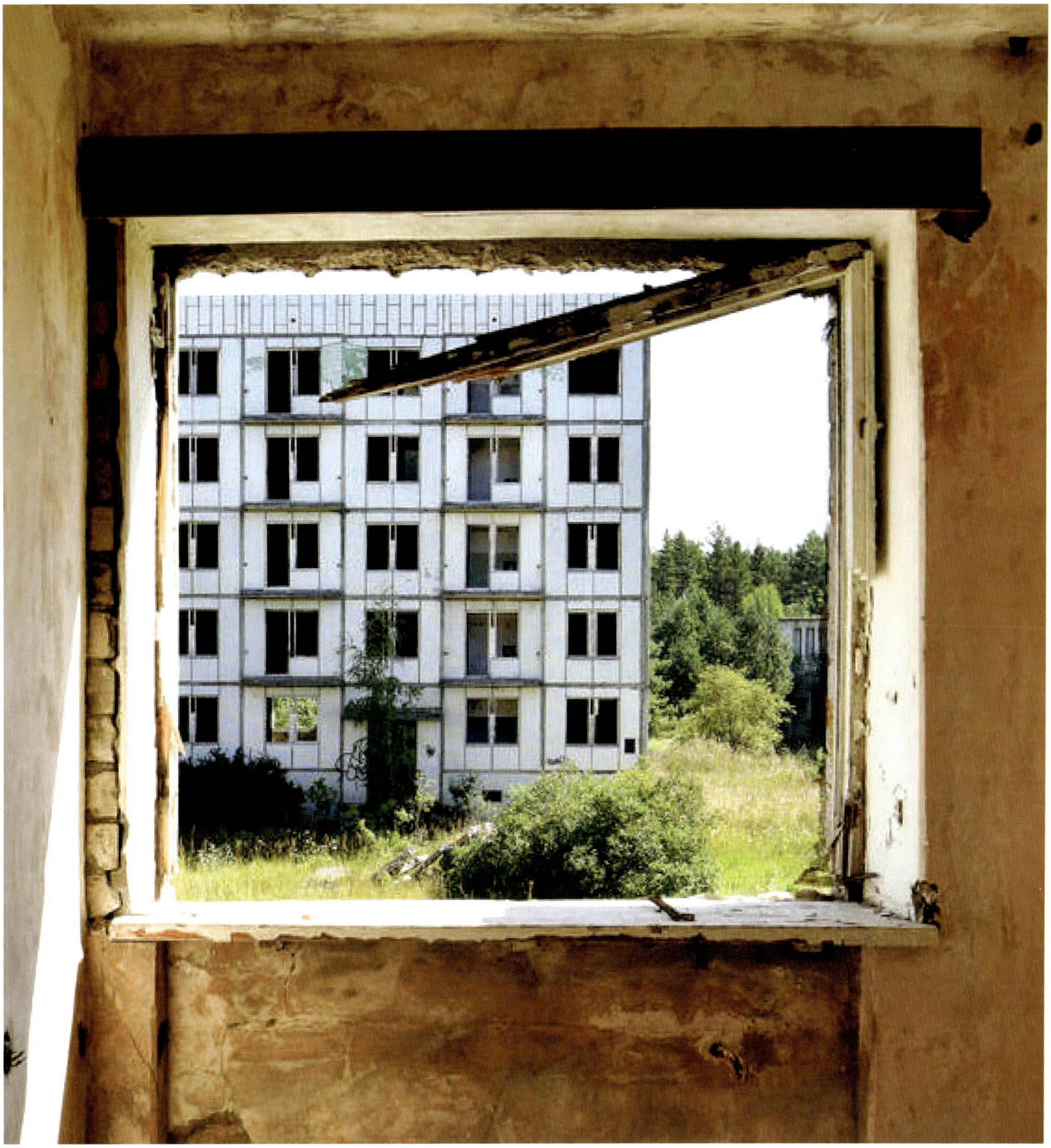

OPPOSITE AND ABOVE:
Former Soviet Listening Station, Irbene, Latvia
In 1974 the Soviet Union built a facility in remote woods near Ventspils on the Baltic Sea coast. The base was equipped with a RT-32 32 metre (105ft) telescope alongside two smaller telescopes and a communications centre. It has been alleged that the KGB used it for listening to NATO communications.

The RT-32 telescope was made of over 20,000 components all mounted on a 25 metre (82ft) tower and weighing 80 tonnes. The dish was supposedly built by a Ukrainian naval facility. The base itself was known as Zveolsdoshka ('Little Star').

A small town was built to house the facility staff, garrison and their families. The base was a secret until the end of the Cold War when Latvia became independent again. Although the station itself is now back in use as a radio astronomy centre, the former Soviet housing blocks have been left to decay.

OVERLEAF:
Former R-12U Missile Silo, Tirza, Latvia
Now overrun by nature, underneath the concrete dome once lurked an R-12U nuclear missile capable of reaching targets throughout Europe. The R-12 was the missile that provoked the Cuban Missile Crisis in 1962. In the event of a nuclear launch being authorized, the dome would have opened to allow the missile to launch.

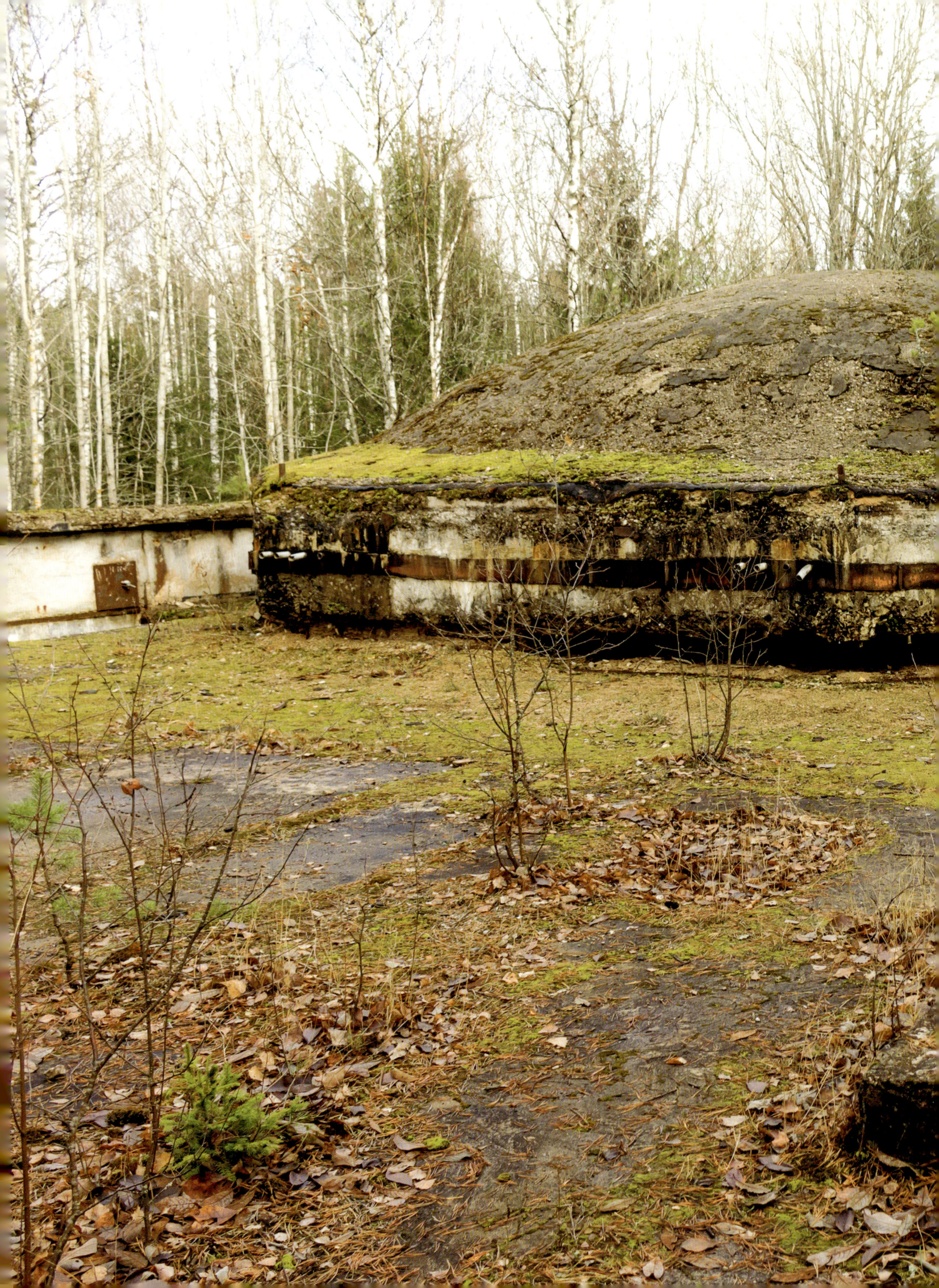

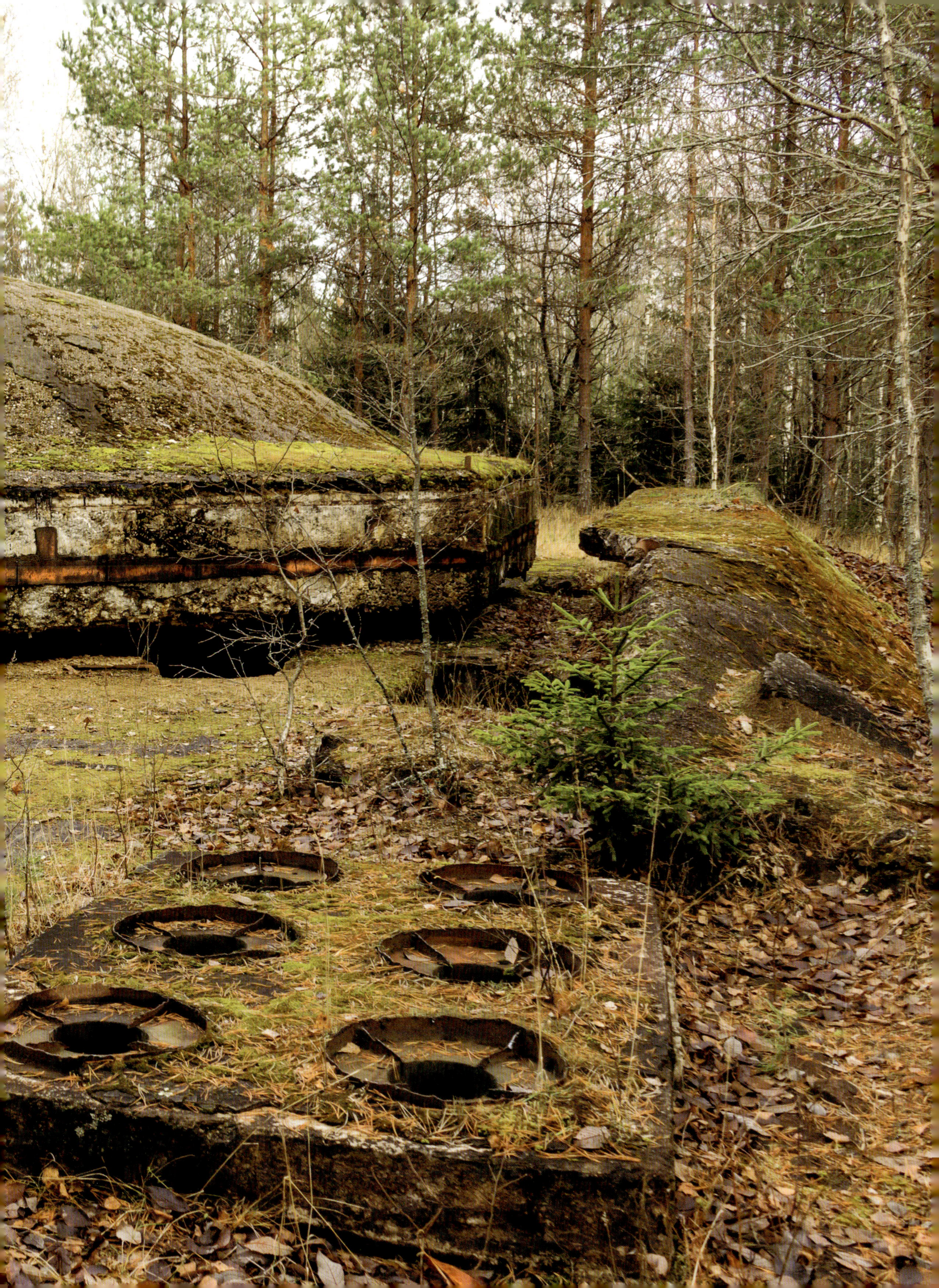

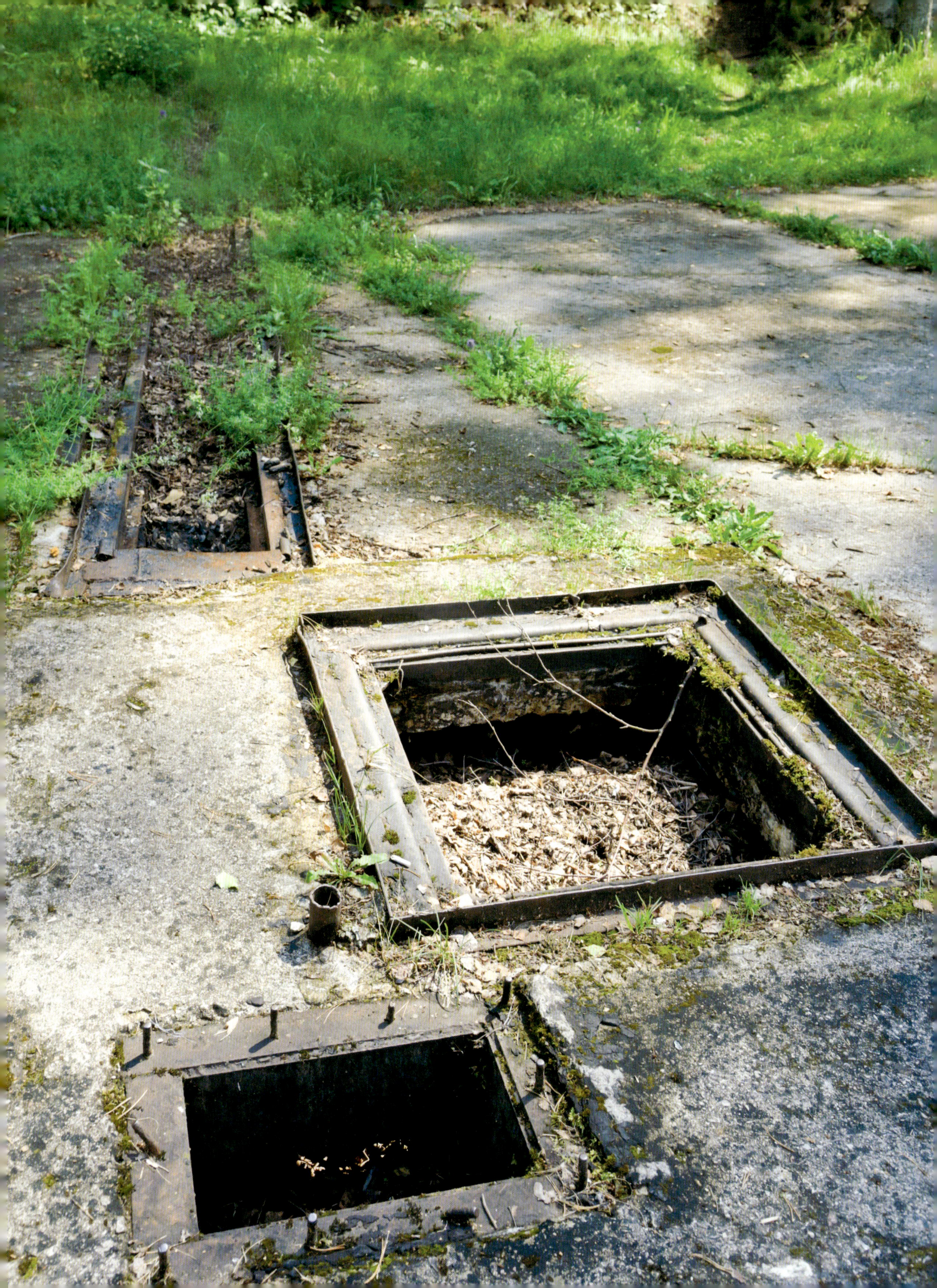

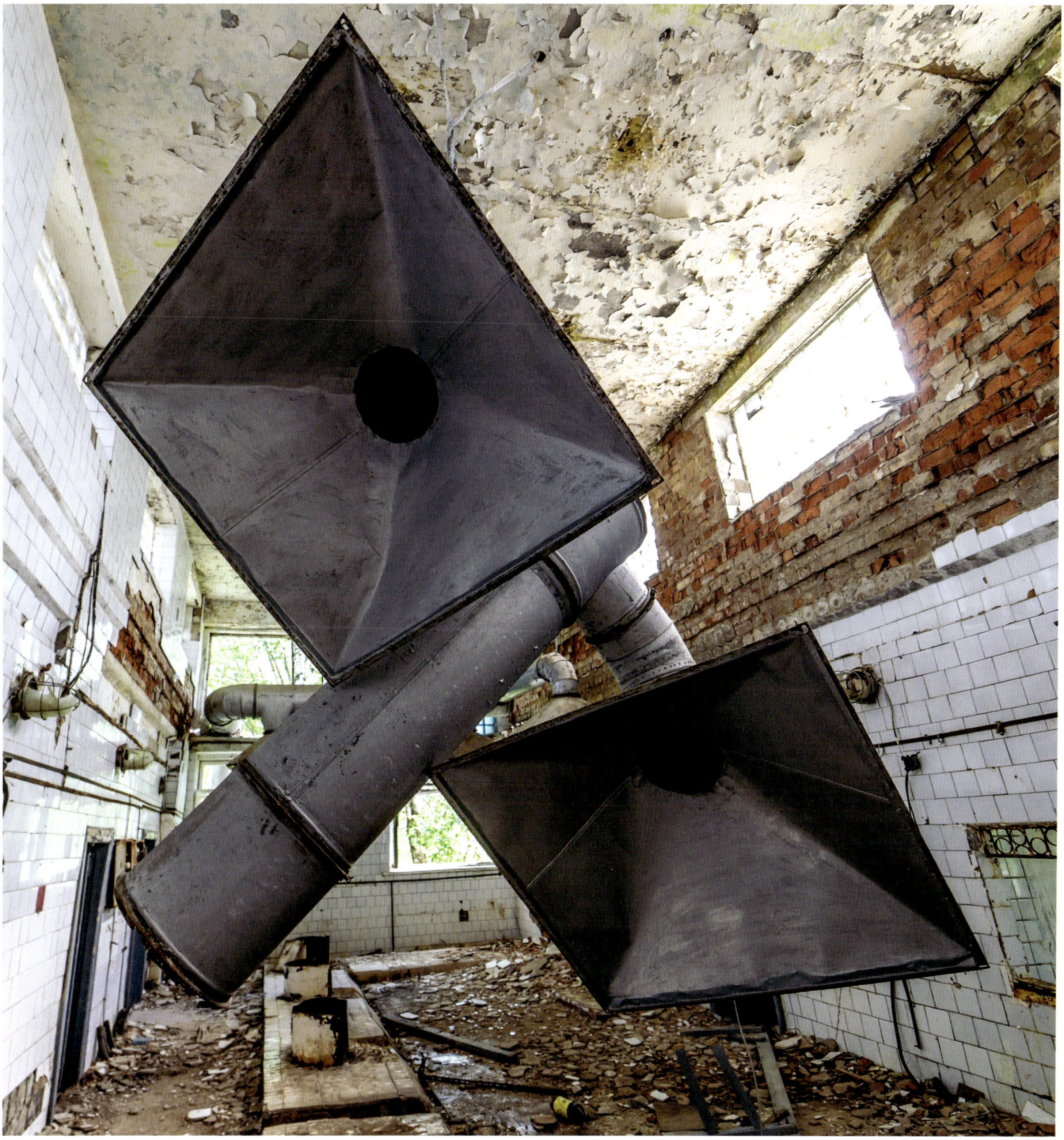

OPPOSITE:
Former R-12 Missile Launch Site, Zeltini, Vidzeme, Latvia
The Zeltini missile site was operational between 1962 and 1984, with eight missiles ready to launch from two surface pads. The standard R-12 warhead was 2.3 megatons. The missile could remain in a fully fuelled state ready to launch for up to one month, with a 30-minute launch preparation time.

ABOVE:
Former Kitchen, Skrunda, Latvia
What appears to be a large extractor fan system lies across this derelict canteen kitchen in Skrunda, a former Dnepr radar base. This 1960s site was a vital part of the Soviet Union's early warning system for tracking ICBM missiles launched from Western Europe. The site finally closed in 1998.

RIGHT:
Former Nuclear Warhead Storage, Plokštine, Estonia
This bunker was built at the Plokštine missile base that became operational at the end of 1962. The base had four R12U missiles ready to launch in silos that had been dug out by hand, with another eight warheads stored in this bunker should further launches be required. The missiles were removed in 1978.

BELOW:
Former Soviet Bunker, Lithuania
A blast door in this disused bunker lies ajar. The wheel in the centre operated the four locking bolts on each corner of the door. The rubber seals that surrounded the door have now perished.

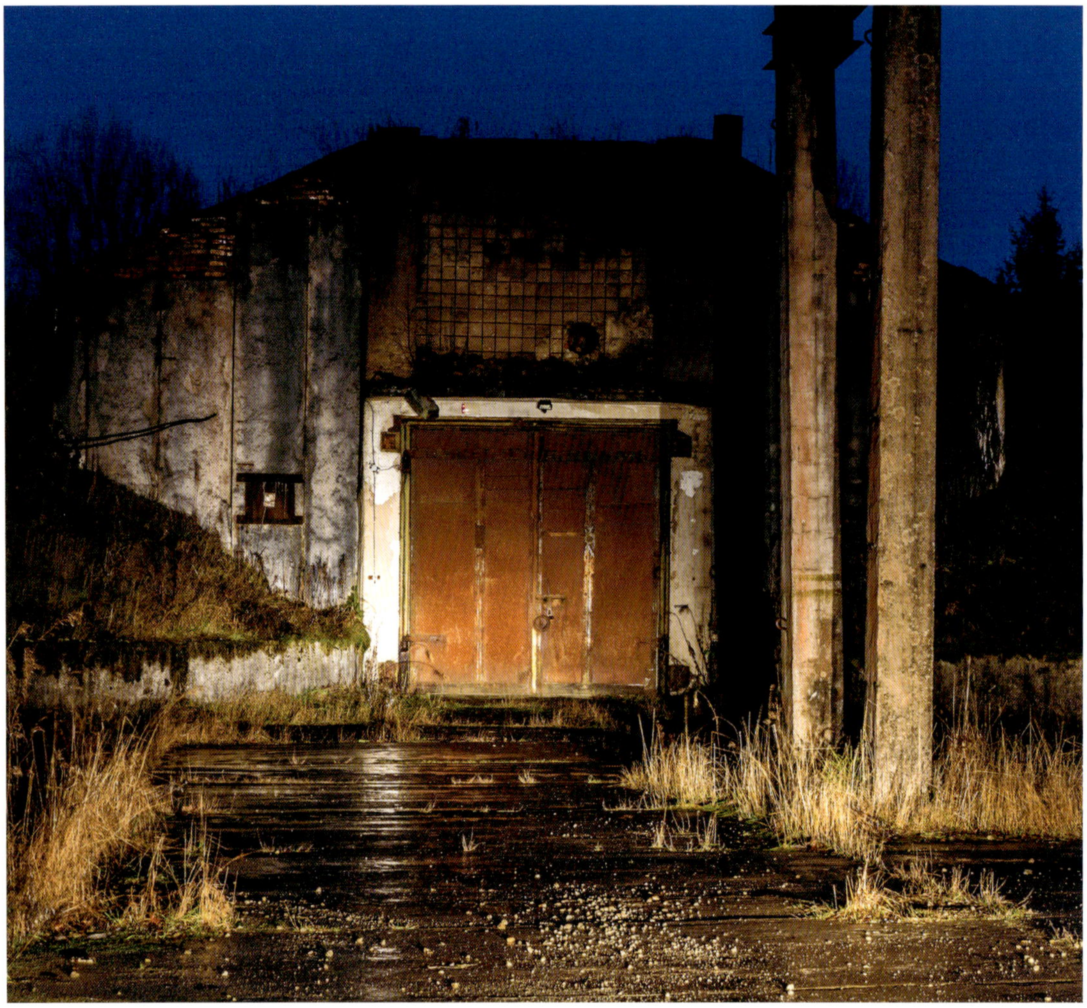

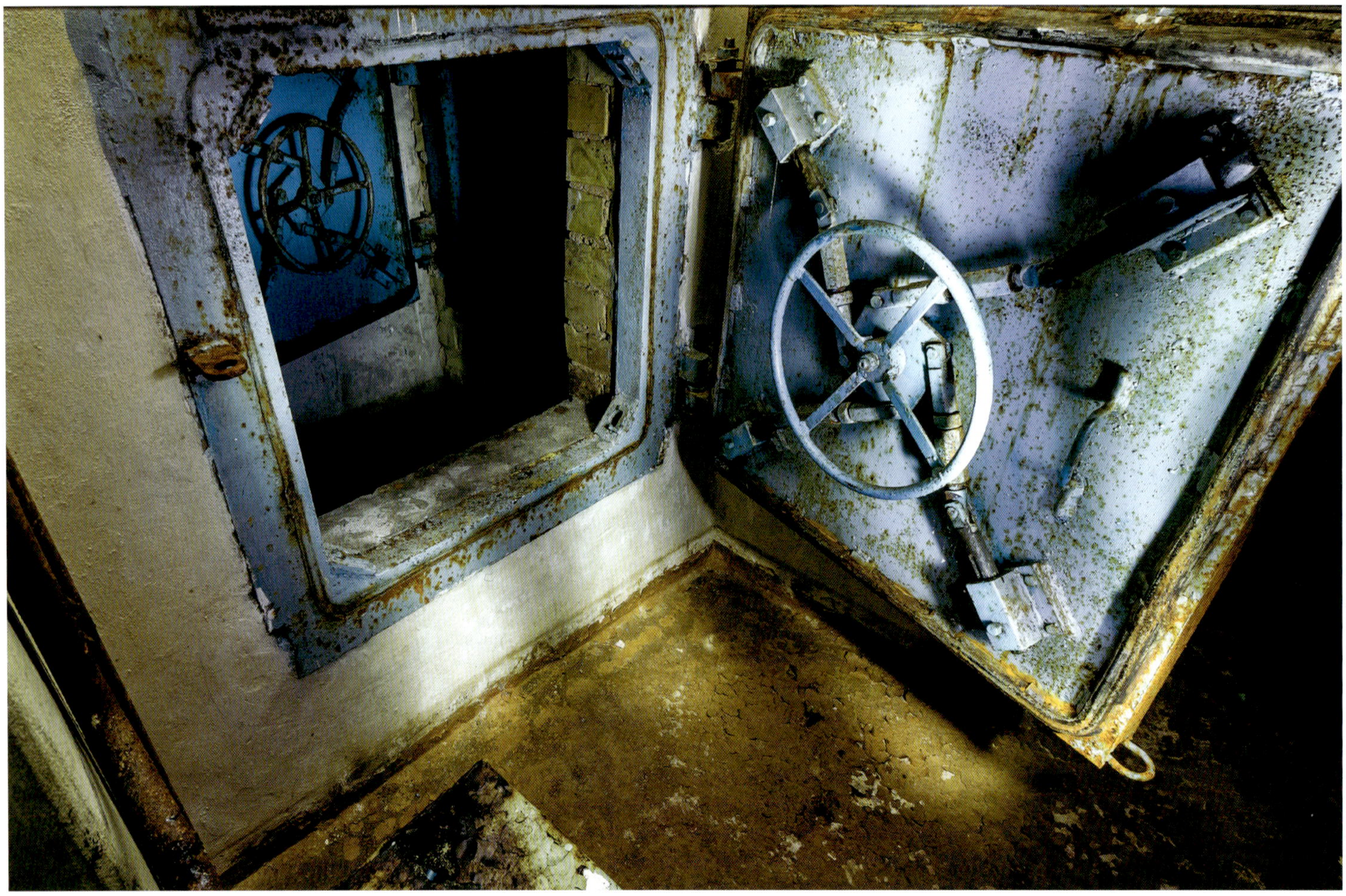

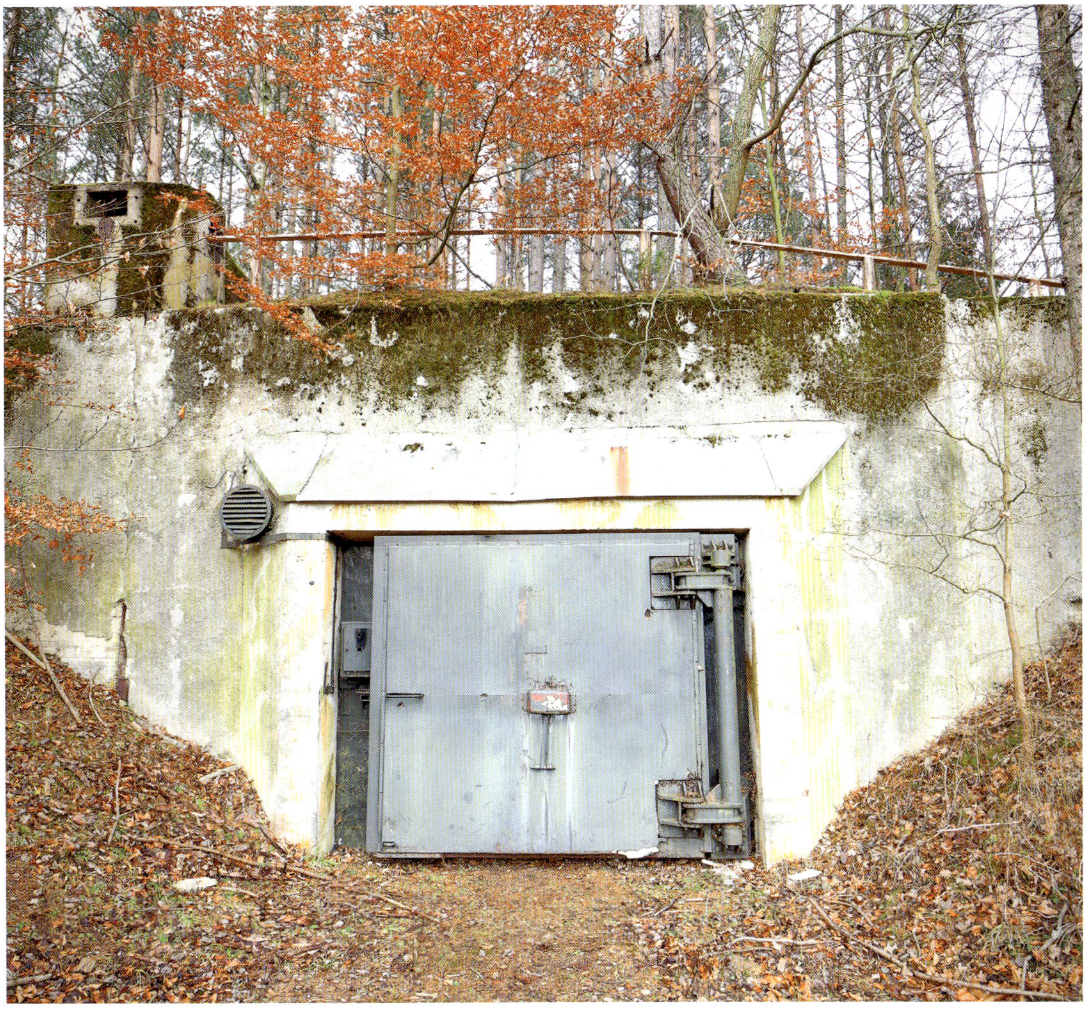

LEFT AND BELOW:
Former Soviet Bunker, Podborsko, Poland
Buried deep in the forest near the village of Podborsko near the Baltic coast is this former Soviet bunker used to store nuclear weapons. Although paid for by the Polish government, this facility was built and manned entirely by Soviet personnel and their families. This blast door is big enough to permit vehicular access to the bunker.

The existence of the nuclear base on Polish soil only became generally known after the Soviet withdrawal in 1991. The loading ramp (pictured below) inside the former Soviet bunker had a crane to lift the warheads from their transport vehicles.

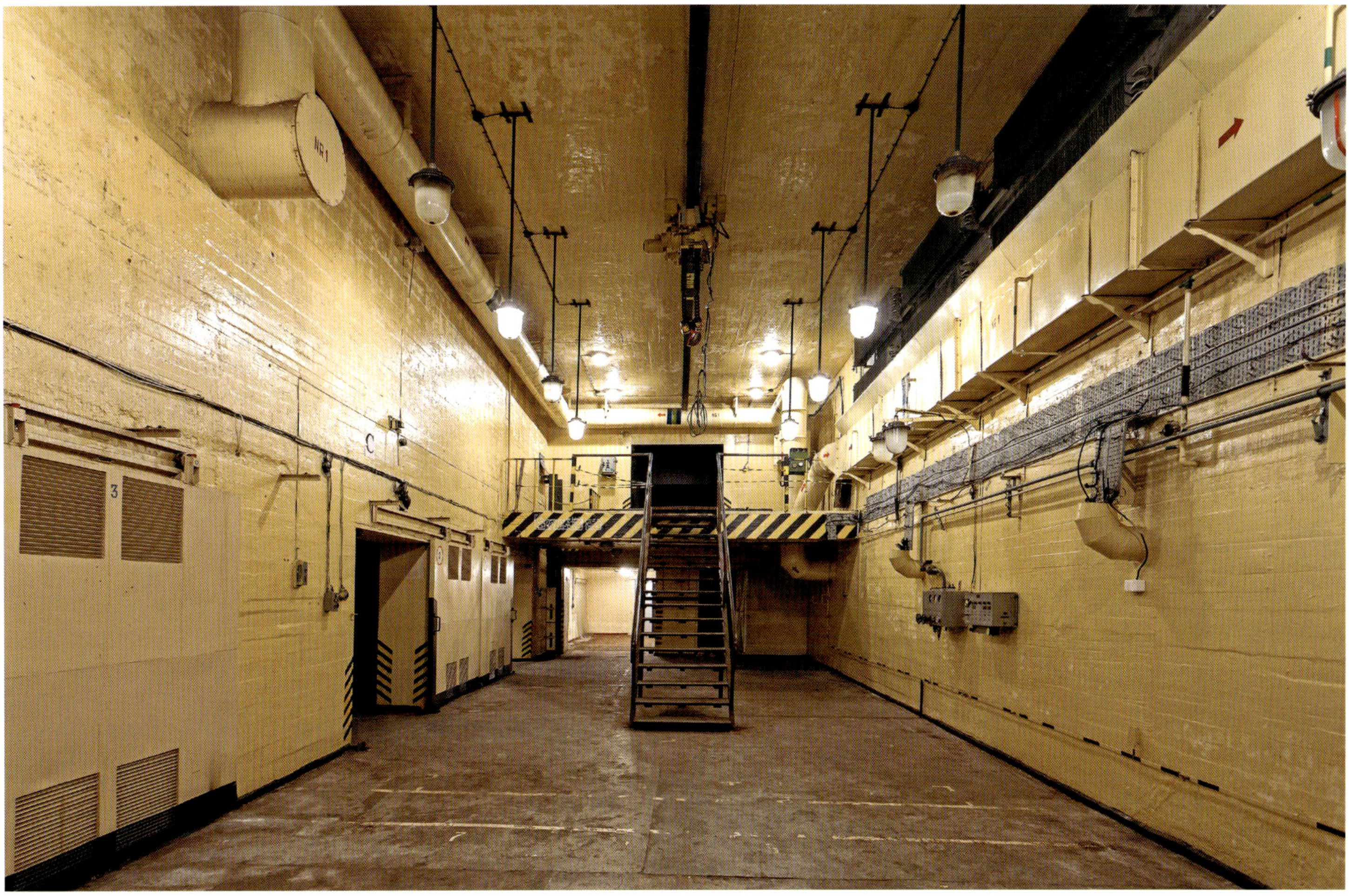

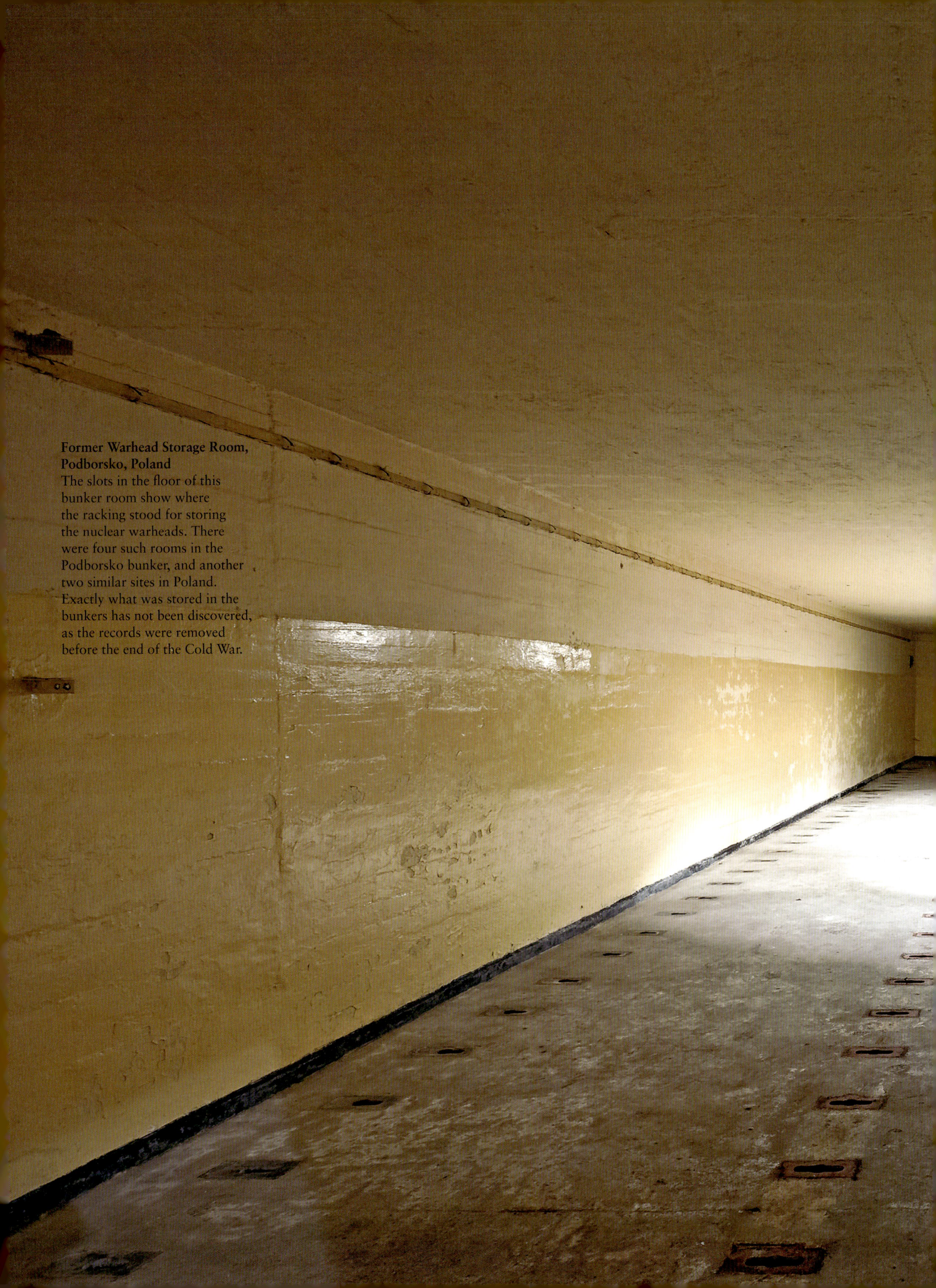

Former Warhead Storage Room, Podborsko, Poland
The slots in the floor of this bunker room show where the racking stood for storing the nuclear warheads. There were four such rooms in the Podborsko bunker, and another two similar sites in Poland. Exactly what was stored in the bunkers has not been discovered, as the records were removed before the end of the Cold War.

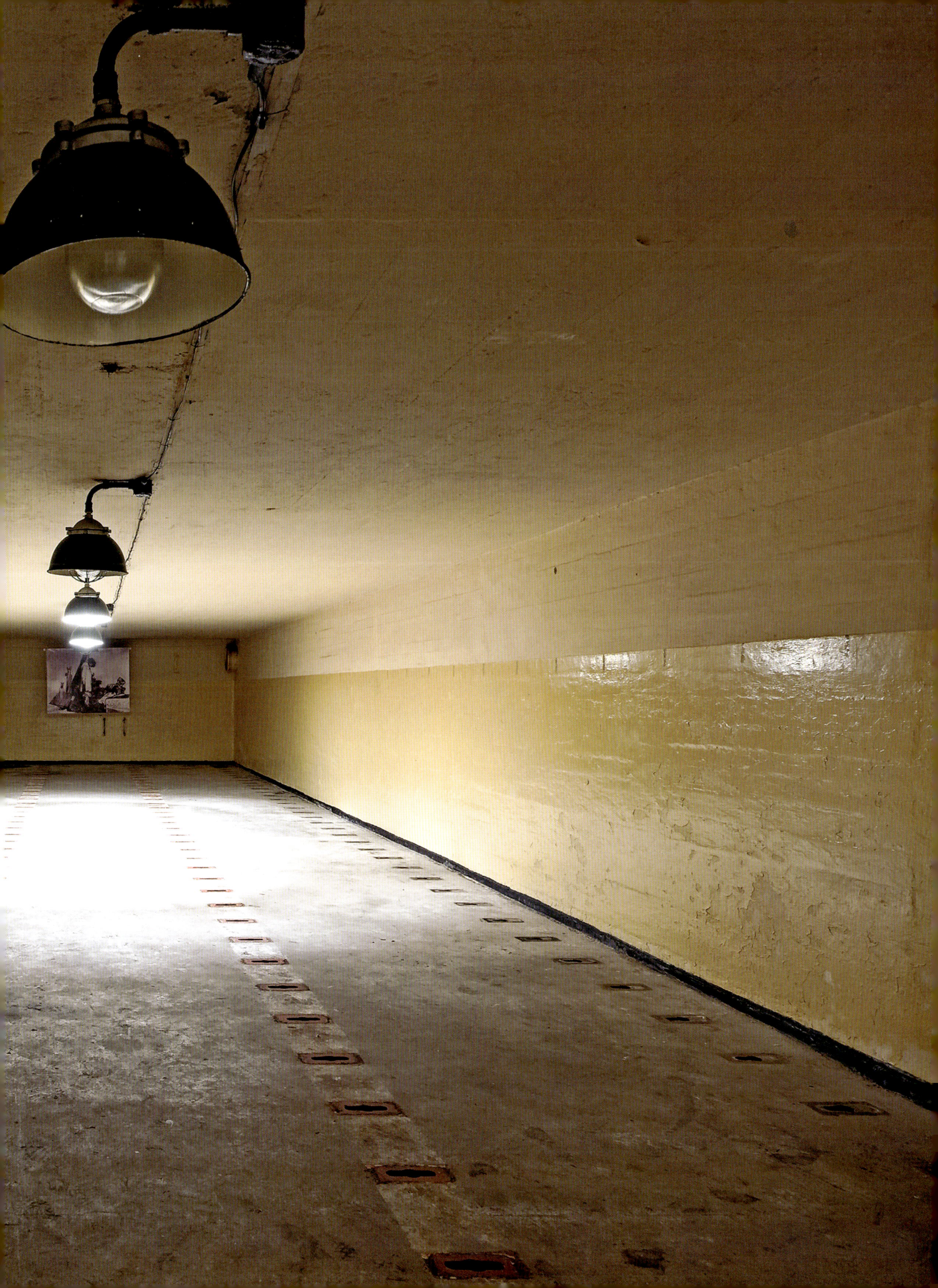

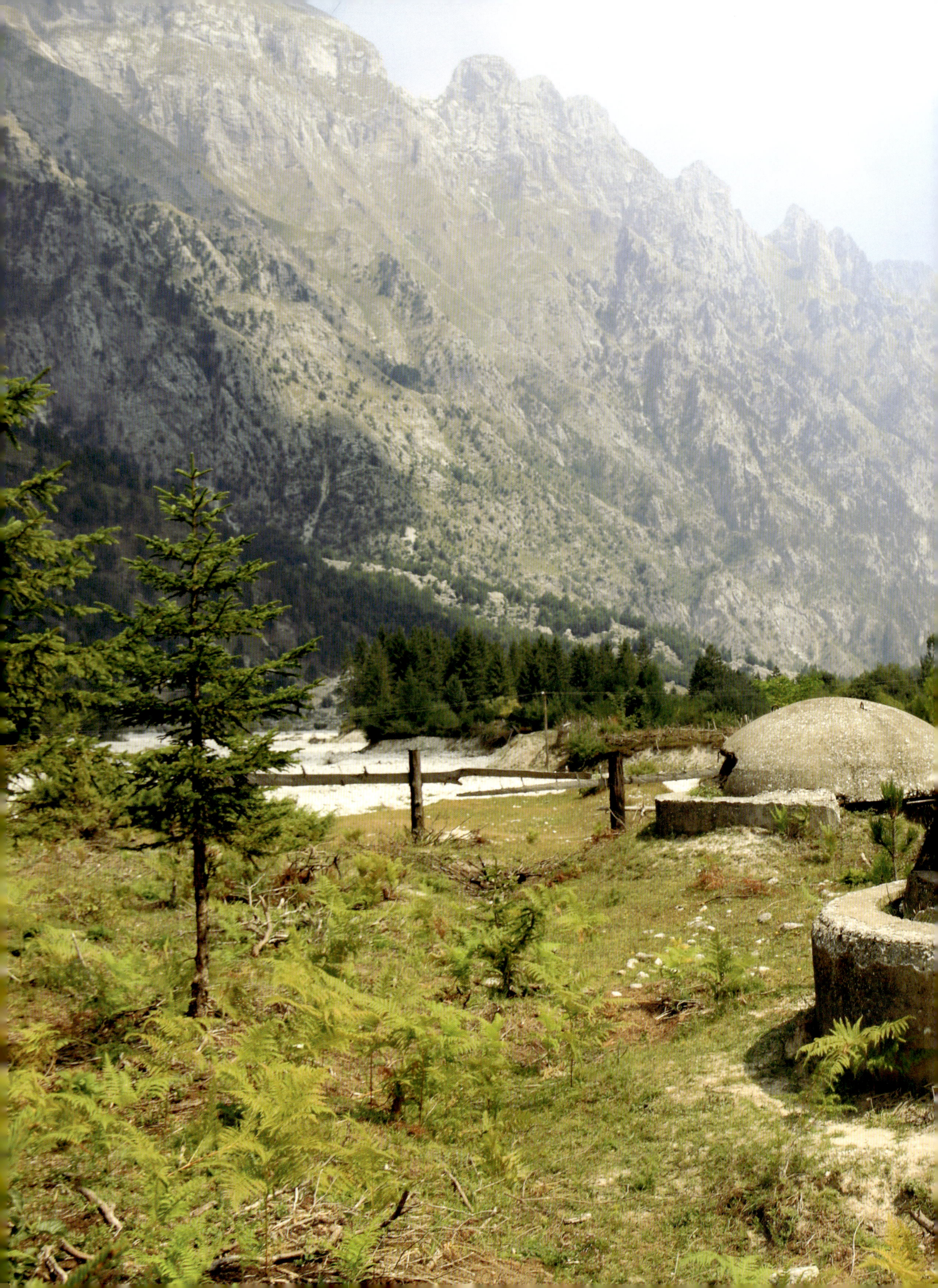

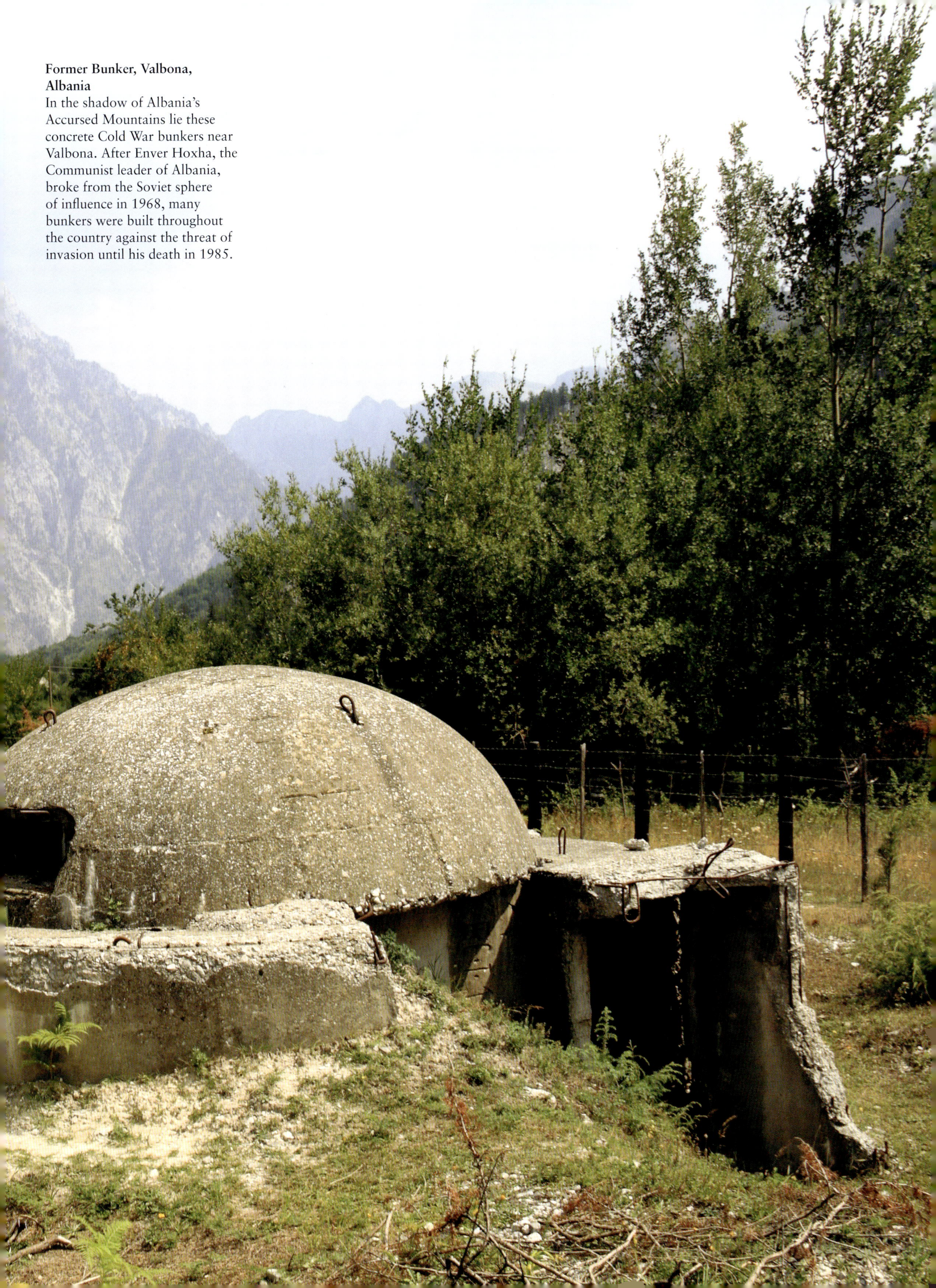

Former Bunker, Valbona, Albania
In the shadow of Albania's Accursed Mountains lie these concrete Cold War bunkers near Valbona. After Enver Hoxha, the Communist leader of Albania, broke from the Soviet sphere of influence in 1968, many bunkers were built throughout the country against the threat of invasion until his death in 1985.

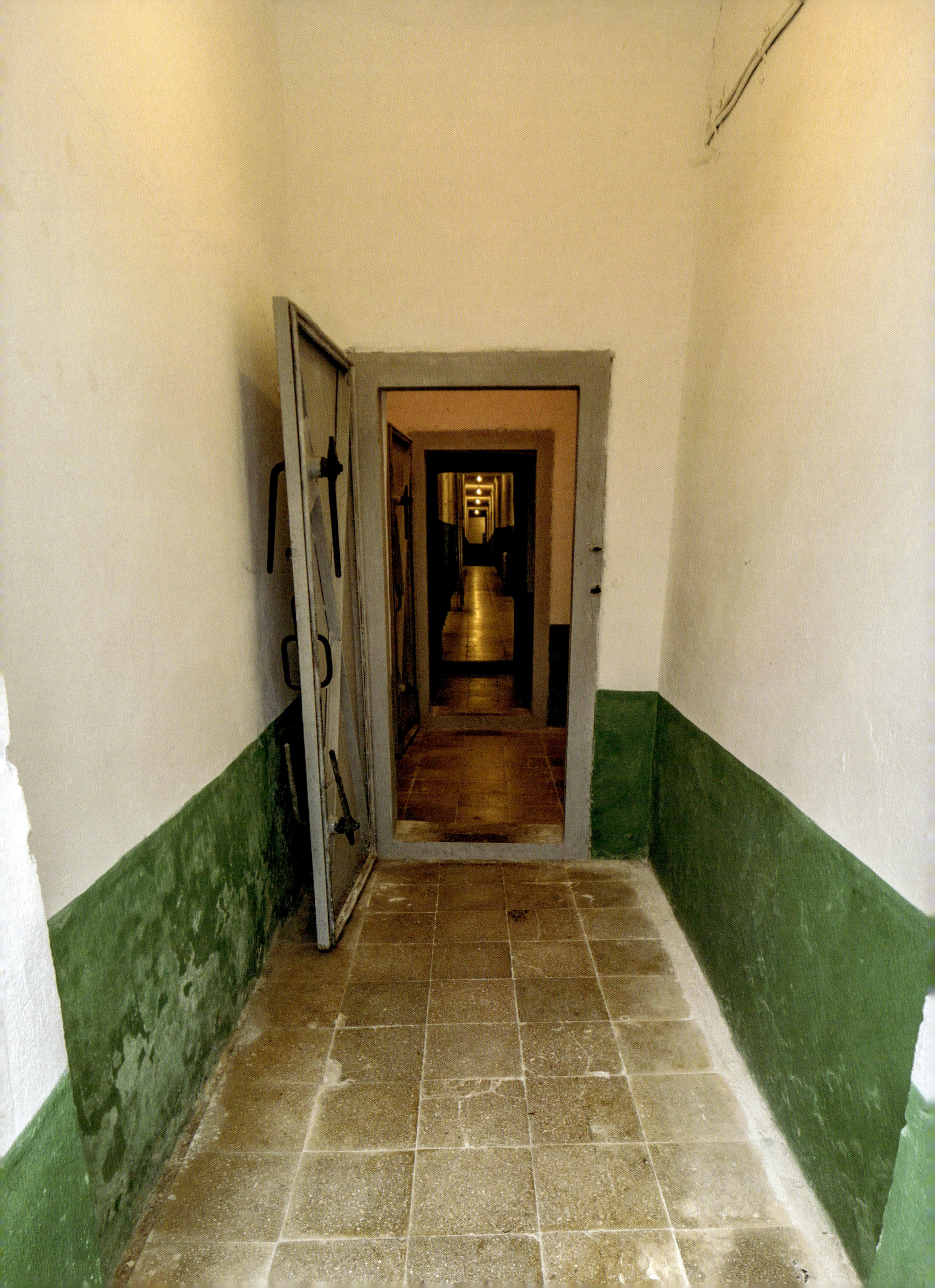

OPPOSITE:
Former Nuclear Bunker, Tirana, Albania
Under the streets of Tirana, the Albanian capital, lies this five-storey former nuclear shelter built for Hoxha himself and his senior government. With tiled floors rather than the usual spartan concrete, it nevertheless still has the steel blast doors in place to protect occupants against nuclear or chemical attack.

BELOW:
ARK D-0 Bunker, Konjic, Bosnia
When first built, this mirrored floor would have been impressive to visitors entering this bunker built for Marshal Tito and the highest echelons of the Yugoslav government, although it is thought that Tito never visited the site. The *Atomska Ratna Komanda* (Atomic War Command) was known as ARK D-0.

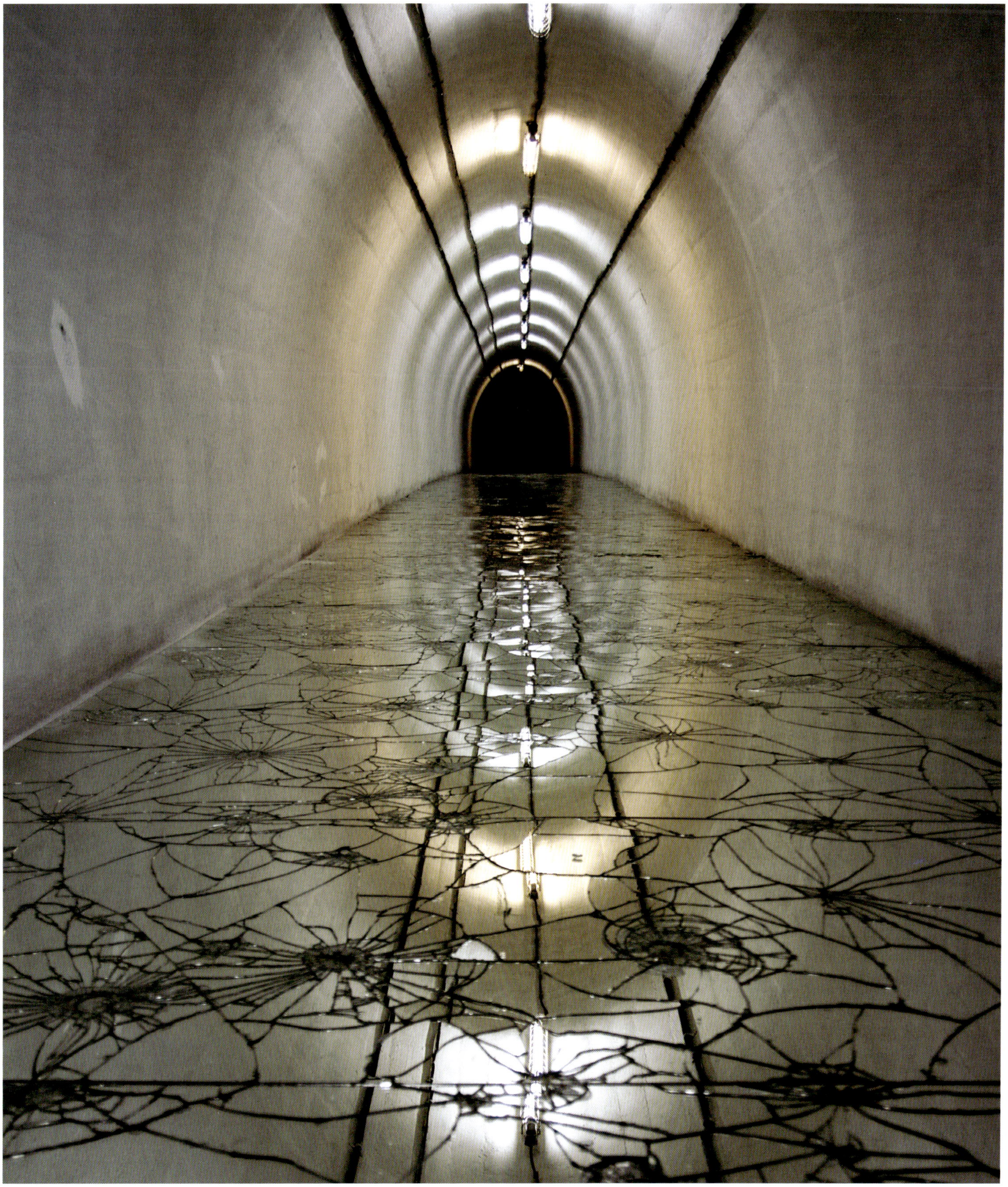

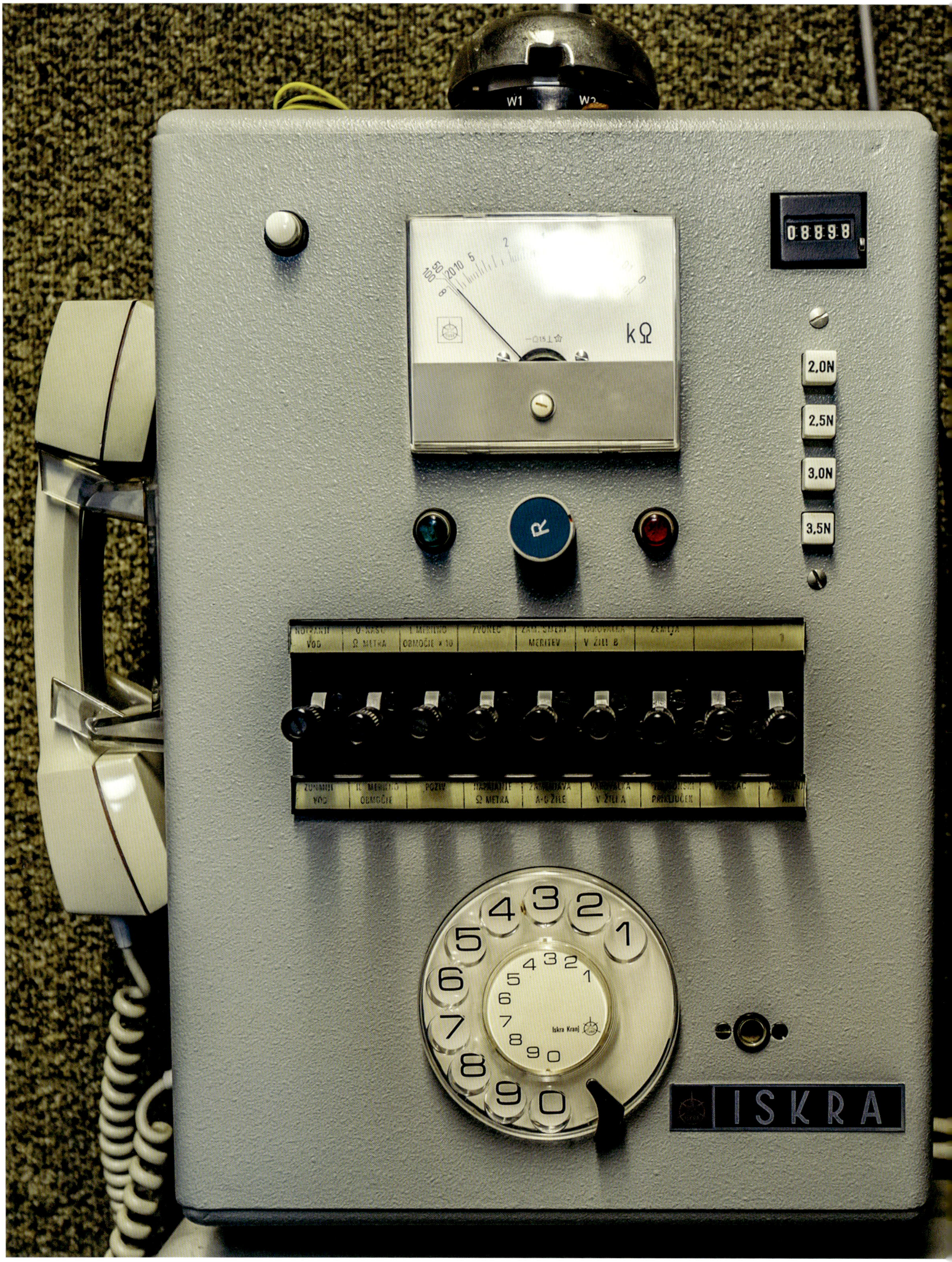
W1
08898
kΩ
2,0N
2,5N
3,0N
3,5N
R
ISKRA
Iskra Kranj

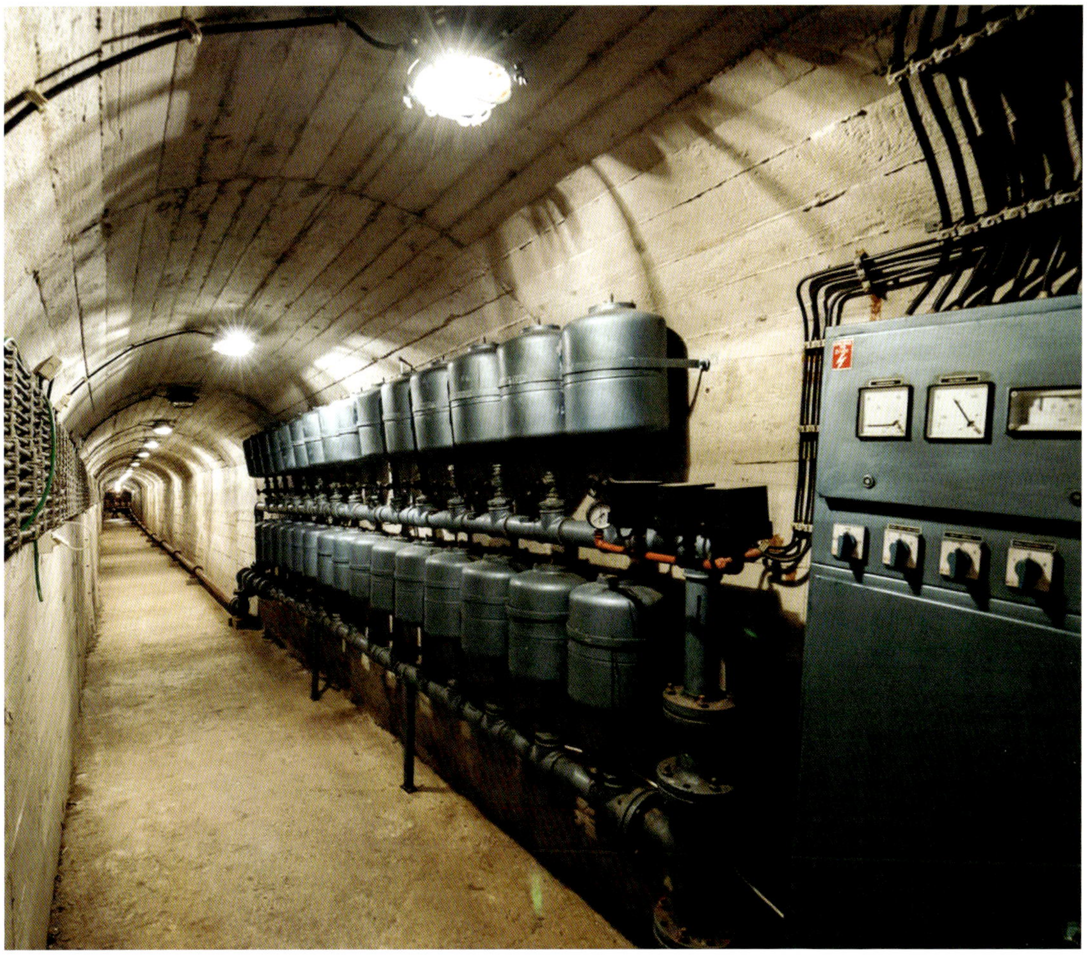

OPPOSITE:
Telecommunications Unit, ARK D-0 Bunker, Konjic, Bosnia
Tito ordered the secret construction of the ARK in what is now Bosnia in 1953. Dug out of a mountain, construction continued until 1979. The entrance was concealed behind what appeared to be an ordinary house. The occupants had supplies to last up to six months, with communications enabling contact with the outside world.

LEFT:
ARK D-0 Bunker, Konjic, Bosnia
The pipework in the tunnel appears to be a system for maintaining pressure. The bunker was constructed in a horseshoe shape, and cost an estimated $4.6 billion. Due to the need to maintain secrecy, workers were escorted to the site in blindfolds.

BELOW:
ARK D-0 Bunker, Konjic, Bosnia
Yellow ventilation ducts with arrows indicate the direction of airflow that runs through the walls and floors of the bunker.

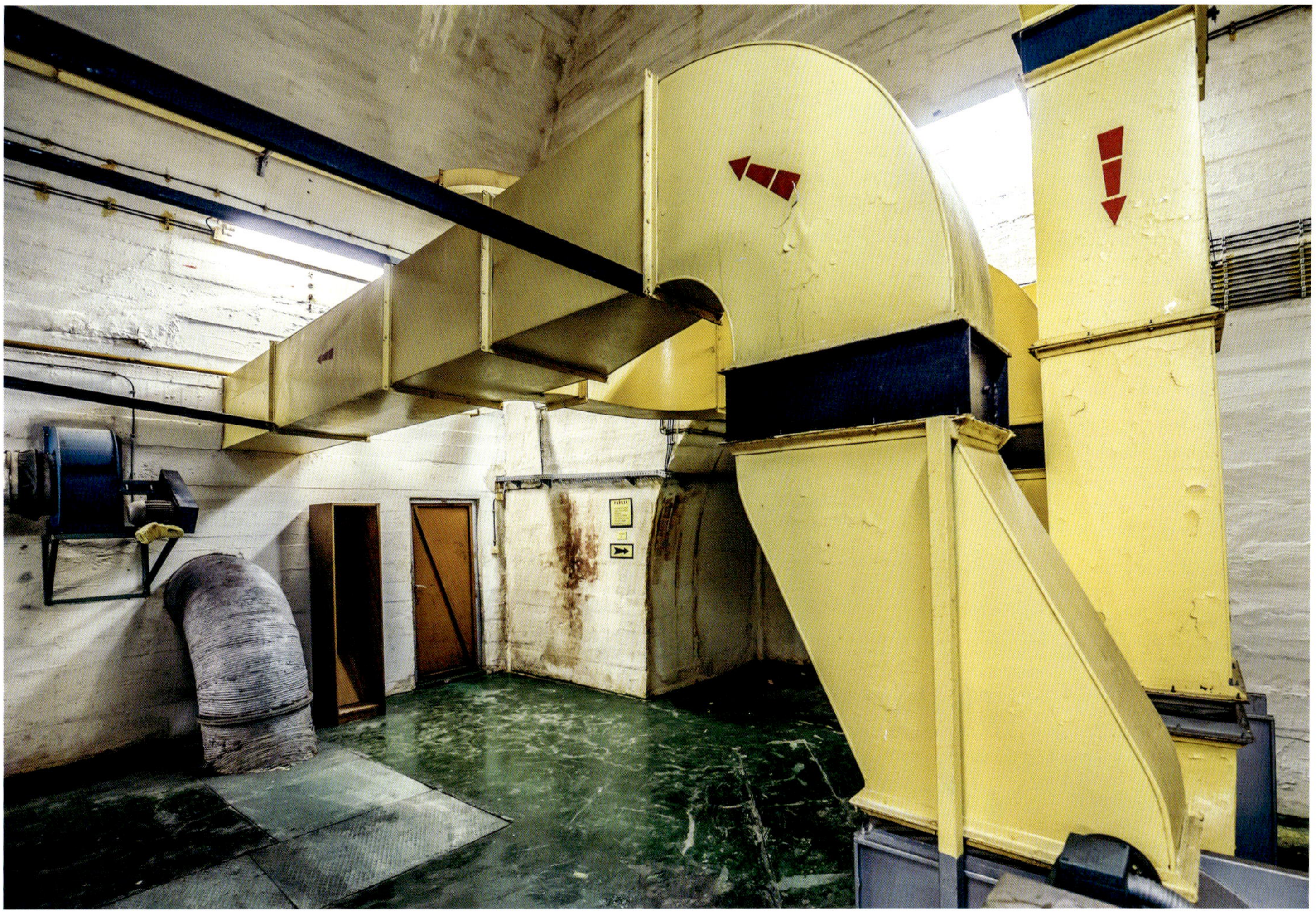

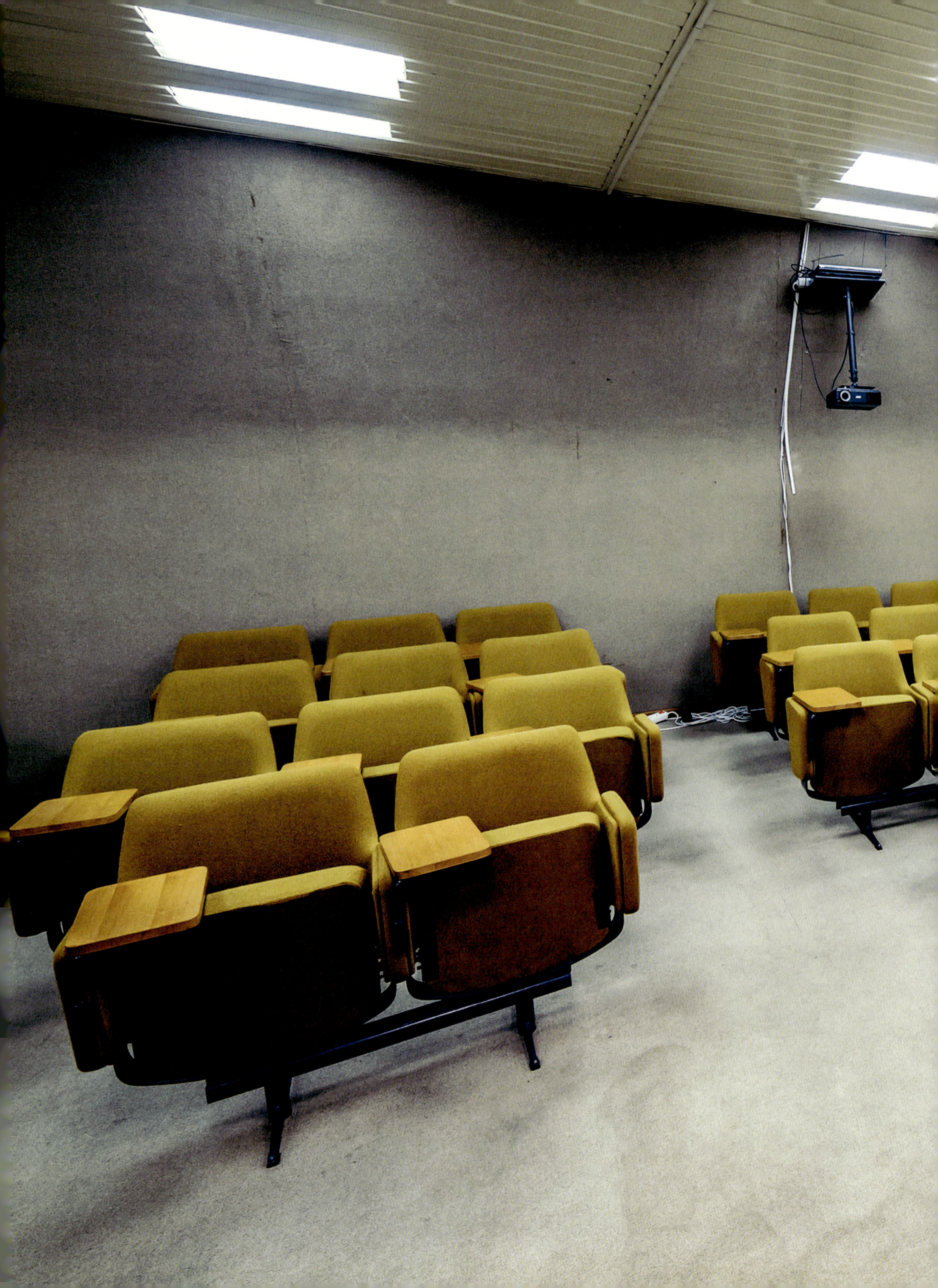

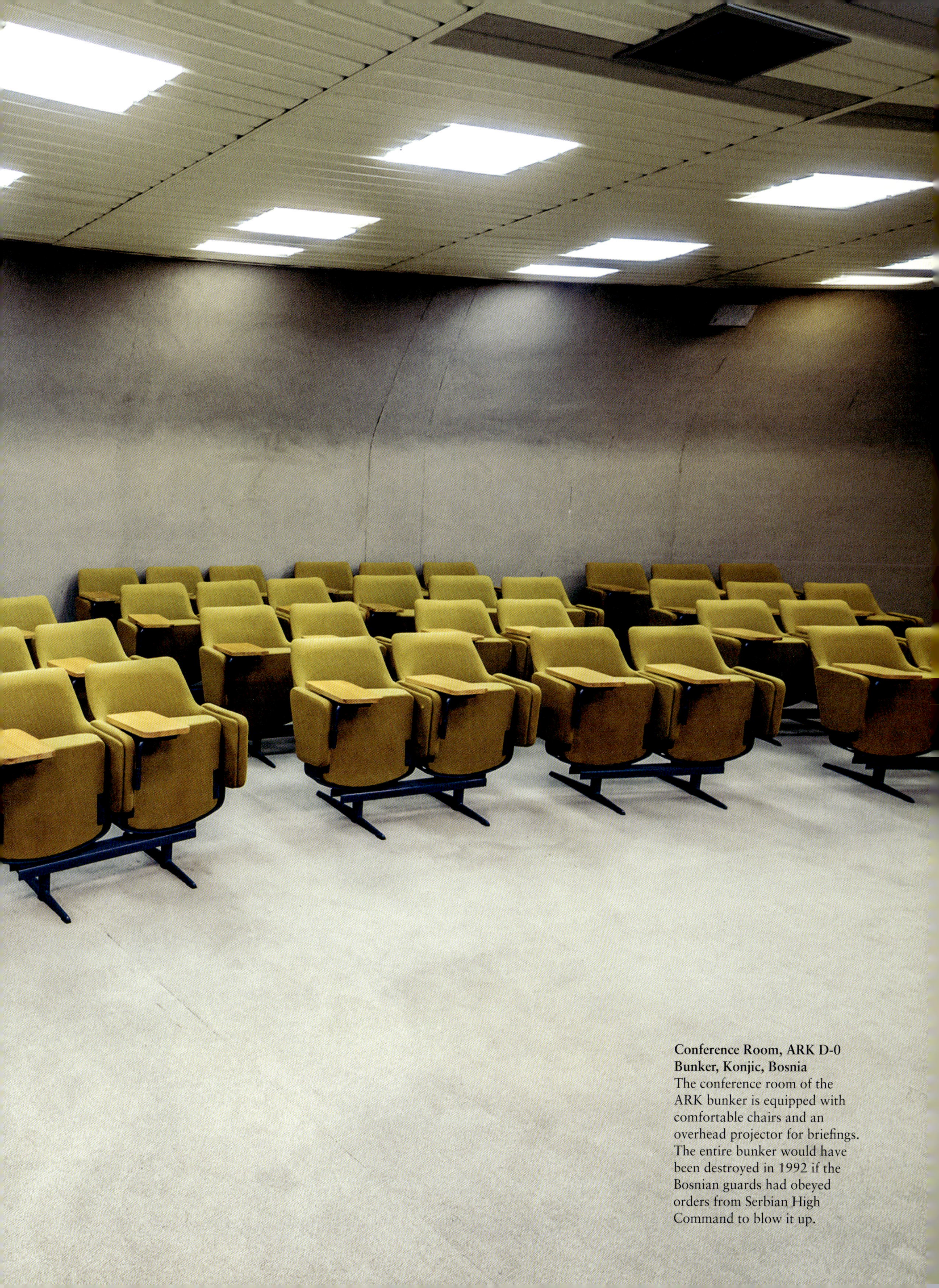

Conference Room, ARK D-0 Bunker, Konjic, Bosnia
The conference room of the ARK bunker is equipped with comfortable chairs and an overhead projector for briefings. The entire bunker would have been destroyed in 1992 if the Bosnian guards had obeyed orders from Serbian High Command to blow it up.

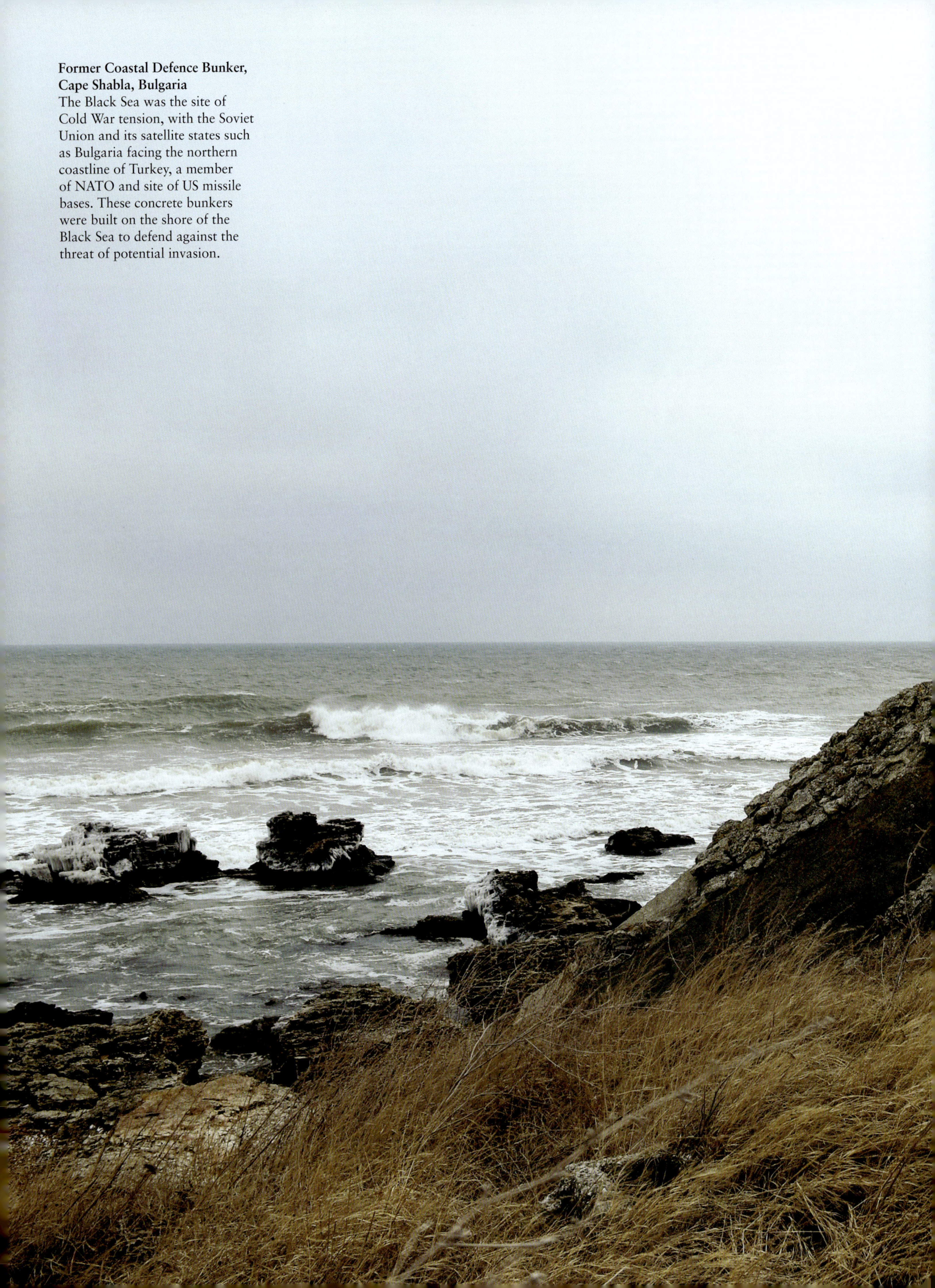

Former Coastal Defence Bunker, Cape Shabla, Bulgaria
The Black Sea was the site of Cold War tension, with the Soviet Union and its satellite states such as Bulgaria facing the northern coastline of Turkey, a member of NATO and site of US missile bases. These concrete bunkers were built on the shore of the Black Sea to defend against the threat of potential invasion.

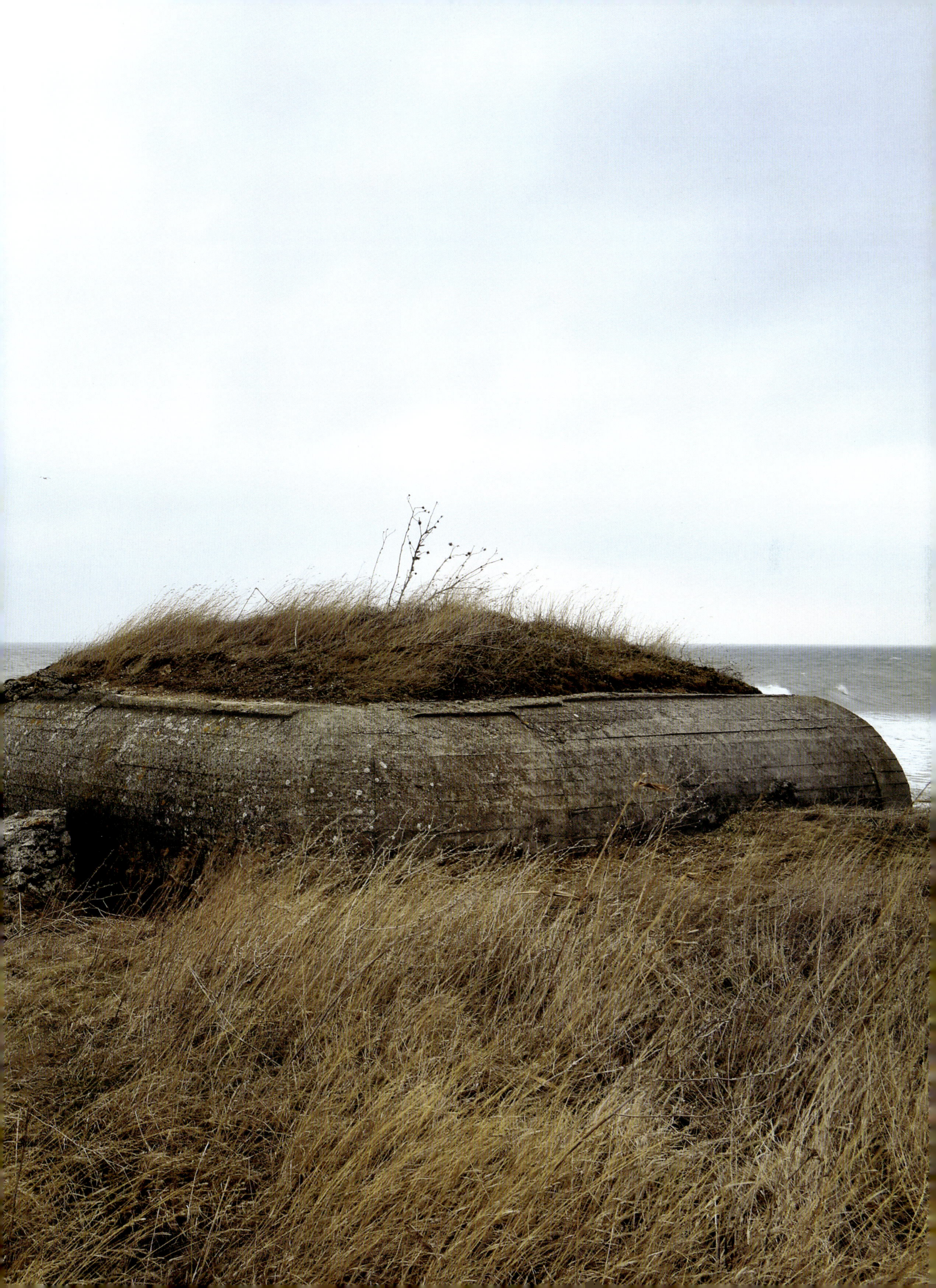

T-62 Tank Wreck, Bulgaria
Bulgaria's armed forces, like those of the other Warsaw Pact nations, operated Soviet-designed equipment, though usually a generation or two behind the types used by the Soviet forces themselves. Some weapons were licence-built in Warsaw Pact factories. Perhaps used for target practice, or as a result of mechanical failure, this T-62 tank has been long abandoned.

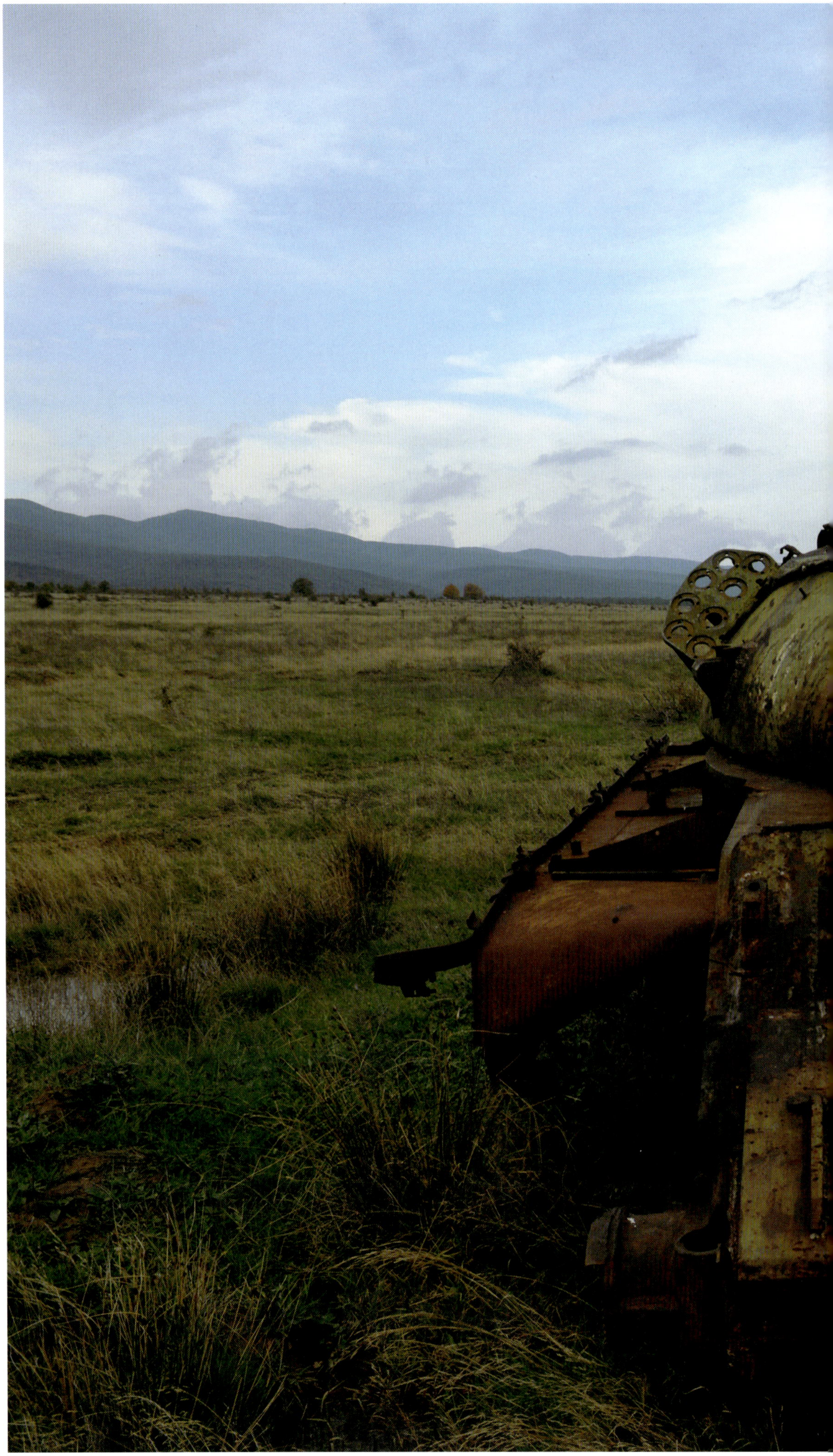

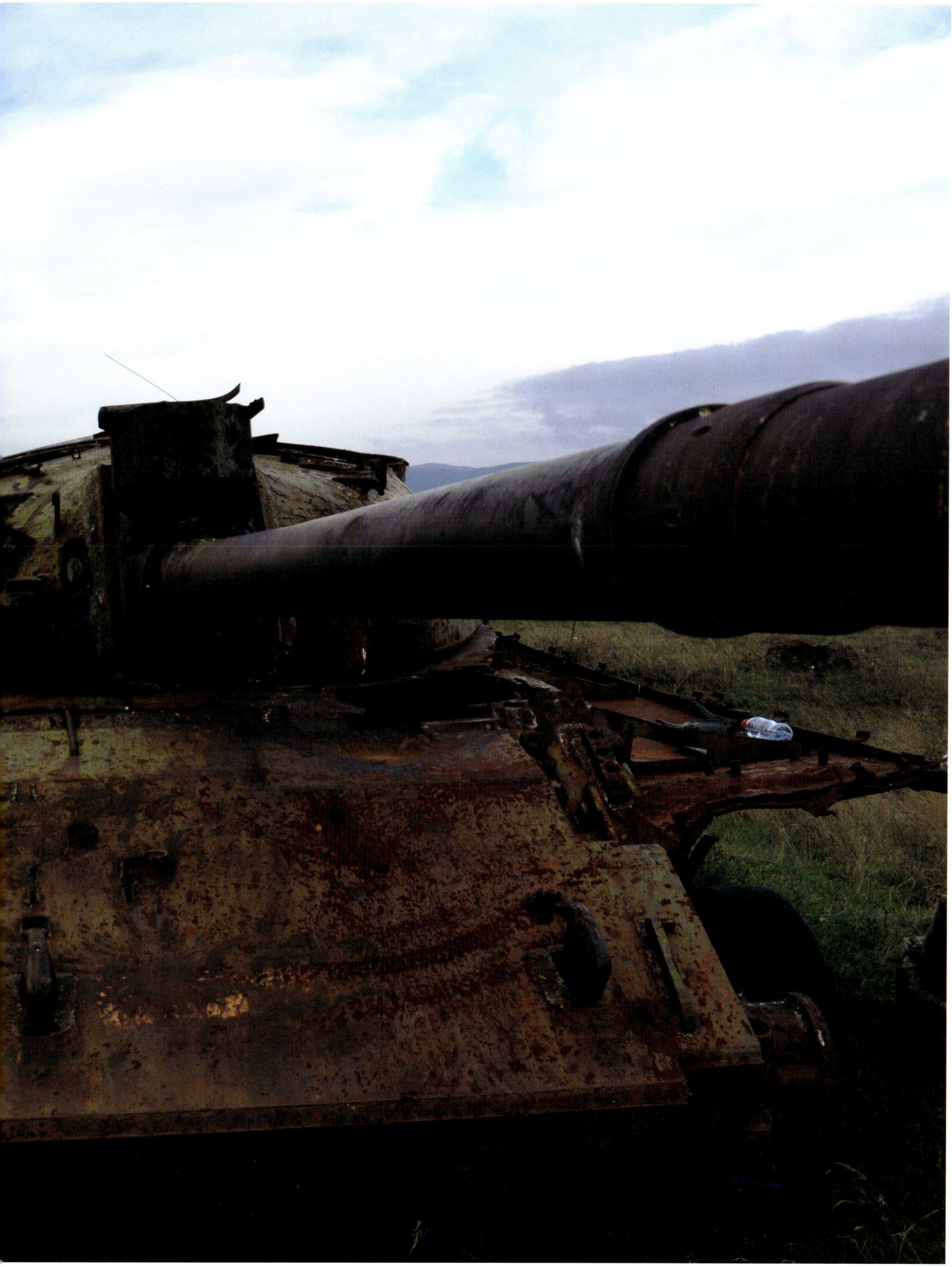

Former Submarine Pen, Vis, Croatia
A small island in the Adriatic Sea, Vis was for a few months during WWII the headquarters of Tito and his Partisans. After the war the island was developed into a major fortified military base, including submarine pens and nuclear bunkers for Tito and his general staff. The Yugoslav Army withdrew in 1992 and it is now a tourist site.

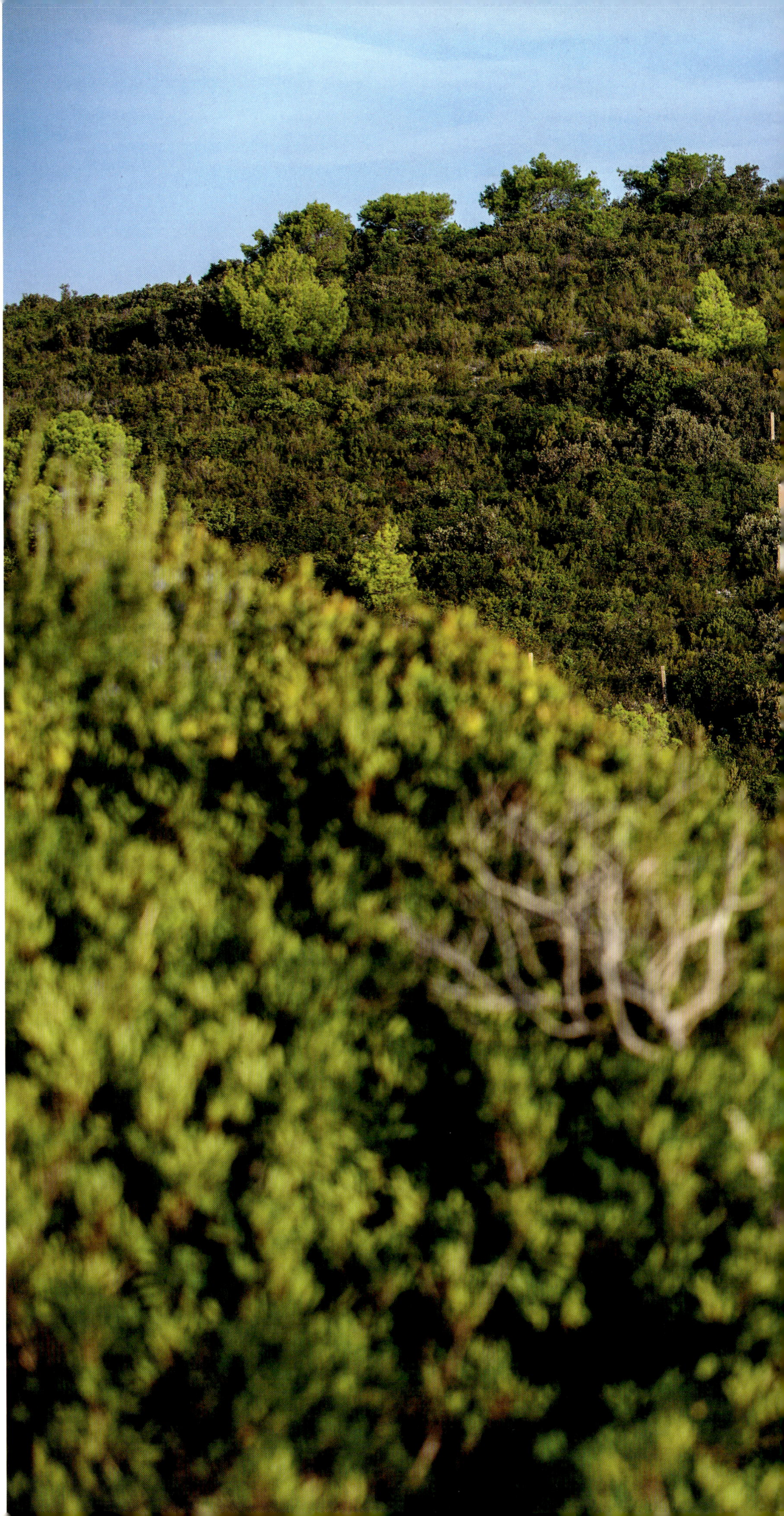

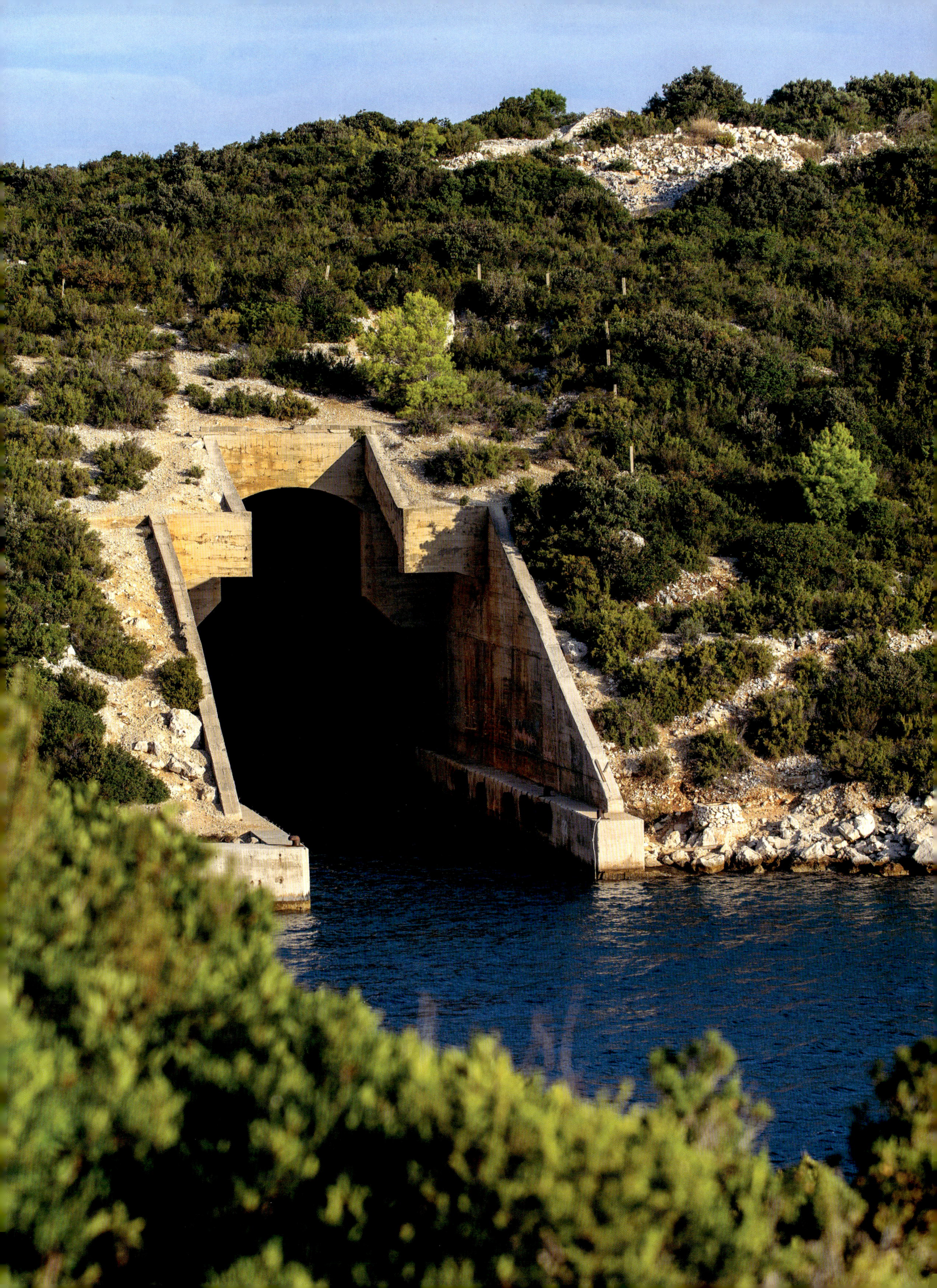

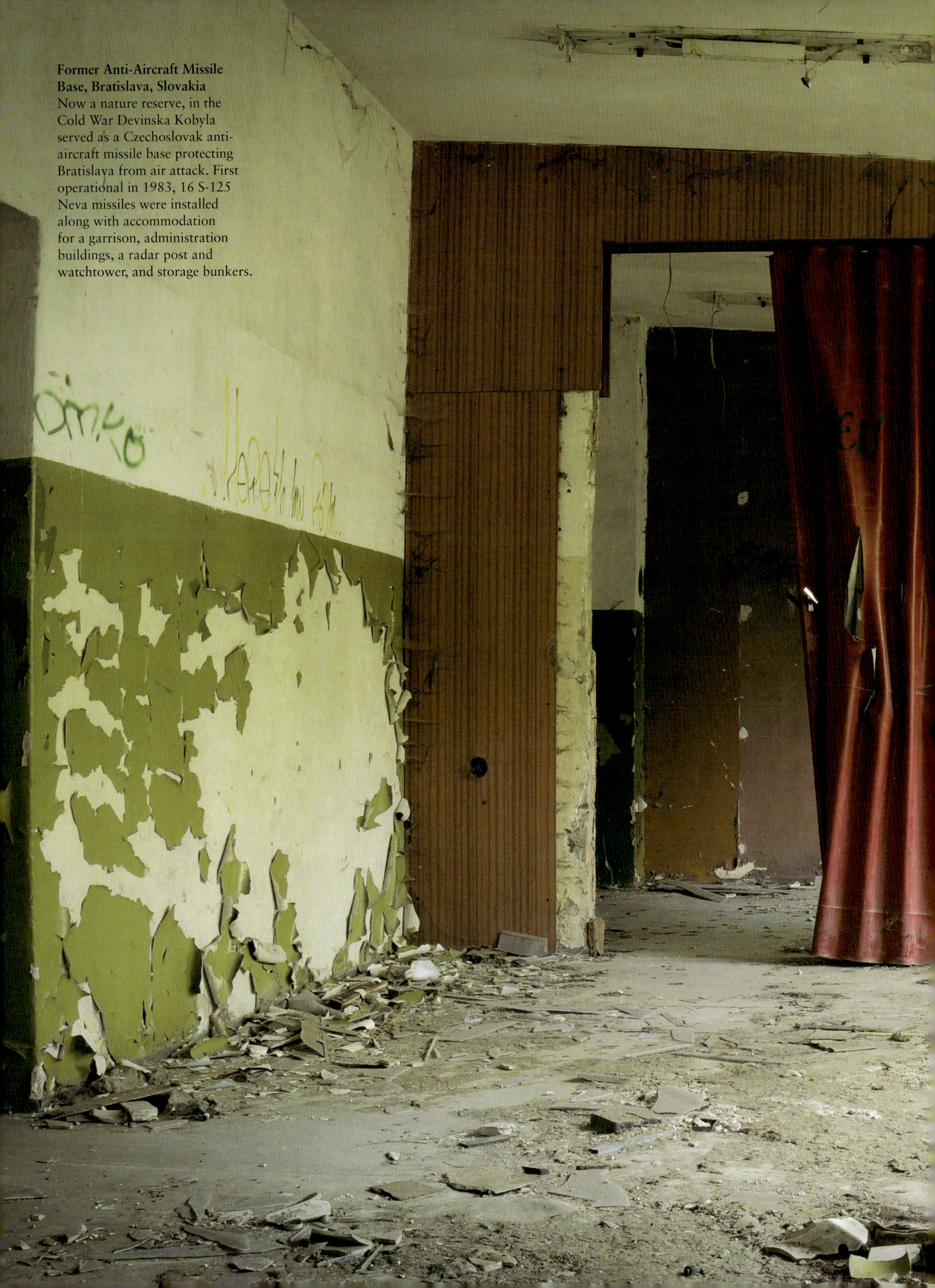

Former Anti-Aircraft Missile Base, Bratislava, Slovakia
Now a nature reserve, in the Cold War Devinska Kobyla served as a Czechoslovak anti-aircraft missile base protecting Bratislava from air attack. First operational in 1983, 16 S-125 Neva missiles were installed along with accommodation for a garrison, administration buildings, a radar post and watchtower, and storage bunkers.

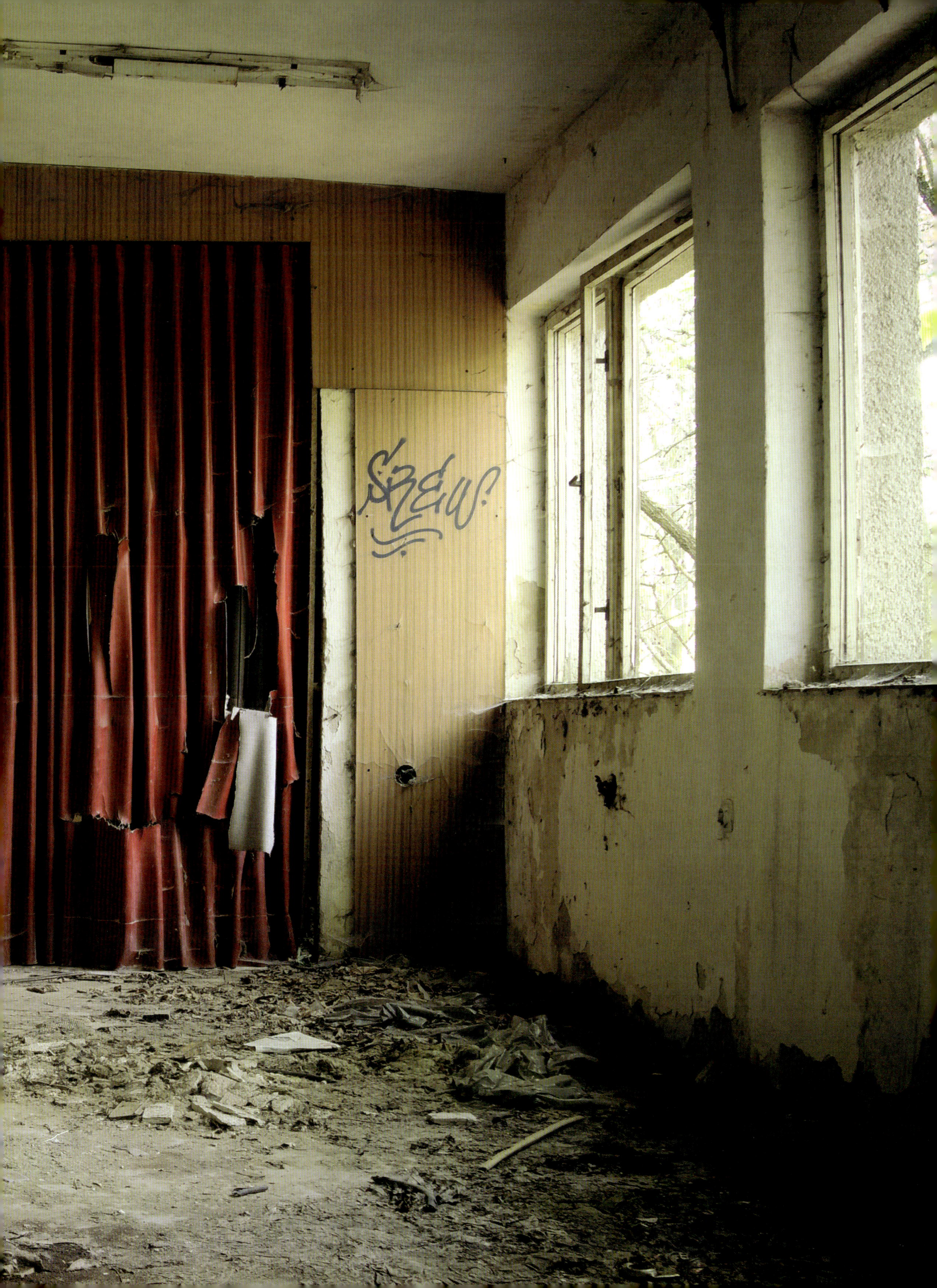

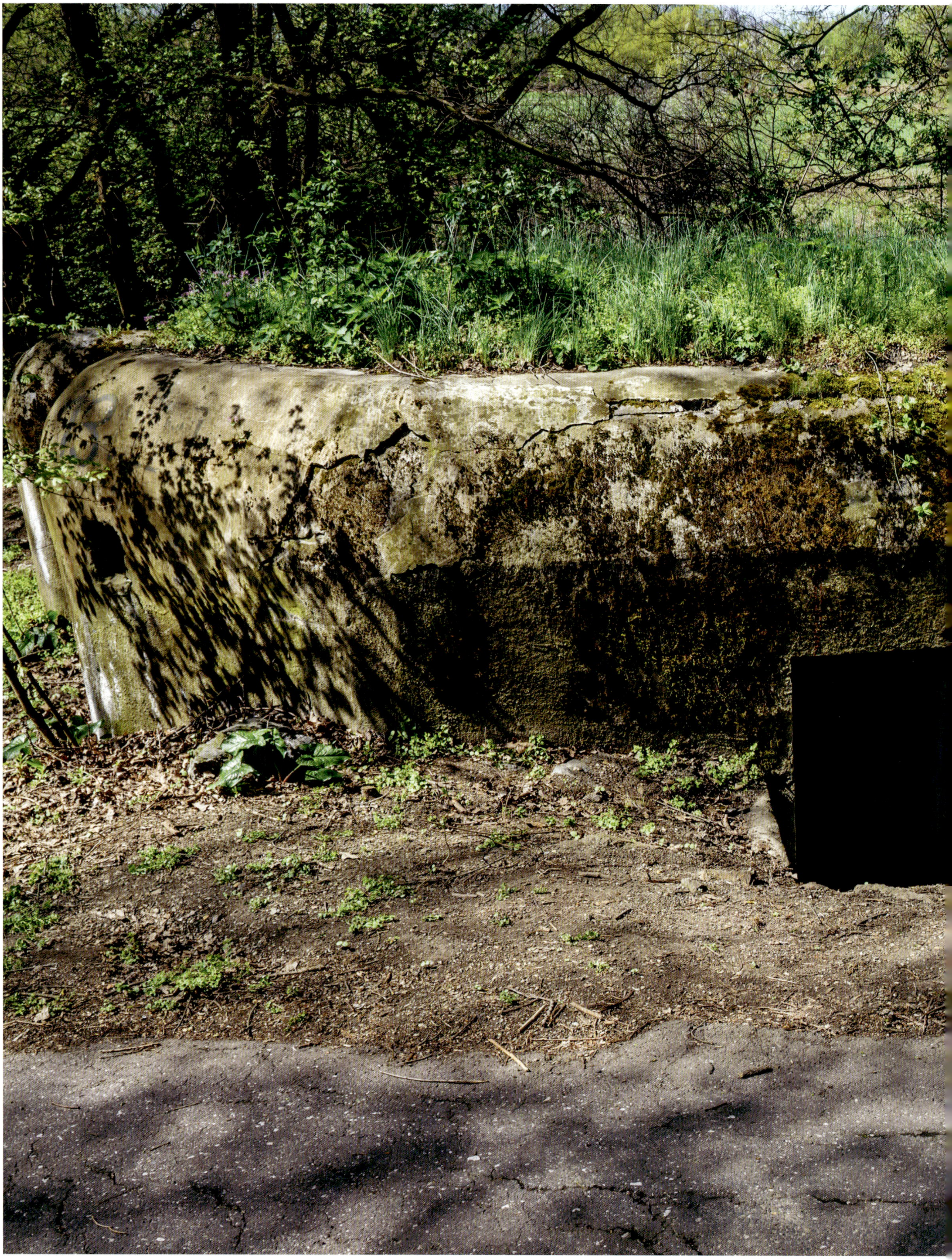

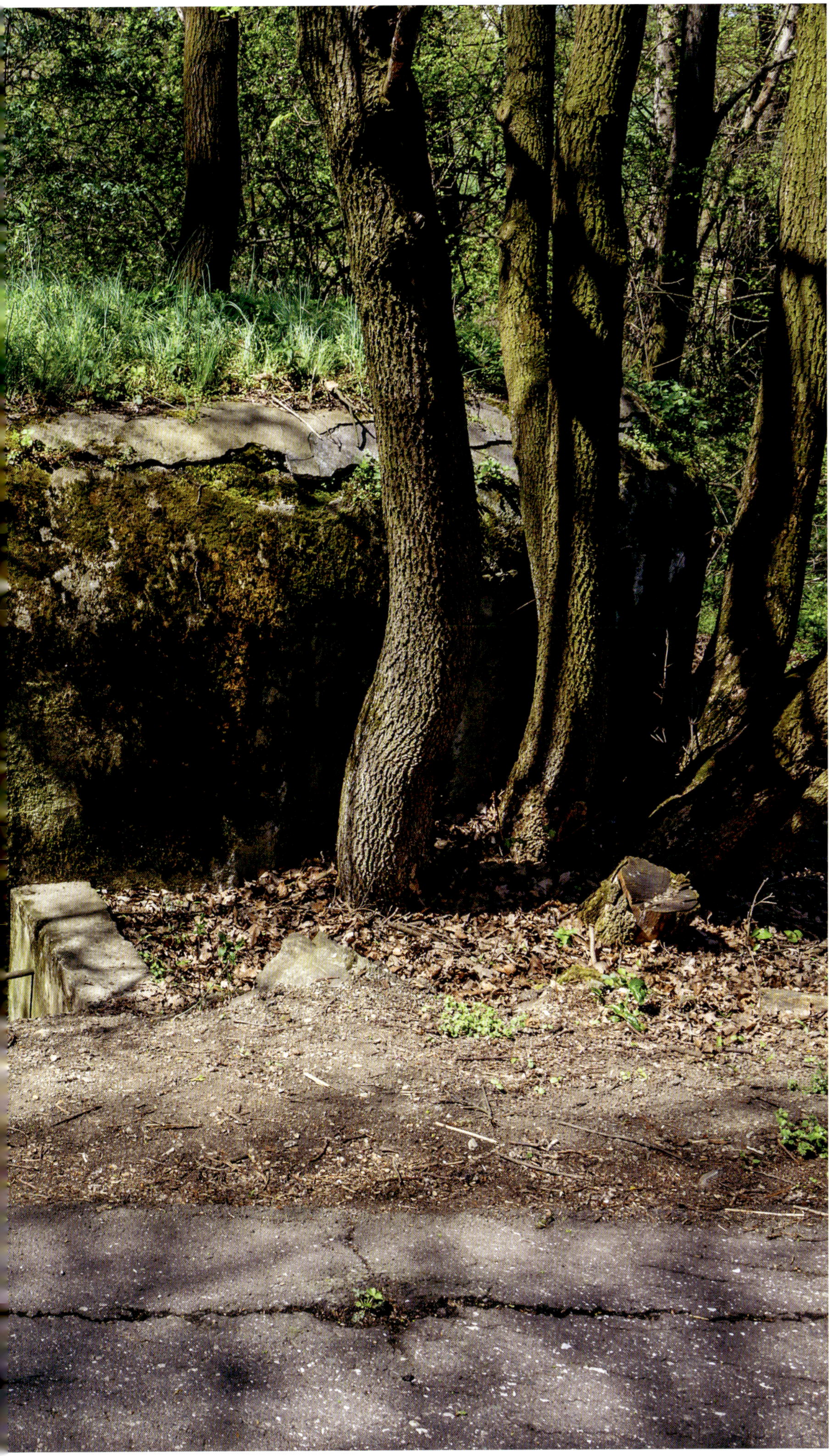

LEFT:
Former Border Bunker Post, Slovakia
The border between East and West, known as the Iron Curtain after Winston Churchill's 1946 US speech in Fulton, ran along the boundary between Austria and the former Czechoslovakia. Many of the concrete bunkers had been originally built to defend against Nazi Germany before World War II, but were adopted by the Warsaw Pact.

OVERLEAF:
Border Defences, Bucina, Czech Republic
This Cold War-era border between Germany and the former Czechoslovakia shows what the Iron Curtain looked like. Barbed wire, electrified fences, a strip of bare earth covering mines, and a line of pyramid-shaped concrete anti-tank traps are all illuminated and overseen by a tall watchtower.

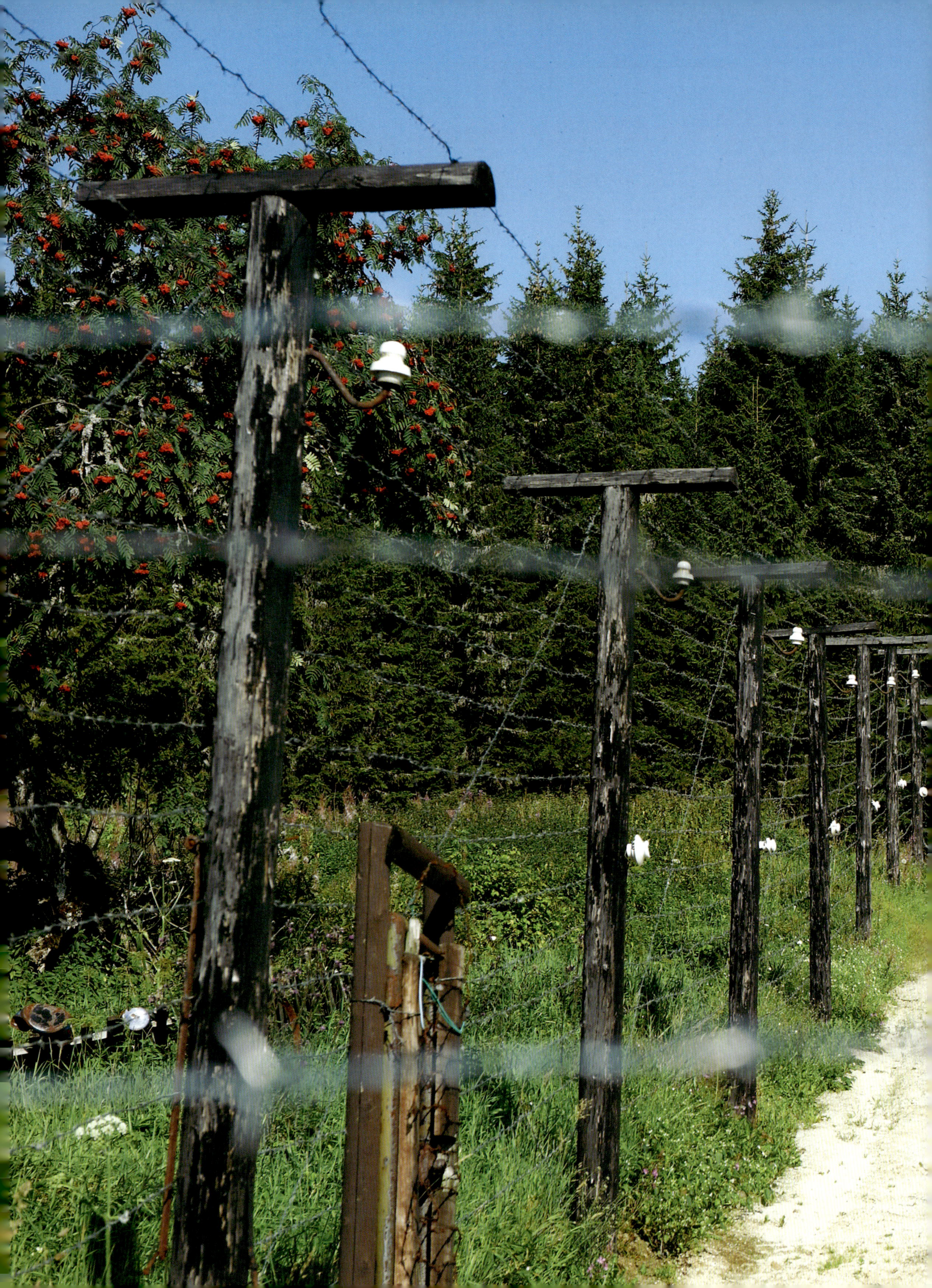

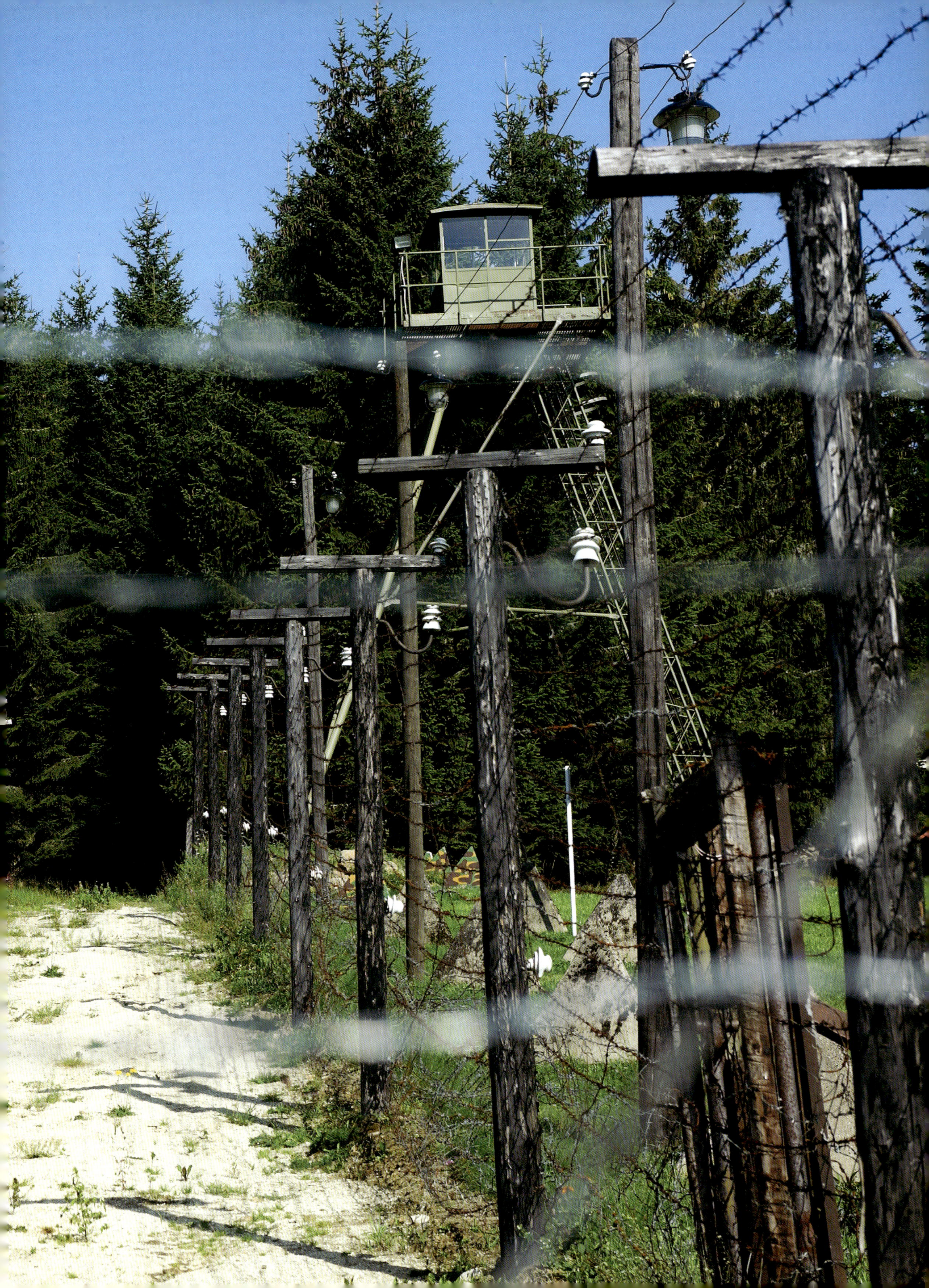

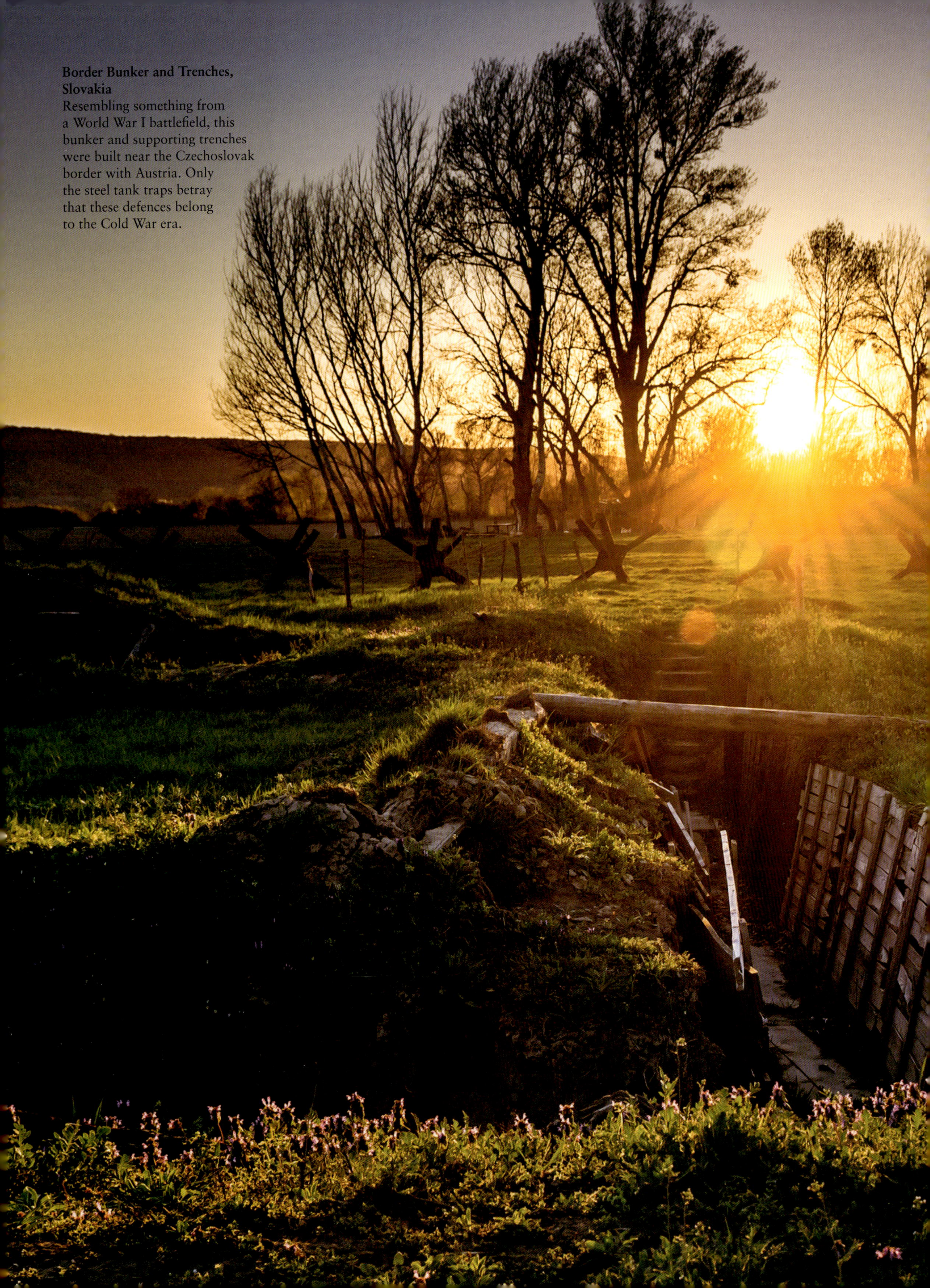

Border Bunker and Trenches, Slovakia
Resembling something from a World War I battlefield, this bunker and supporting trenches were built near the Czechoslovak border with Austria. Only the steel tank traps betray that these defences belong to the Cold War era.

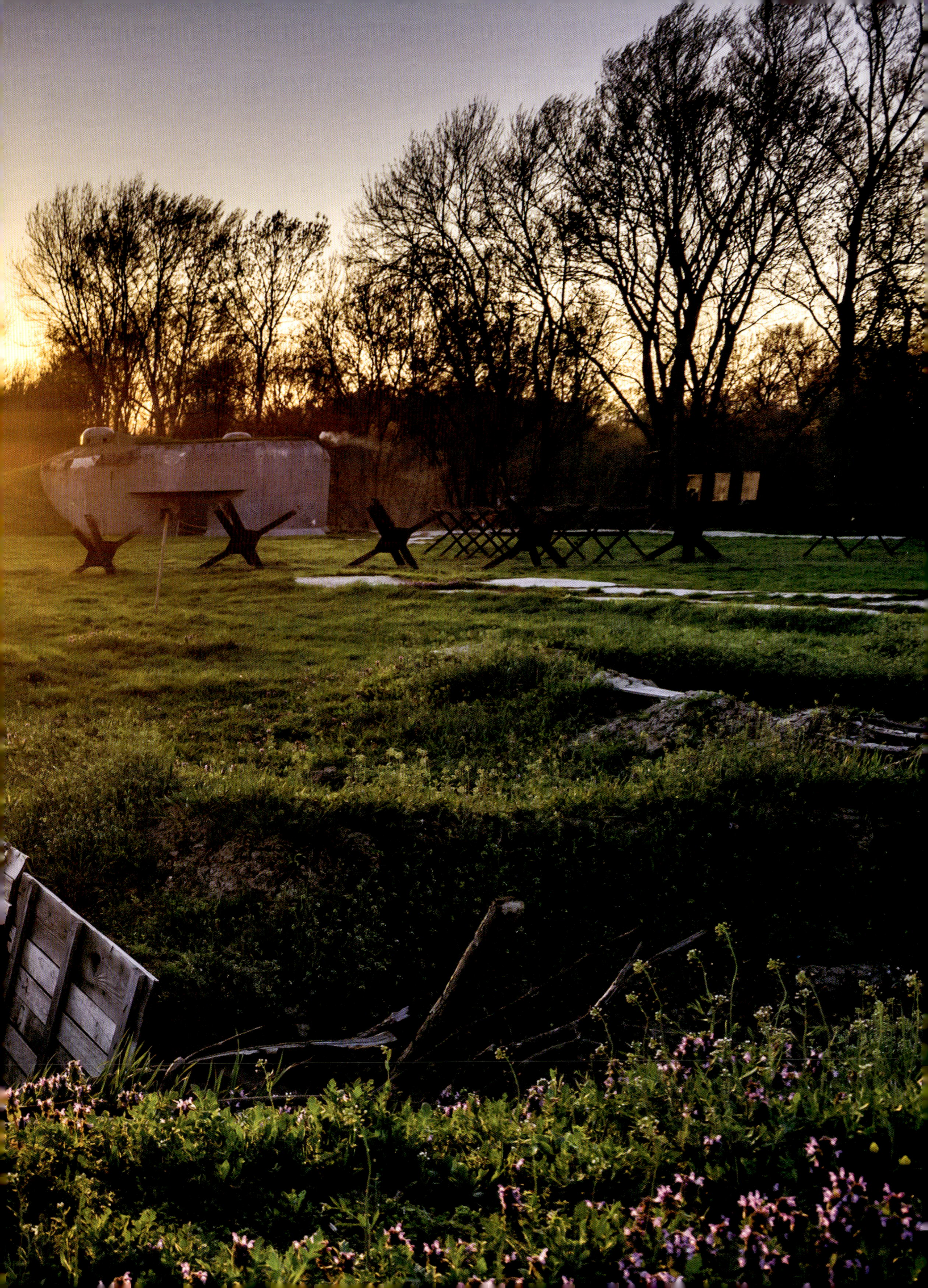

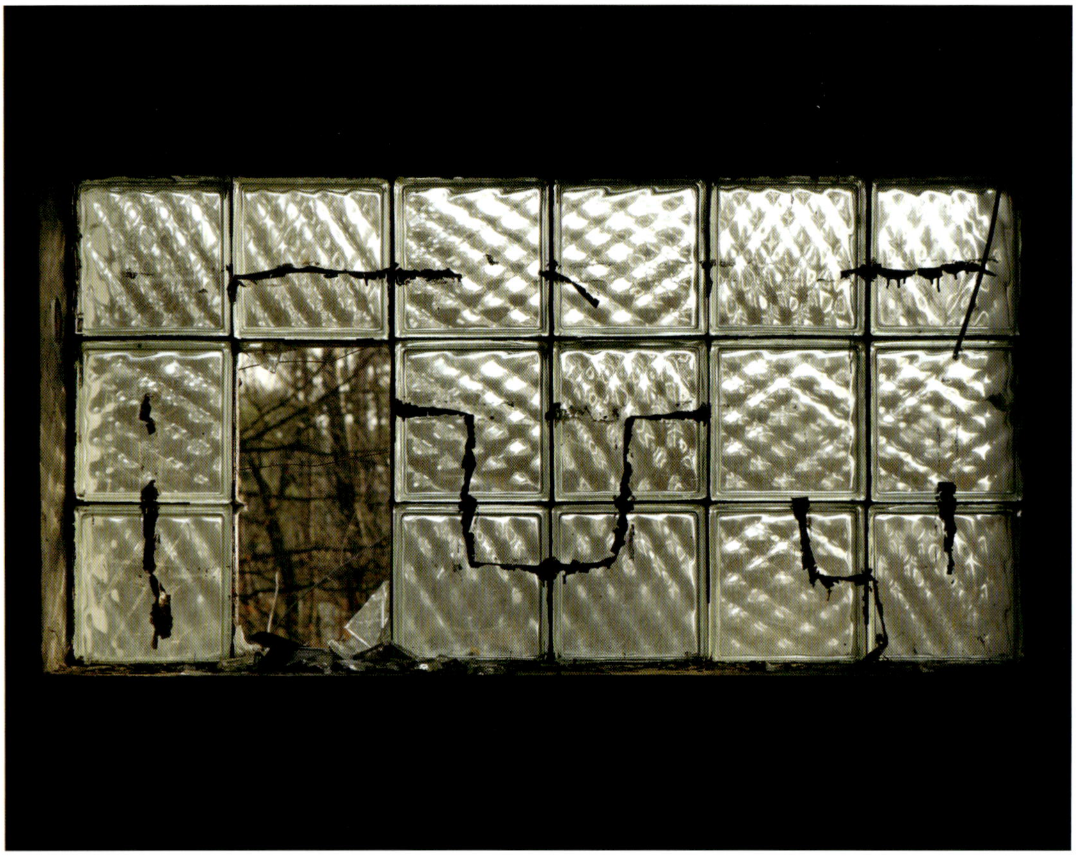

RIGHT, BELOW AND OPPOSITE:
Former Soviet Base, Milovice, Czech Republic
After the 'Prague Spring' unrest of 1968, the Soviet Union established the Soviet Central Group of Forces based at a site north of Prague. Until its closure in 1991, all Soviet armed forces units in what was Czechoslovakia were controlled from this base, including air, land and missile assets.

As well as an underground command centre, administrative offices and other standard military facilities, such as ammunition dumps and firing ranges, the base had accommodation for the families of the Soviet military personnel, such as these abandoned apartments.

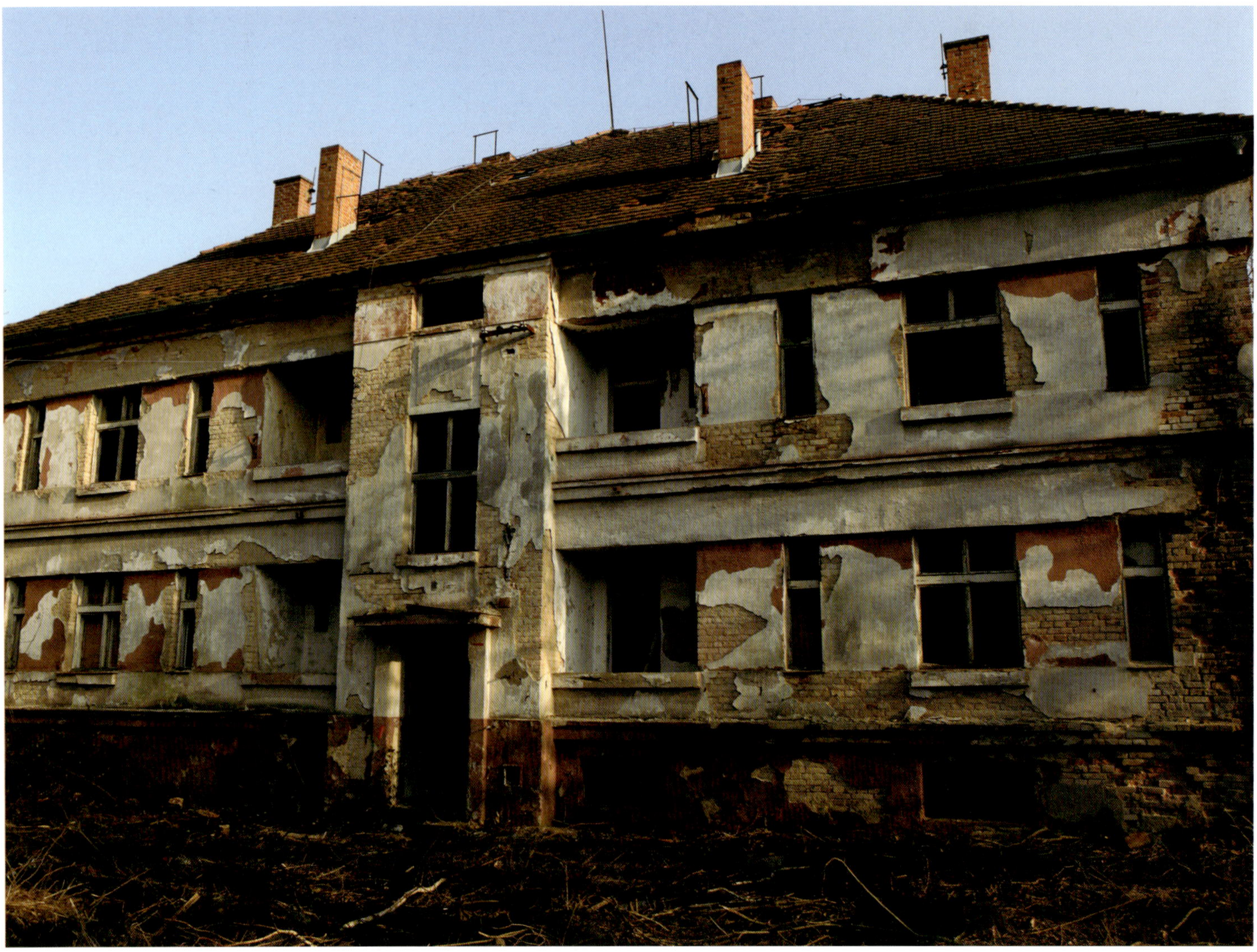

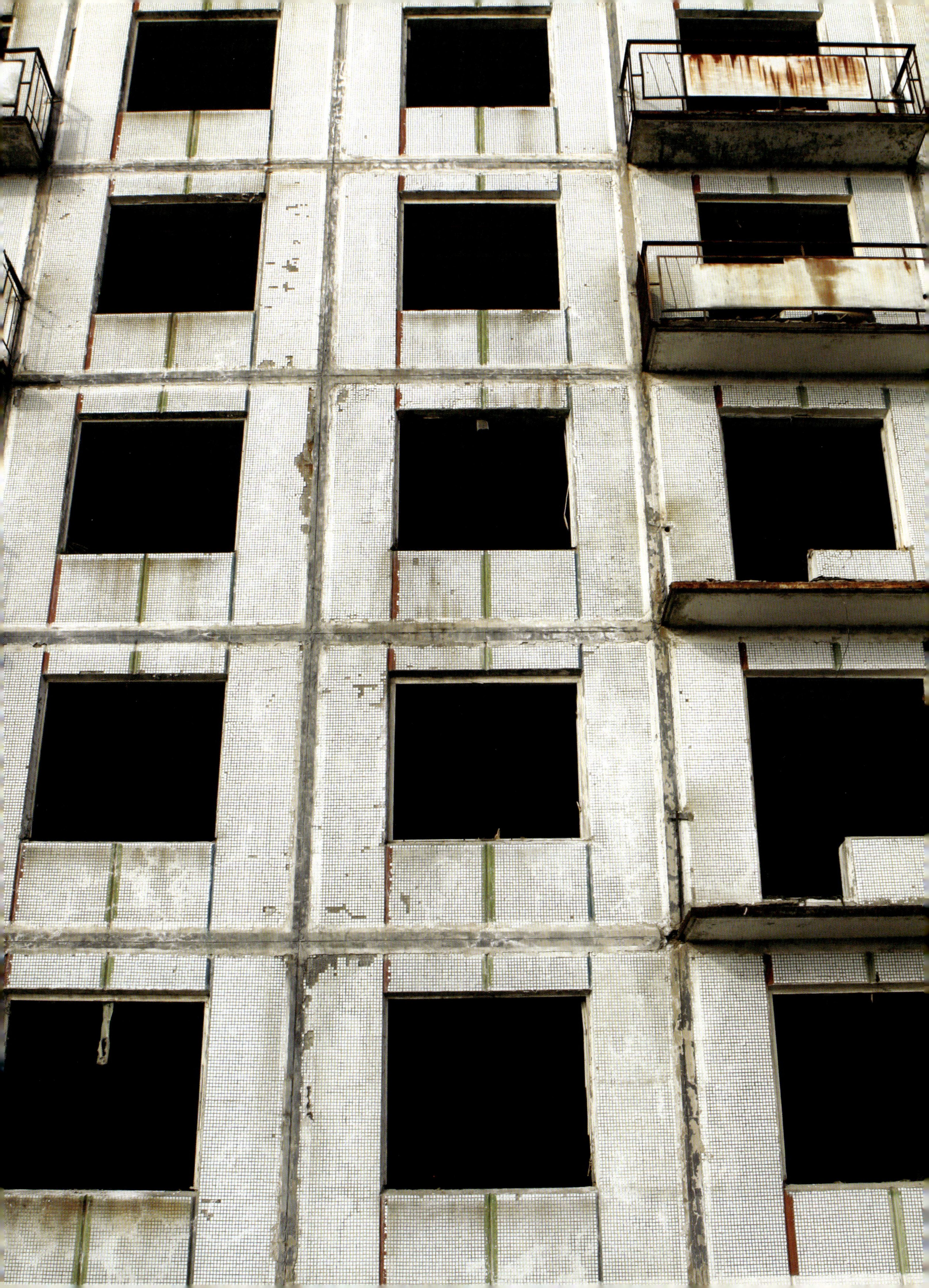

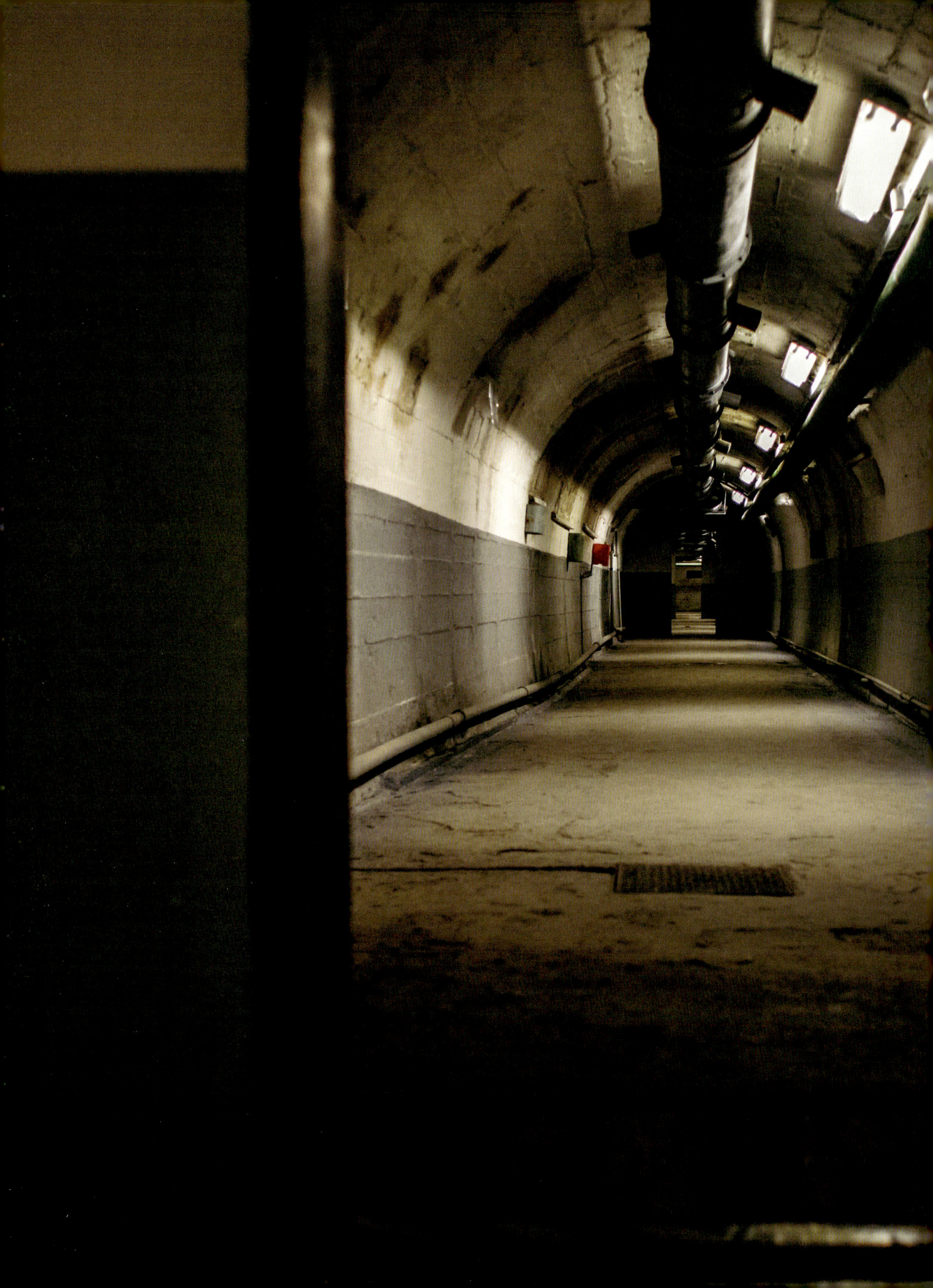

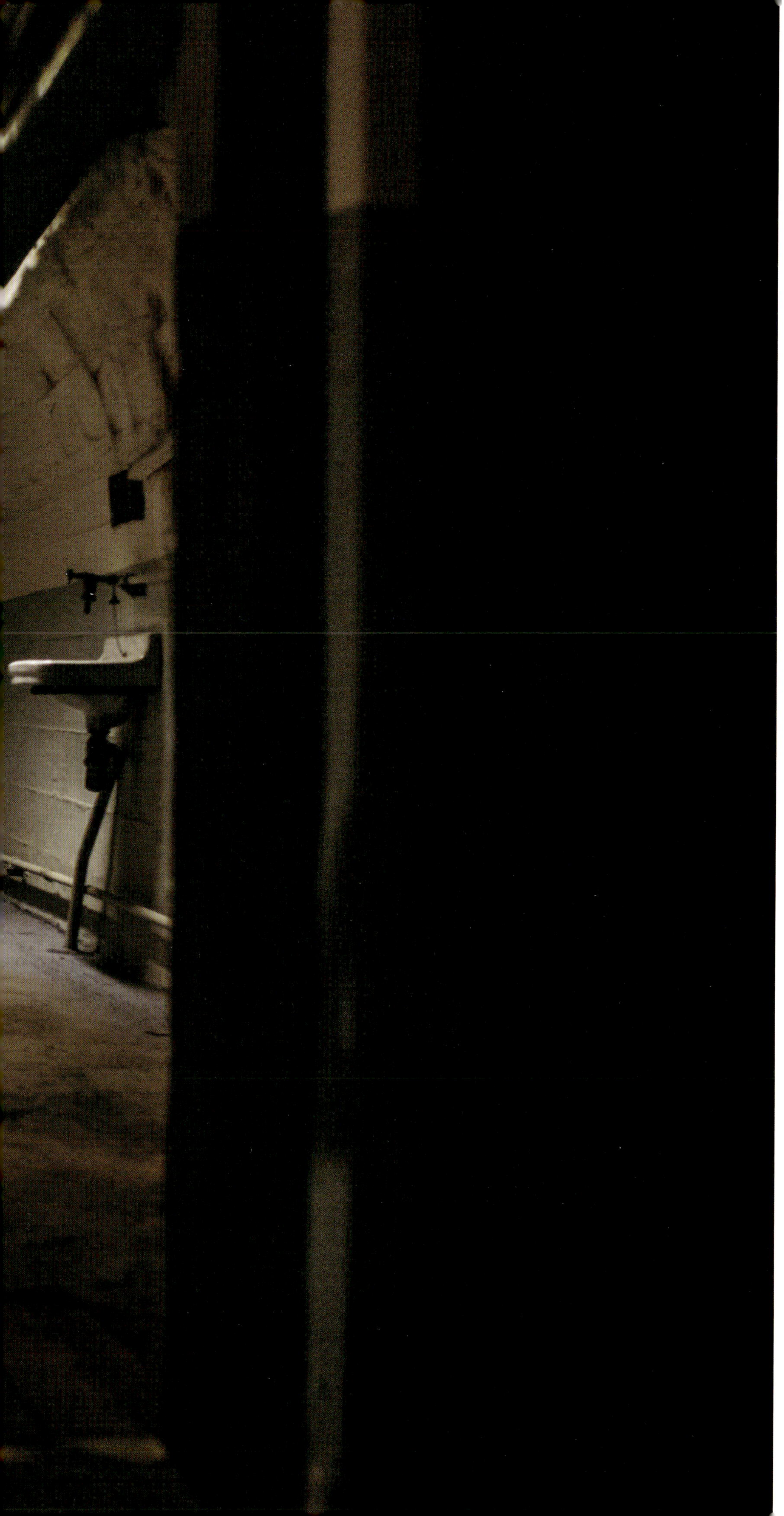

Nuclear Shelter, Prague, Czech Republic
The washbasin and drain cover in this tunnel suggest that this could have been a decontamination area in this shelter built during the 1950s for public use in case of war. It has sufficient capacity to house 1300 people for a few days. Prague has more than 800 underground shelters able to hold over half a million people.

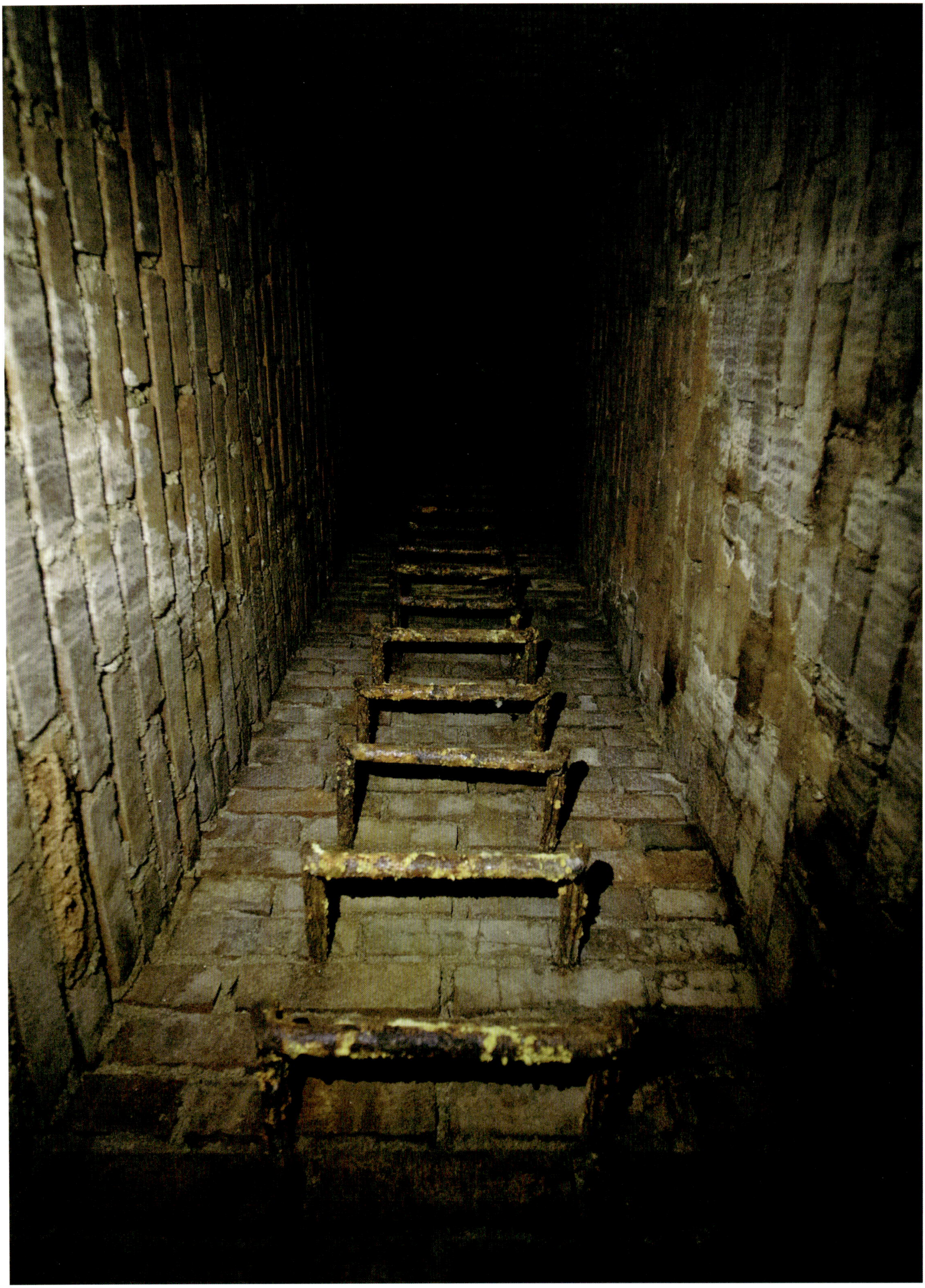

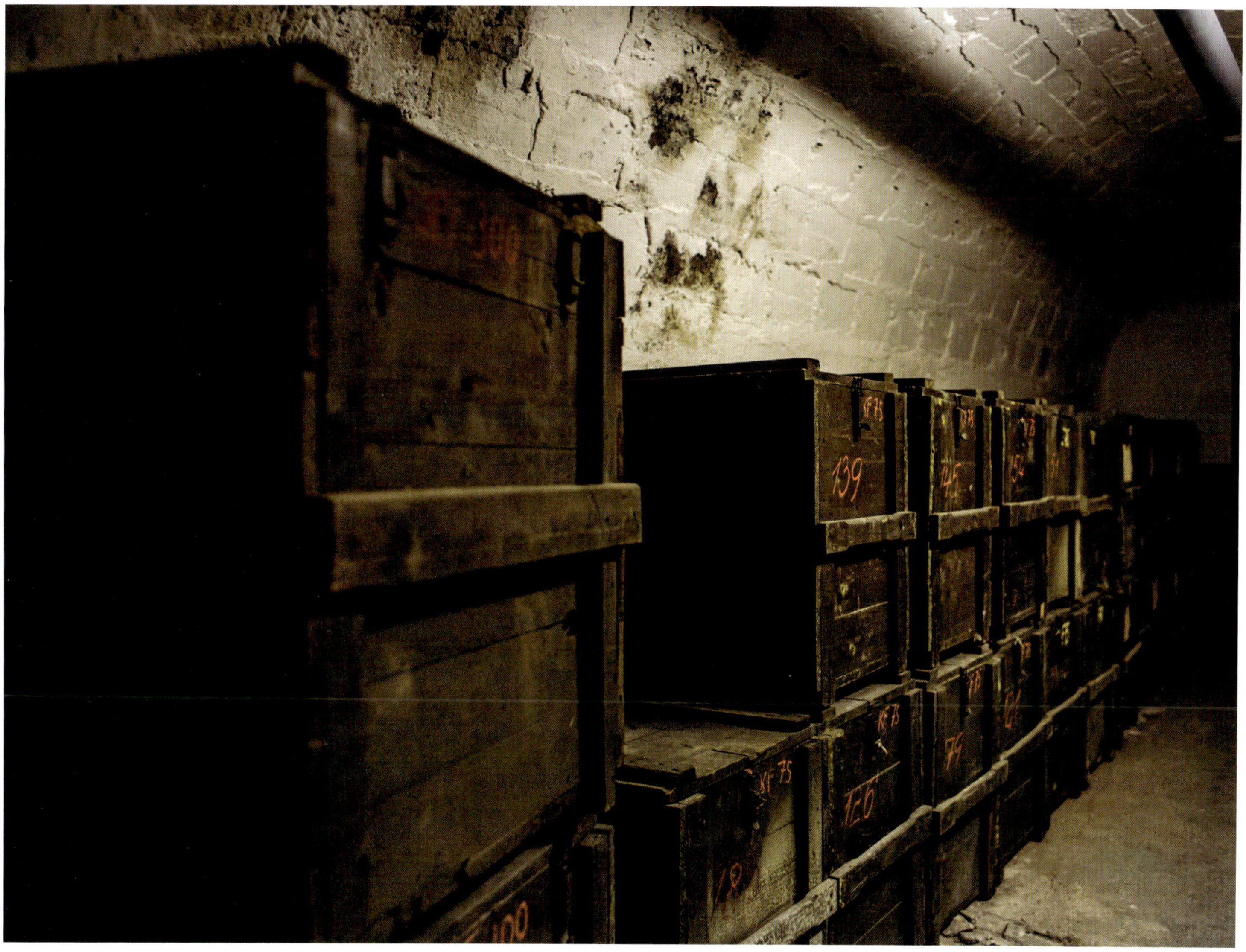

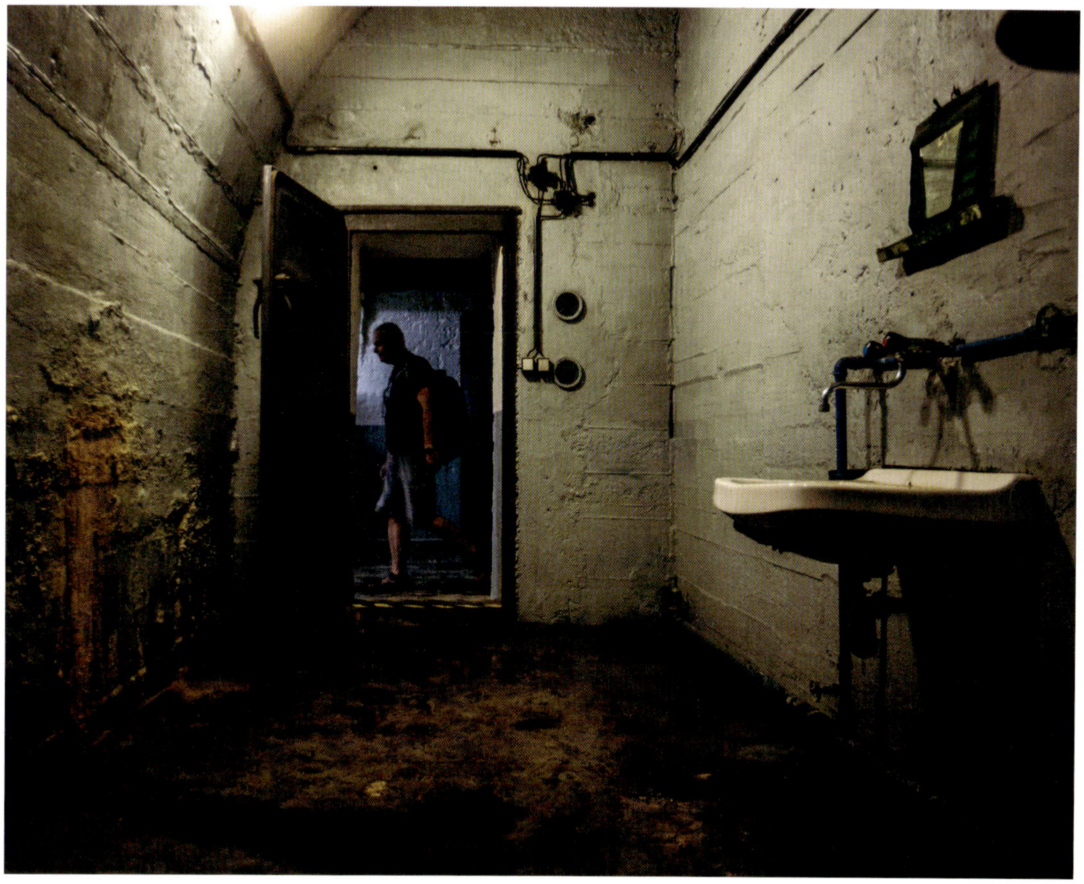

ABOVE:
Filter Boxes, Nuclear Shelter, Prague, Czech Republic
Gas mask filters are stacked in boxes ready for emergency use at this public nuclear shelter beneath Prague. The masks would require regular filter changes to remain effective under sustained exposure to chemical attack or radioactive dust.

RIGHT:
Former Washroom, Nuclear Shelter, Prague, Czech Republic
A primitive mirror has been attached above this washbasin in a room off the main tunnel.

OPPOSITE:
Escape Ladder, Hotel Jalta Shelter, Prague, Czech Republic
The rusty rungs of this ladder lead up an escape shaft to an emergency exit on the surface. The shelter was built in the 1950s under the Hotel Jalta on Wenceslas Square, but it was also used as a listening post to spy on the hotel's guests.

Former Soviet Submarine Repair Facility, Balaklava, Crimea
This former Soviet submarine repair base in the Crimea would have been a likely target for a NATO nuclear missile, or certainly been affected by any fallout from an attack on the major Soviet naval base at Sevastapol nearby. Workers at the base would have been required to wear full protective equipment against nuclear or chemical attack, such as the suits shown, which would have seriously inhibited their ability to work.

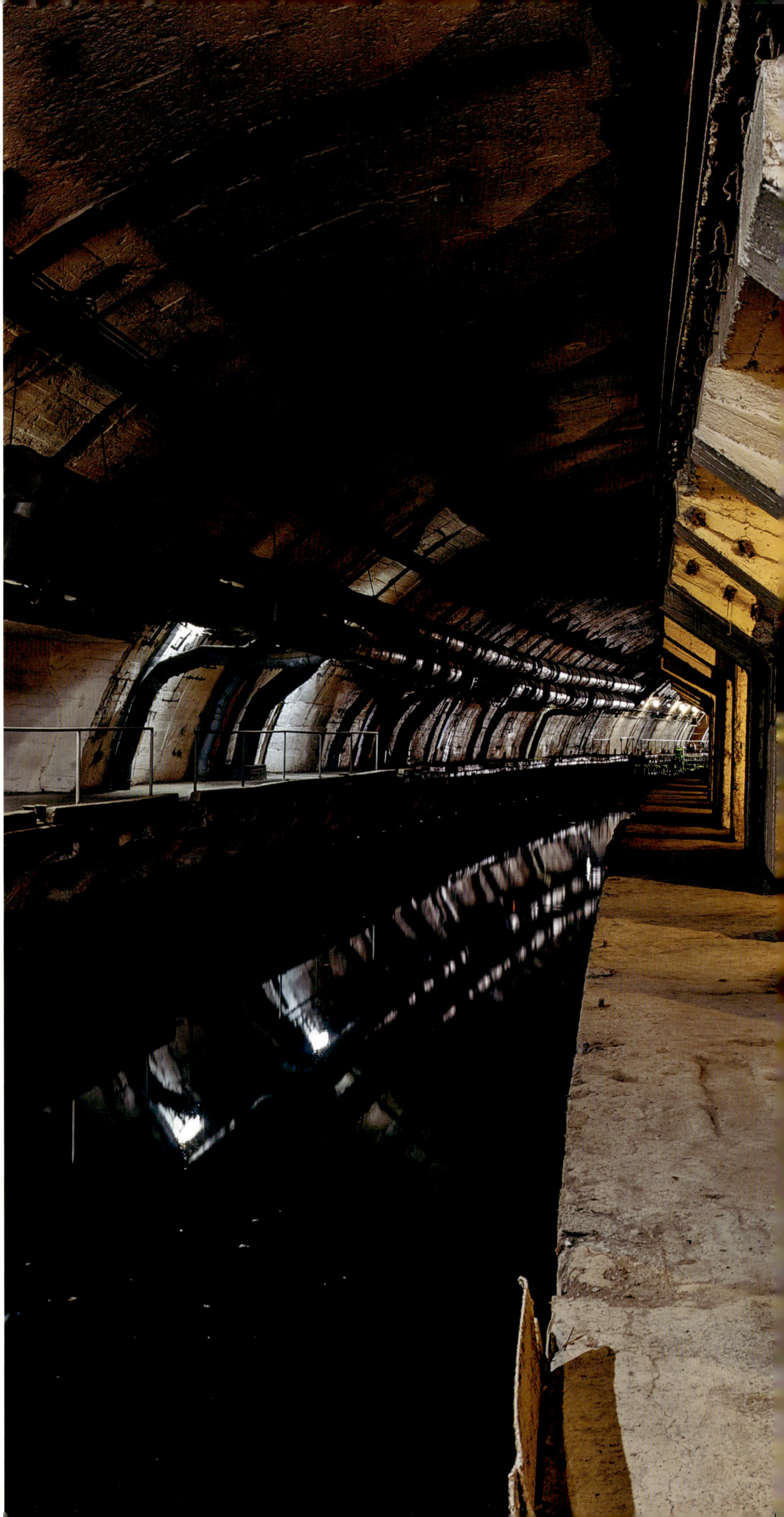

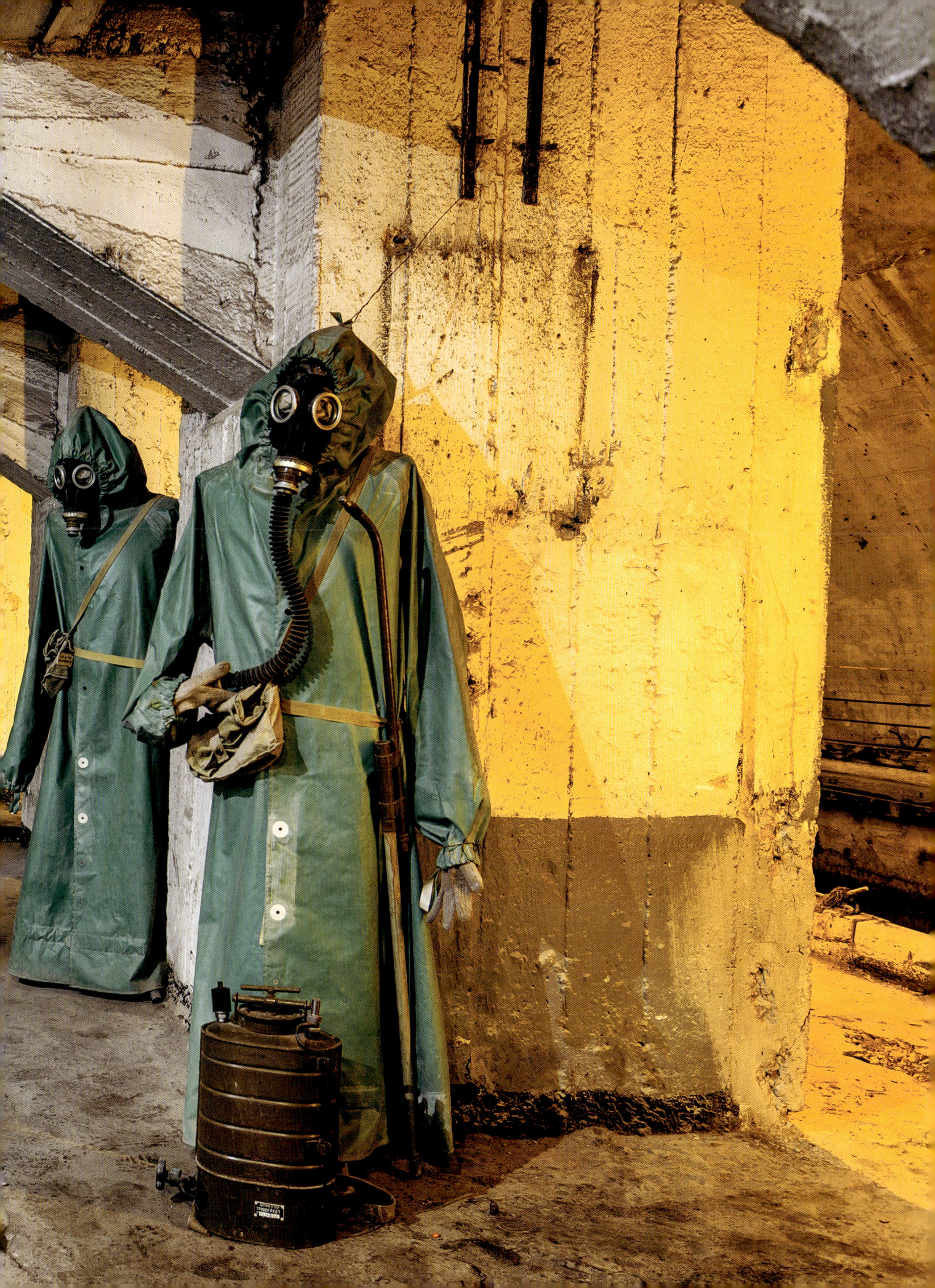

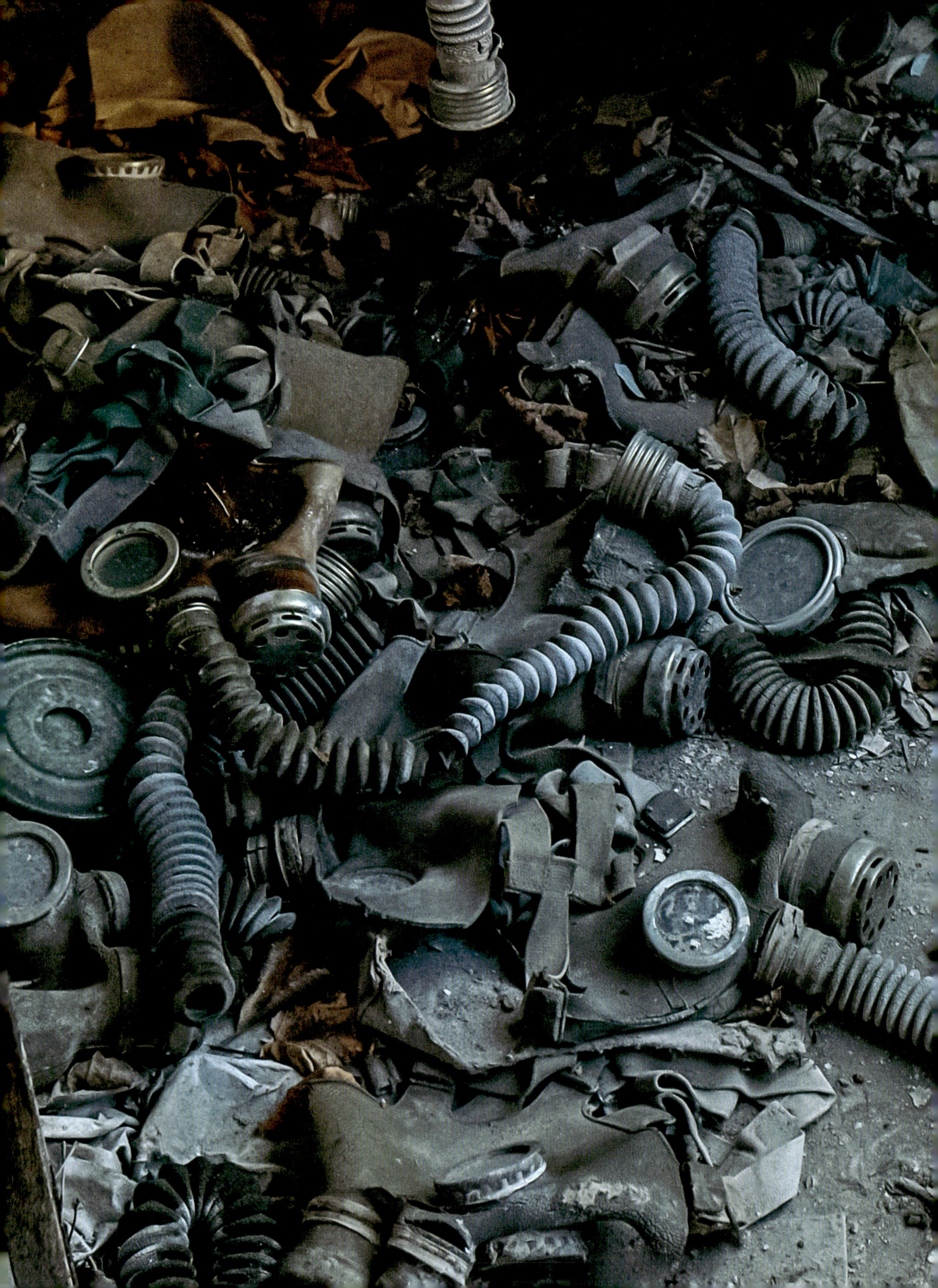

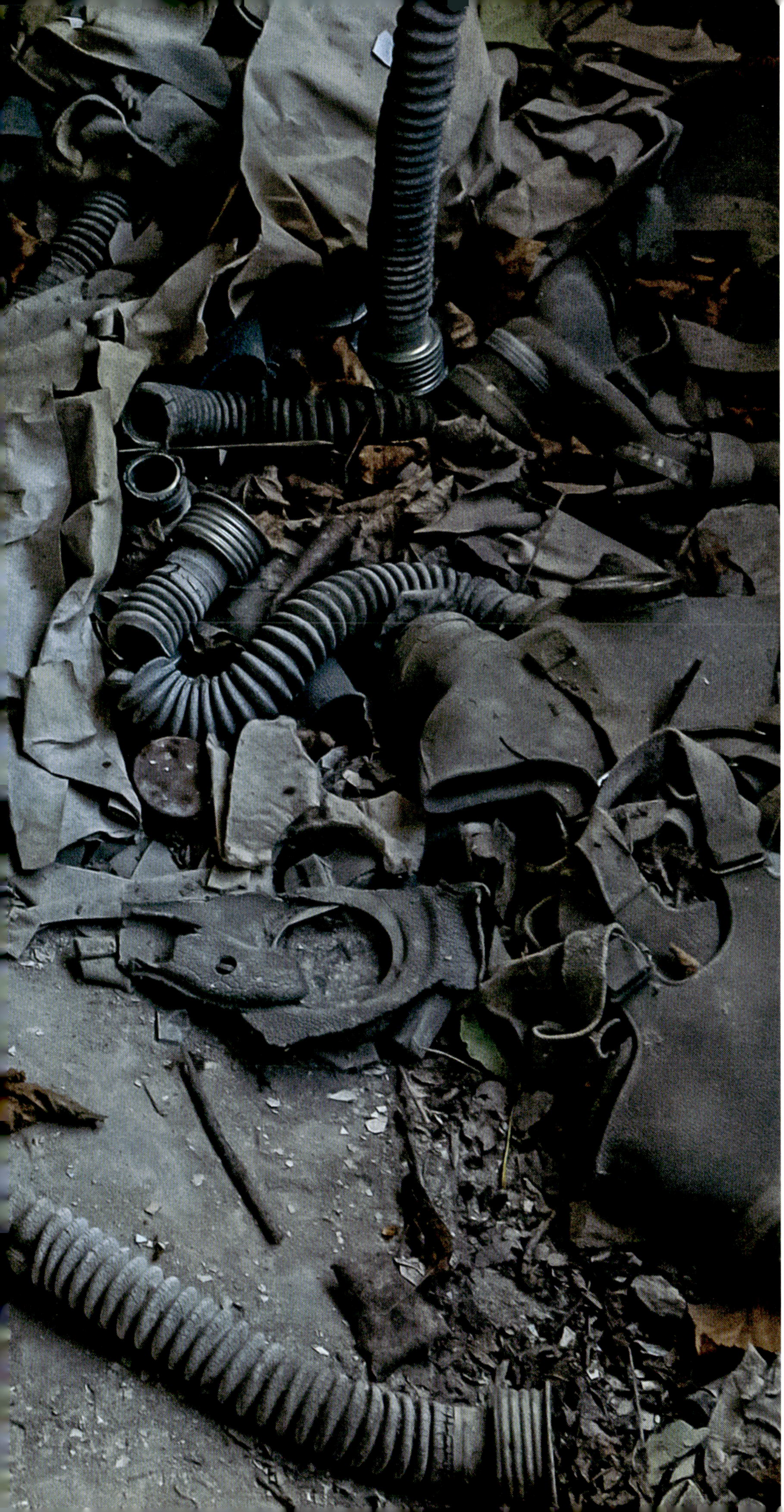

Children's Gas Masks, Pripyat, Ukraine
These masks have been strewn over the floor of this abandoned classroom in Pripyat, a town that was only evacuated a day after the reactor explosion at the nearby Chernobyl nuclear plant on 26 April 1986. Tragically many of the inhabitants had already suffered exposure to high levels of radiation.

Abandoned Building, Chernobyl, Ukraine
This housing block still bears the hammer and sickle of the Soviet Union on its roof. The explosion led to a mass evacuation of the local population and the establishment of a permanent exclusion zone due to the high levels of radiation that remained. The absence of humans has allowed wild animals and trees to reclaim the city.

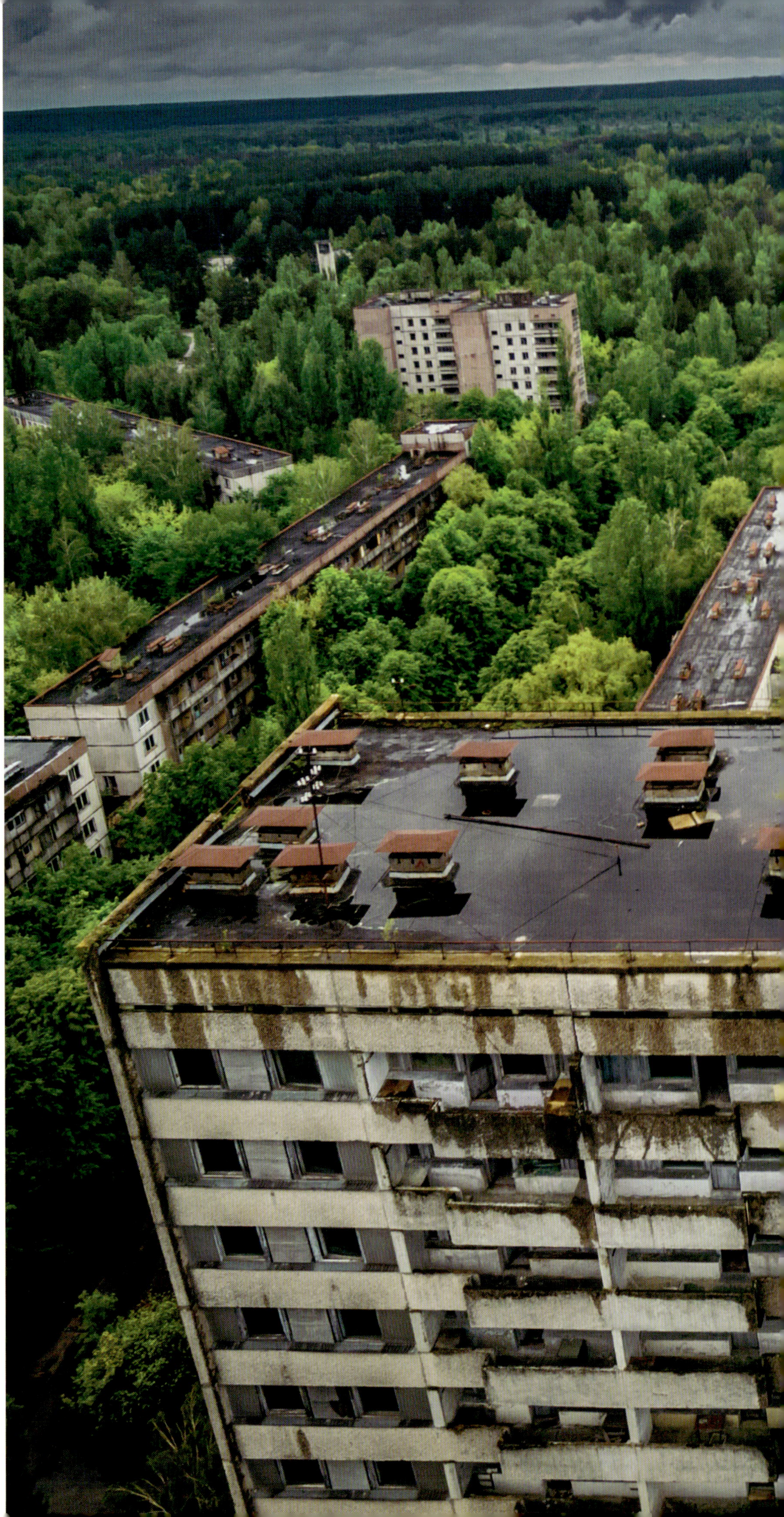

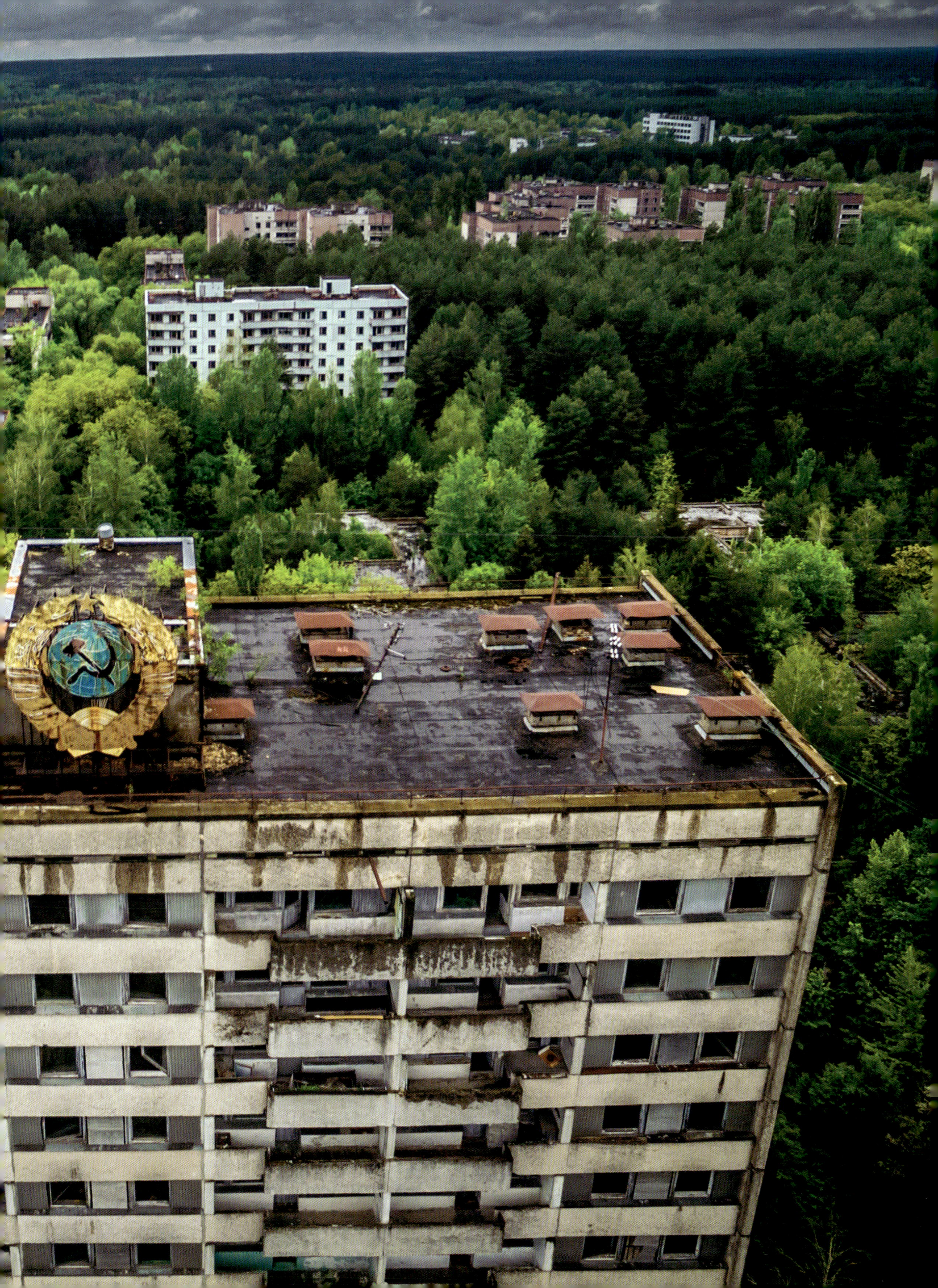

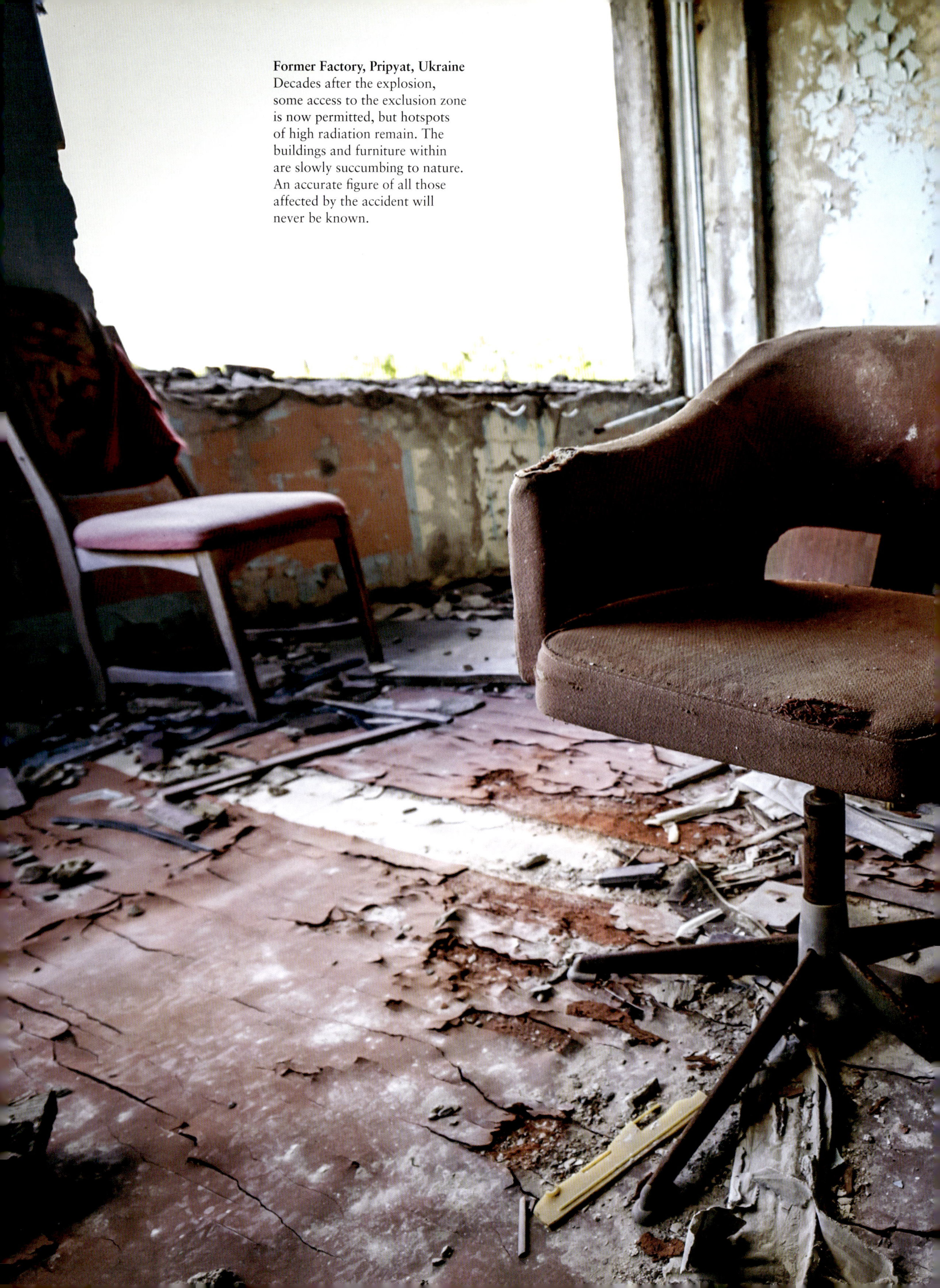

Former Factory, Pripyat, Ukraine
Decades after the explosion, some access to the exclusion zone is now permitted, but hotspots of high radiation remain. The buildings and furniture within are slowly succumbing to nature. An accurate figure of all those affected by the accident will never be known.

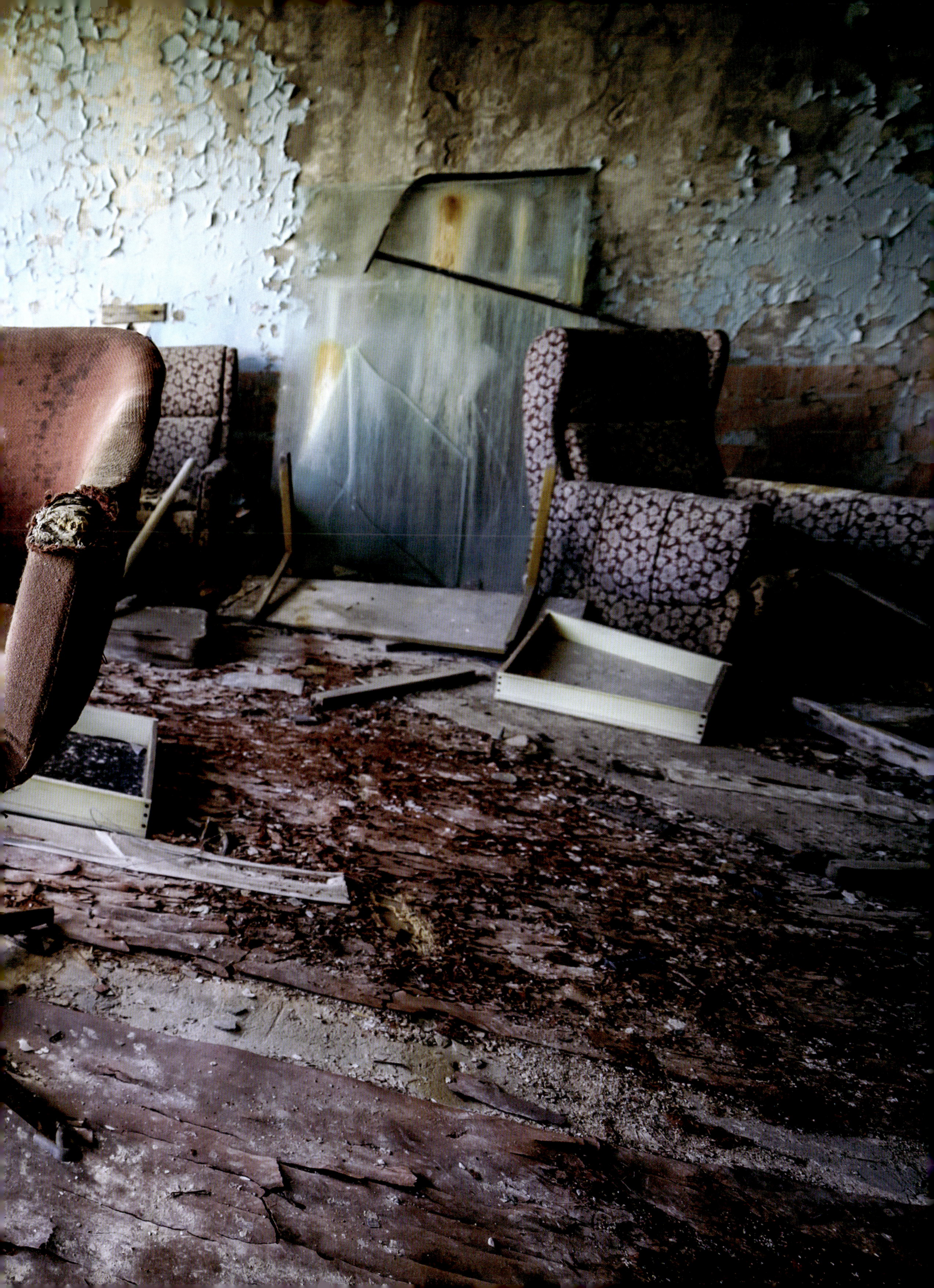

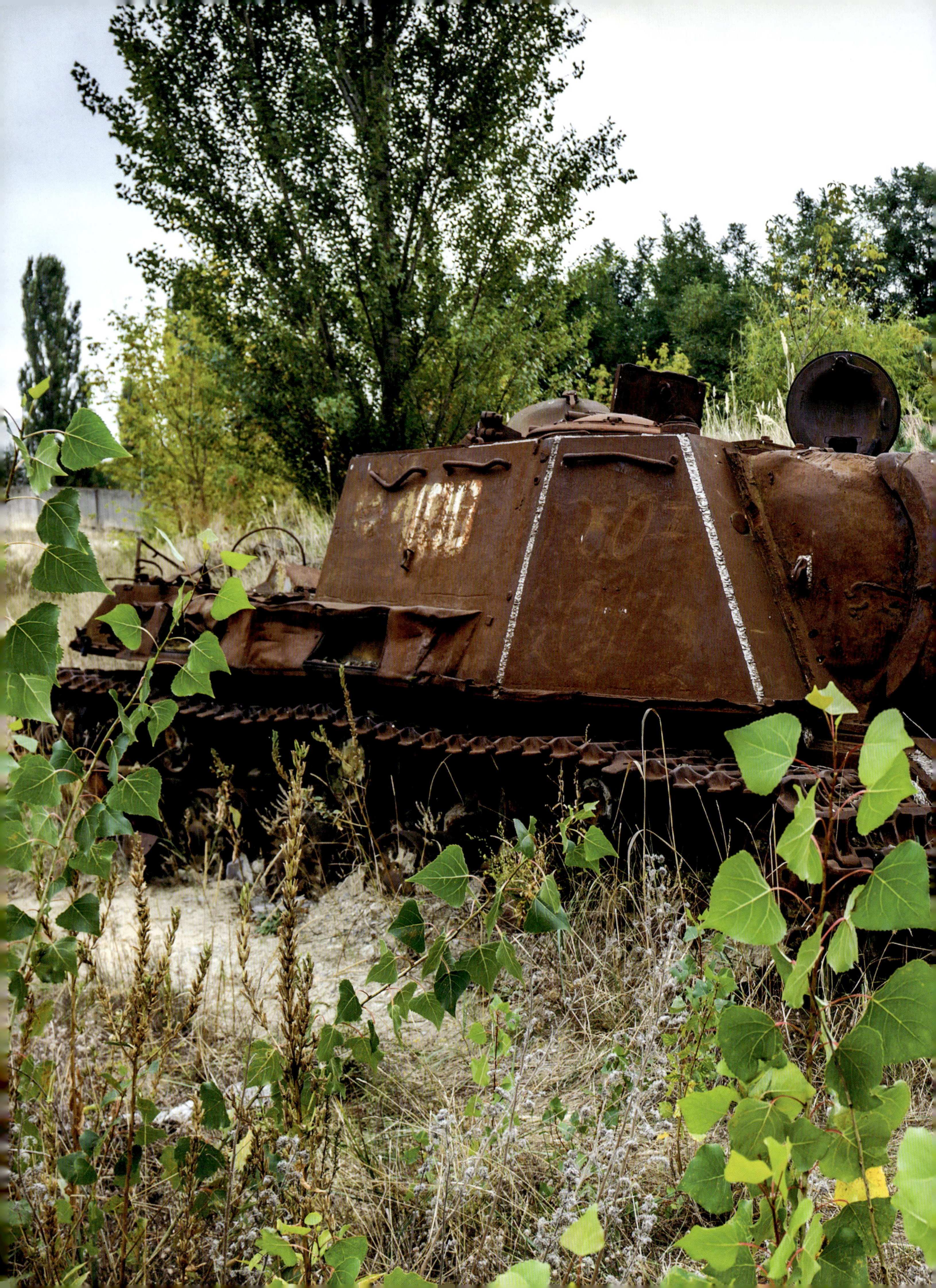

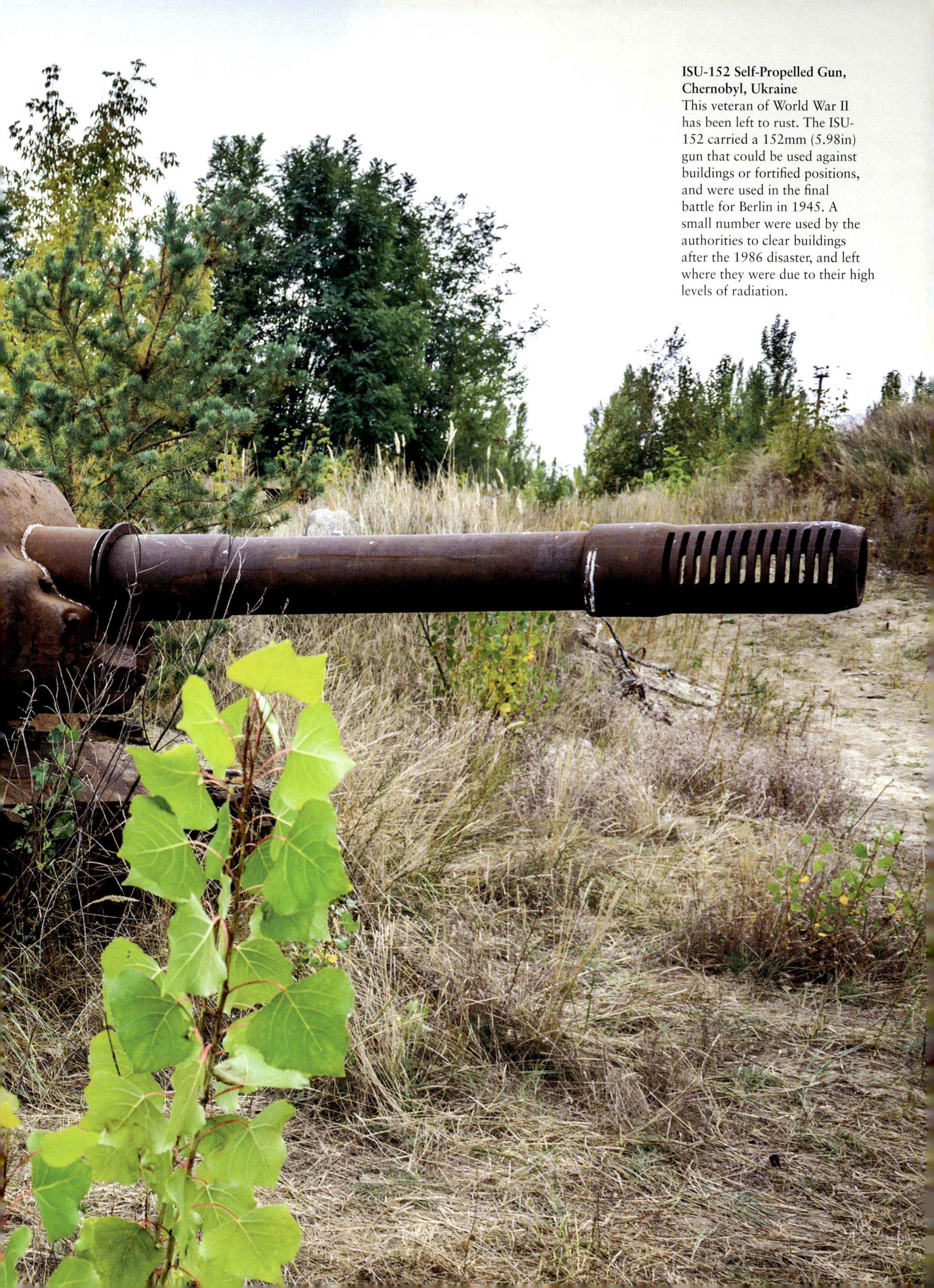

ISU-152 Self-Propelled Gun, Chernobyl, Ukraine
This veteran of World War II has been left to rust. The ISU-152 carried a 152mm (5.98in) gun that could be used against buildings or fortified positions, and were used in the final battle for Berlin in 1945. A small number were used by the authorities to clear buildings after the 1986 disaster, and left where they were due to their high levels of radiation.

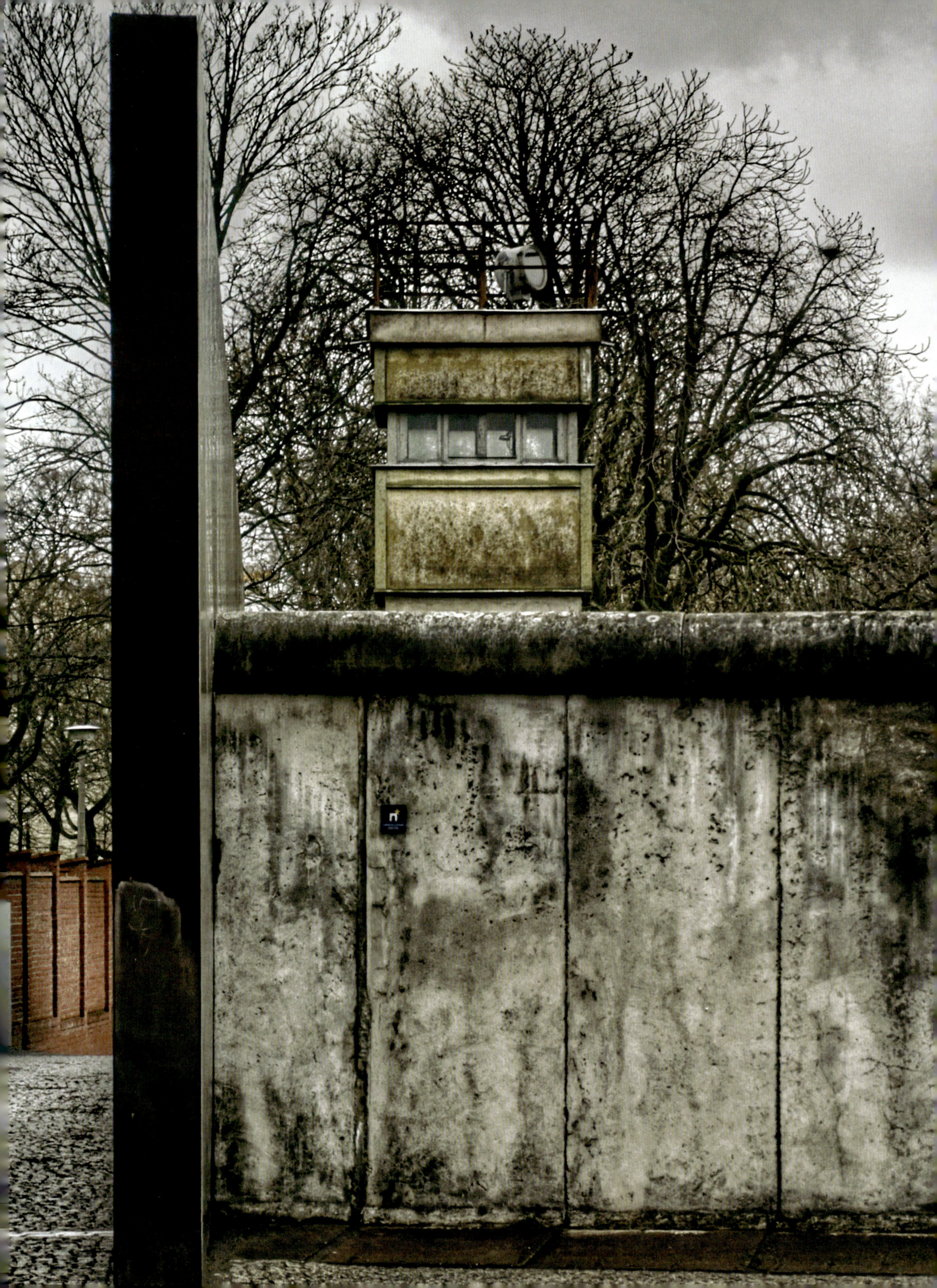

Western Europe

When it became clear that the Soviet Union was not going to withdraw its occupying forces from Eastern Europe, and that puppet Communist governments were being installed in their zone of control, the US drawdown of forces after World War II was reversed, and new plans made for a new potential conflict in Europe. Attitudes hardened and tensions rose during the Berlin crisis of 1948–9, when, in an attempt to have the Western Allies withdraw from the German capital, the Soviet military closed all access to the city except for the three air corridors. The subsequent airlift of supplies required a significant effort from the air forces of the US, UK, France, Canada, Australia, New Zealand and South Africa.

Although the Soviet blockade ended in May 1949, the previous month saw the US signing a treaty creating NATO to defend Western Europe against the threat of Soviet attack. As a consequence new US bases were established on the continent that would remain in place until the end of the Cold War. Even longstanding neutral countries did not escape the prospect of nuclear hostilities. Both Switzerland and Sweden invested heavily in bunkers to protect their citizens from a nuclear war.

OPPOSITE:
Watchtower, Berlin Wall, Berlin, Germany
By 1961 3.5 million East Germans had left for West Germany. To prevent the continued loss of its population, from midnight on 12 August the border was closed, with construction of a more permanent concrete barrier around West Berlin officially known as the Anti-Fascist Protection Rampart starting shortly afterwards.

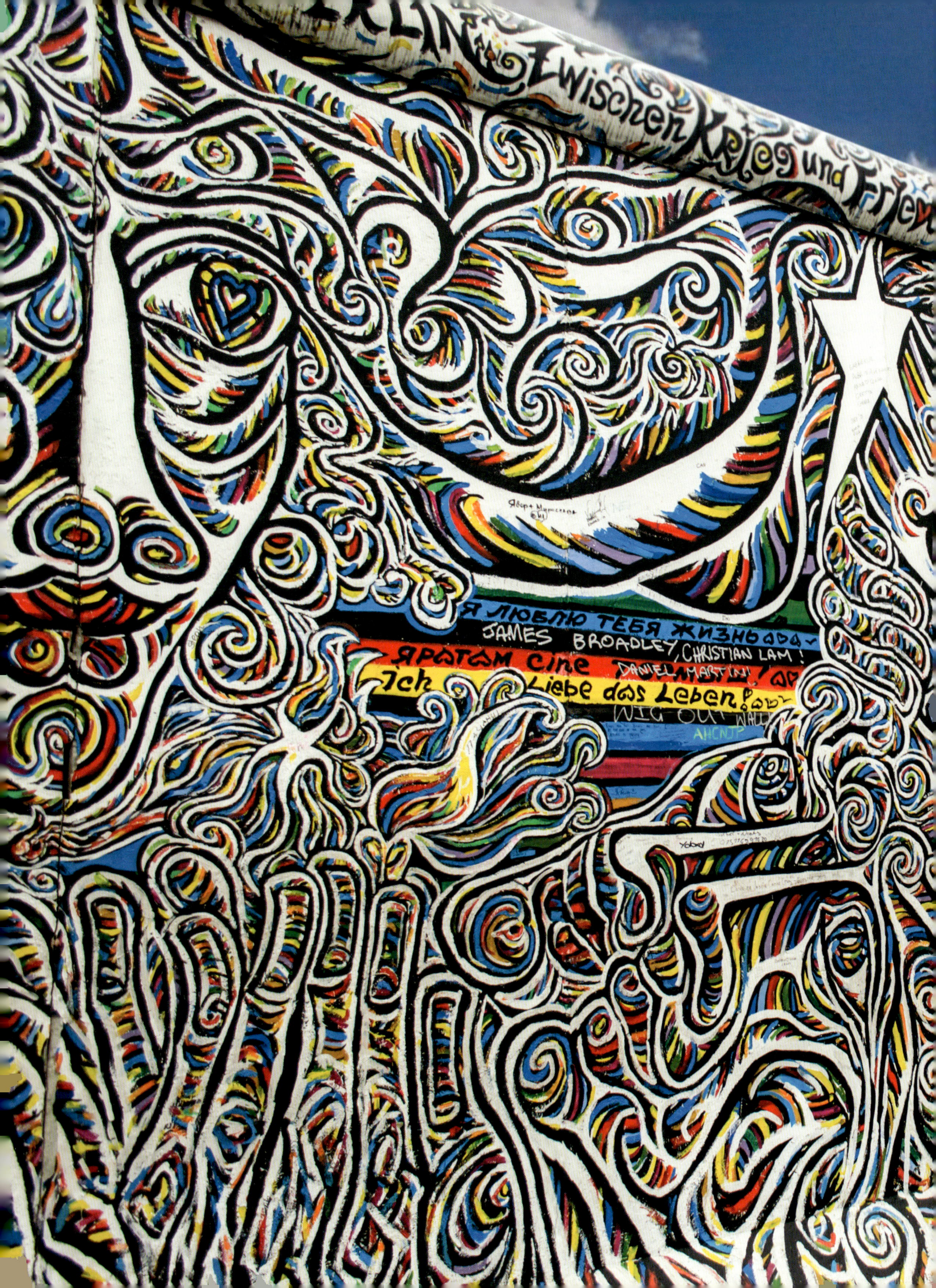
Zwischen Krieg und Frie
Я ЛЮБЛЮ ТЕБЯ ЖИЗНЬ
JAMES BROADLEY, CHRISTIAN LAM!
DANIEL MARTIN
Ich liebe das Leben!

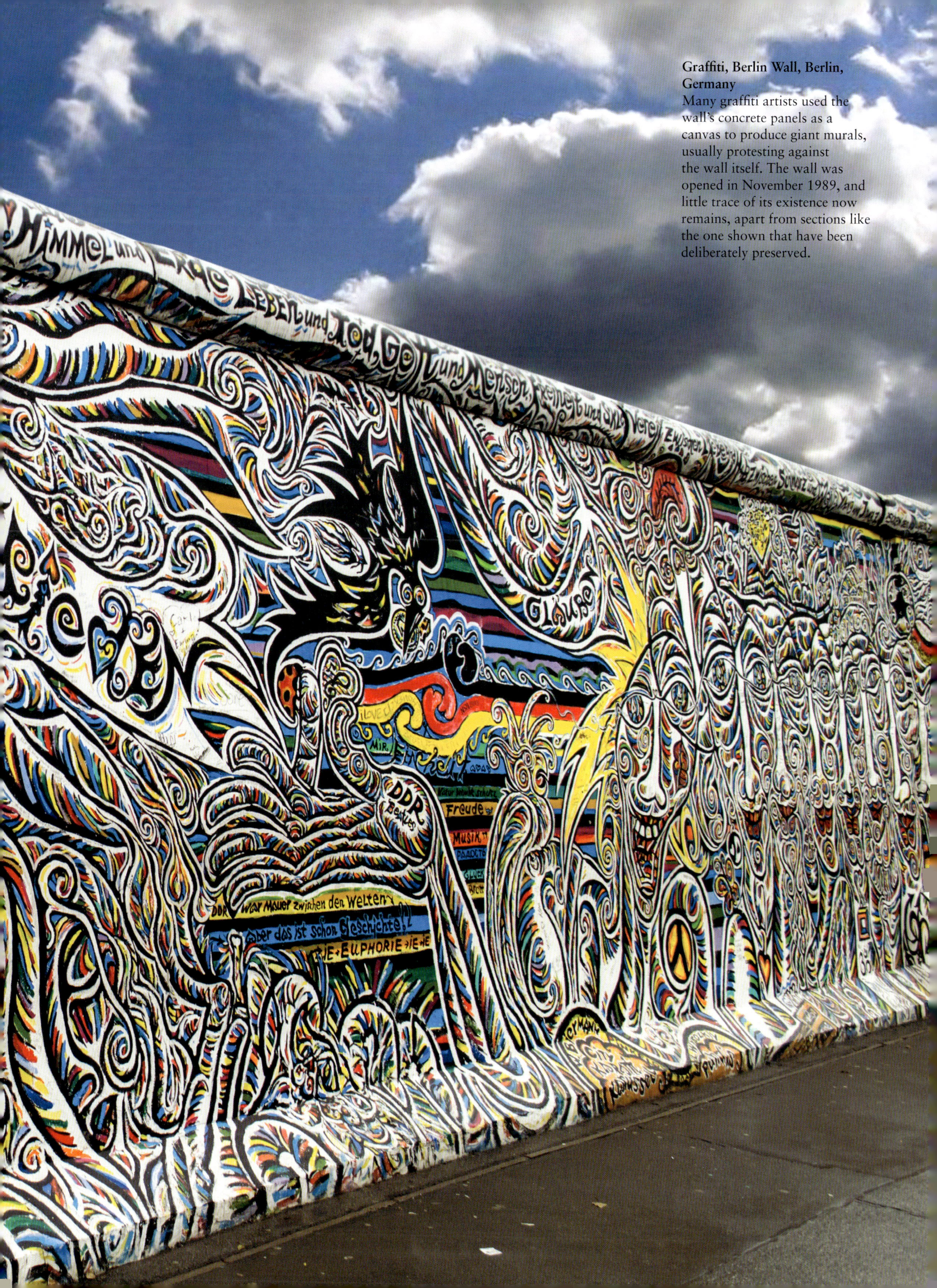

Graffiti, Berlin Wall, Berlin, Germany
Many graffiti artists used the wall's concrete panels as a canvas to produce giant murals, usually protesting against the wall itself. The wall was opened in November 1989, and little trace of its existence now remains, apart from sections like the one shown that have been deliberately preserved.

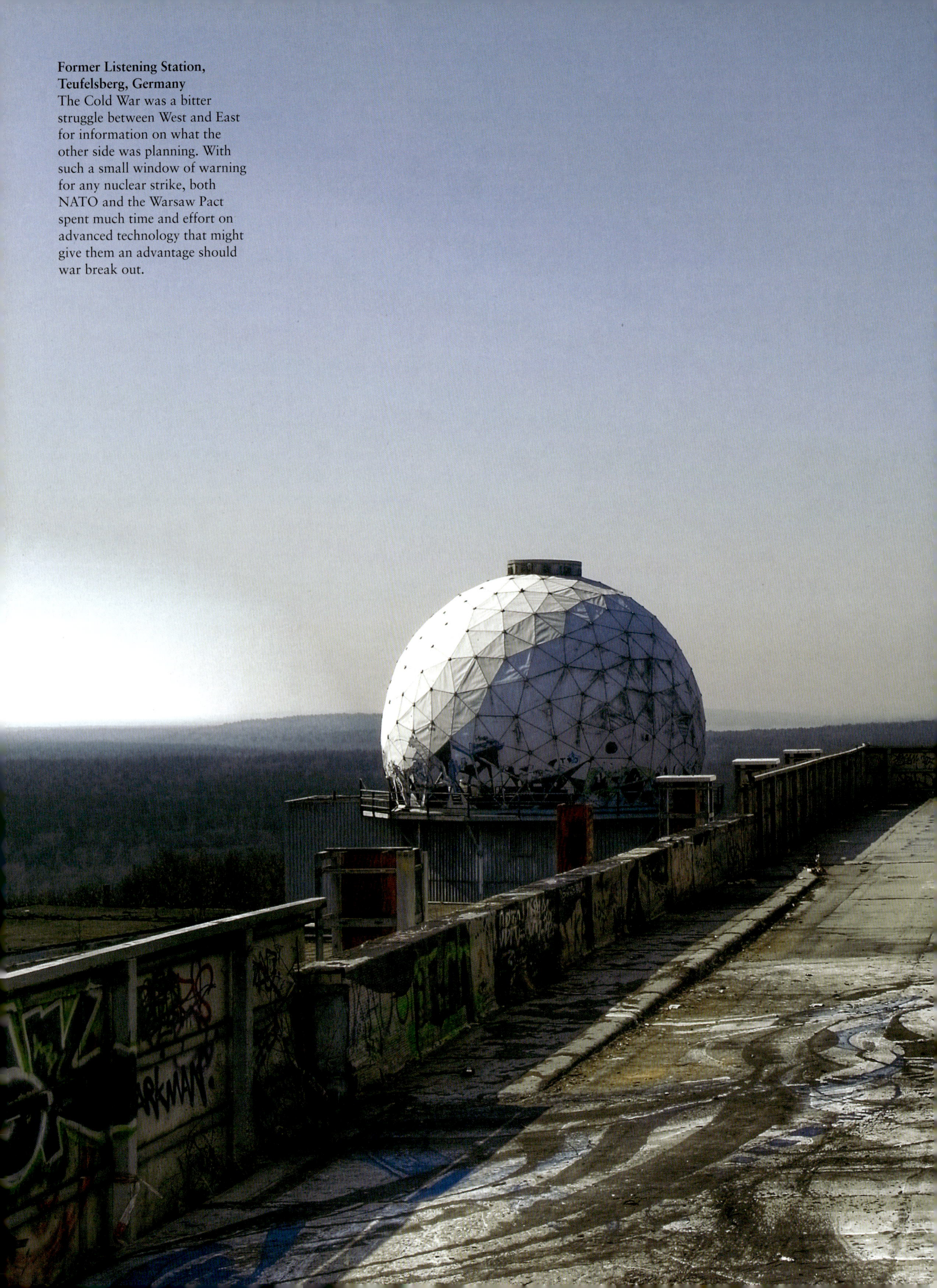

Former Listening Station, Teufelsberg, Germany
The Cold War was a bitter struggle between West and East for information on what the other side was planning. With such a small window of warning for any nuclear strike, both NATO and the Warsaw Pact spent much time and effort on advanced technology that might give them an advantage should war break out.

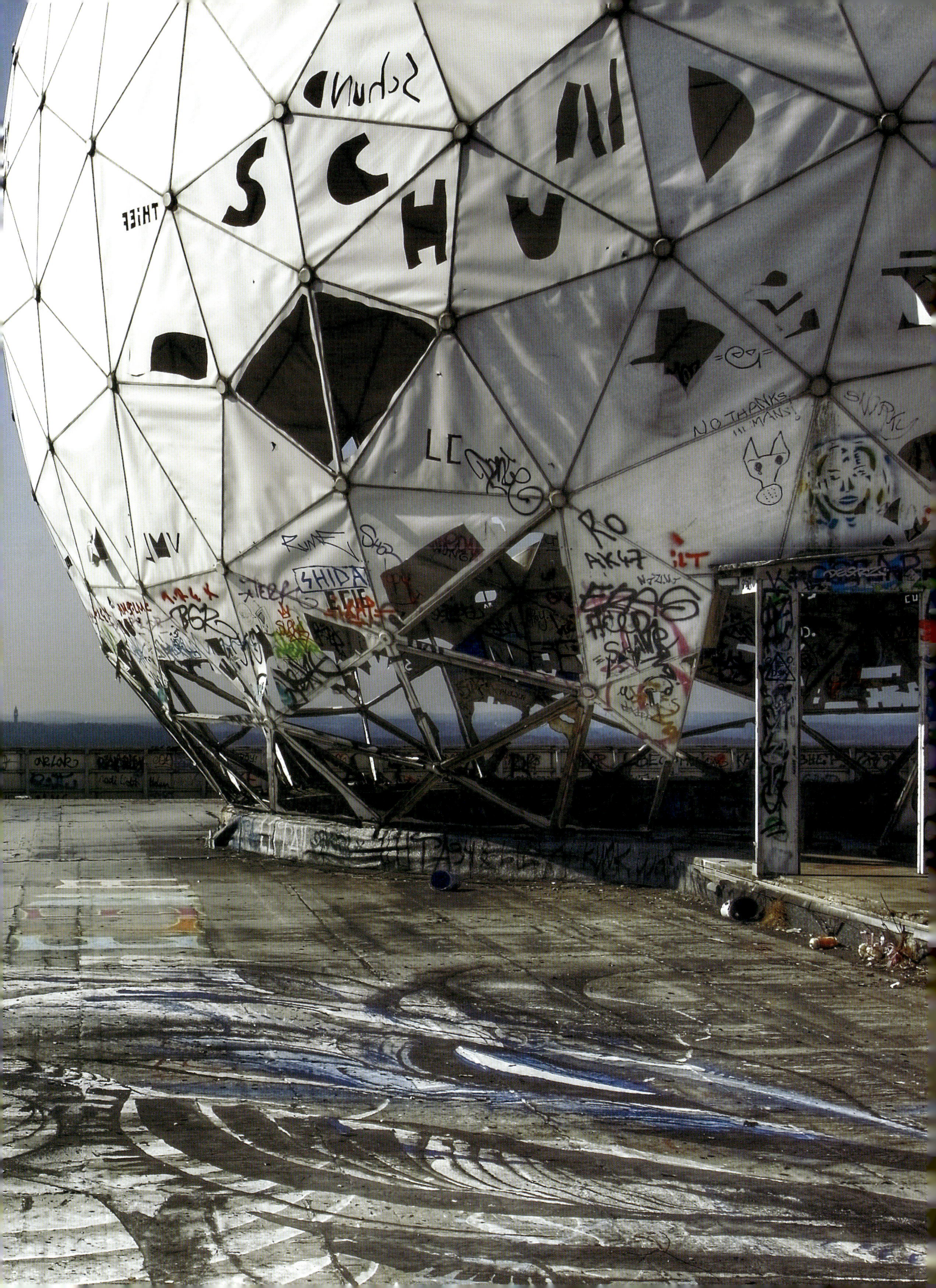
SCHUND
NO THANKS

RIGHT:

Checkpoint Charlie, Friedrichstadt, Berlin, Germany

There were a number of border crossing points from West Berlin to the East, but the busiest and most notorious was Checkpoint Charlie, located in the American zone of Berlin. A café nearby allowed curious Westerners to view East Berliner life while enjoying their coffee.

BELOW:

Fuel Pumps, Tempelhof Airport, Berlin, Germany

Built before World War II as the main airport for the German capital, Tempelhof played a vital role in the Berlin Airlift after the Soviets sealed off the land routes into West Berlin between June 1948 and May 1949. At the height of the airlift, aircraft were landing every 45 seconds, equivalent to peak rates at London's Heathrow airport today. The airlift pilots dropped sweets to children watching the aircraft, earning themselves the nickname *Rosinenbomber* (Raisin Bombers).

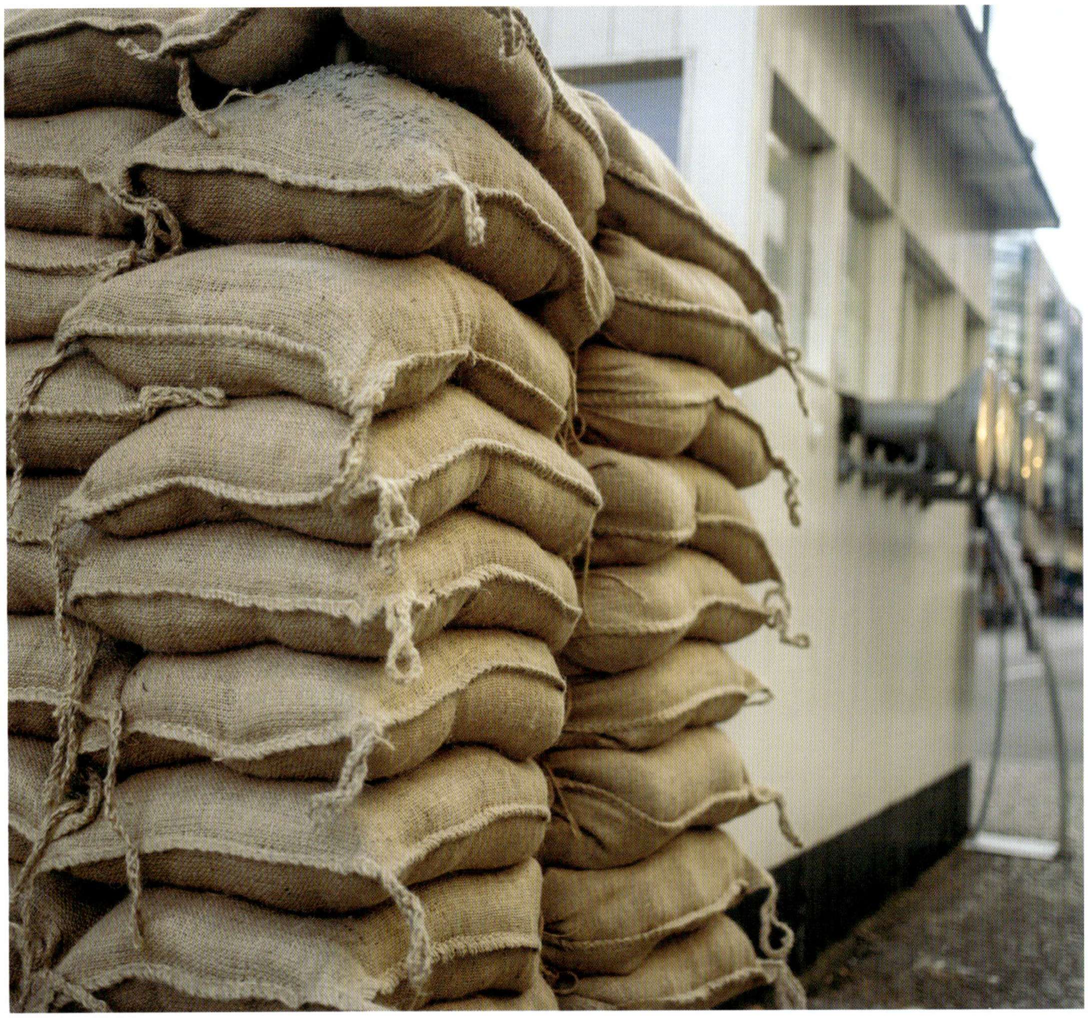

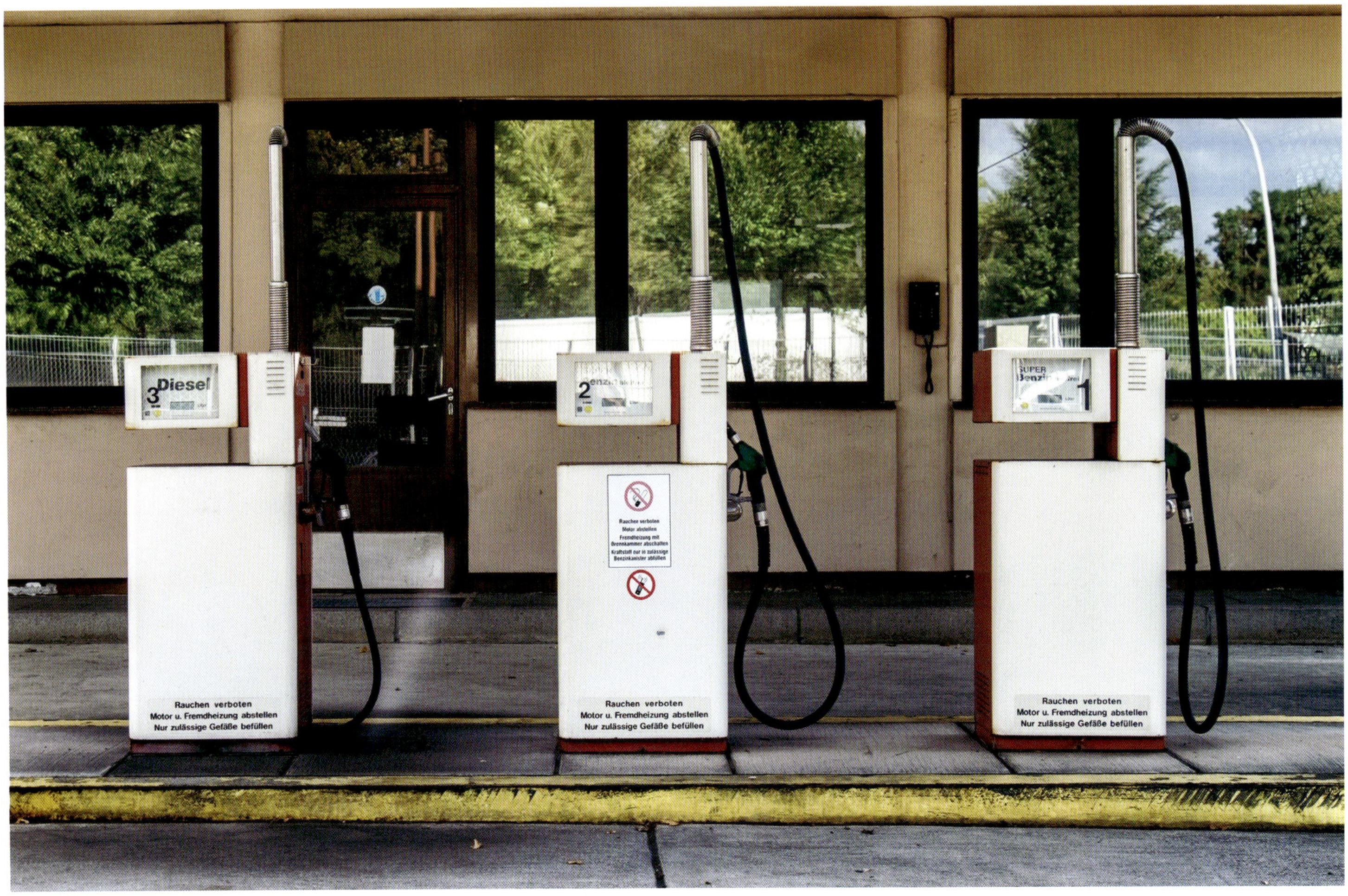

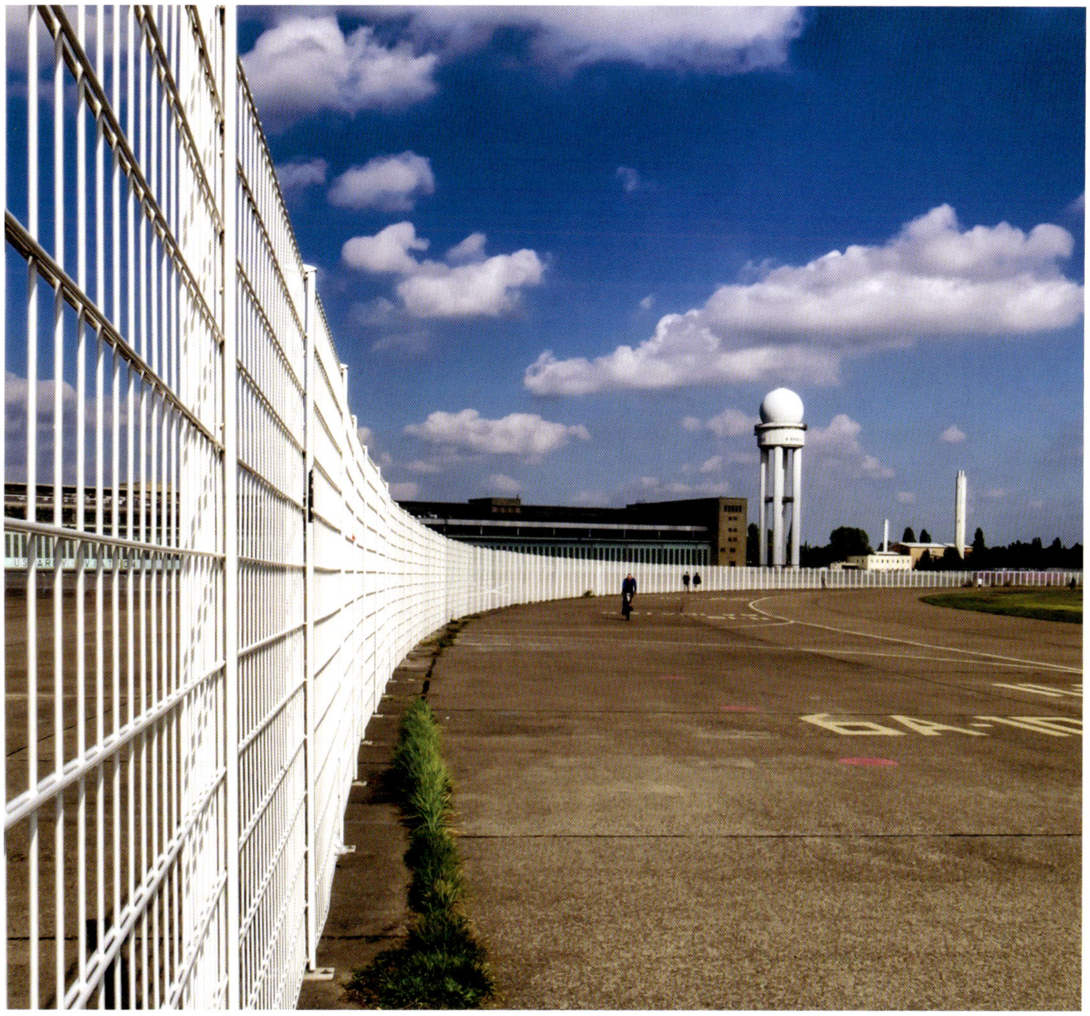

LEFT AND BELOW:
Radar Tower And Former US Army Aviation Hangars, Tempelhof Airport, Berlin, Germany
The airport's terminal building has been called the 'mother of all airports' by Norman Foster. It was one of the largest buildings in the world when completed, with integral hangars and shelter from the elements for passengers disembarking aircraft. The airport was closed in 2008.

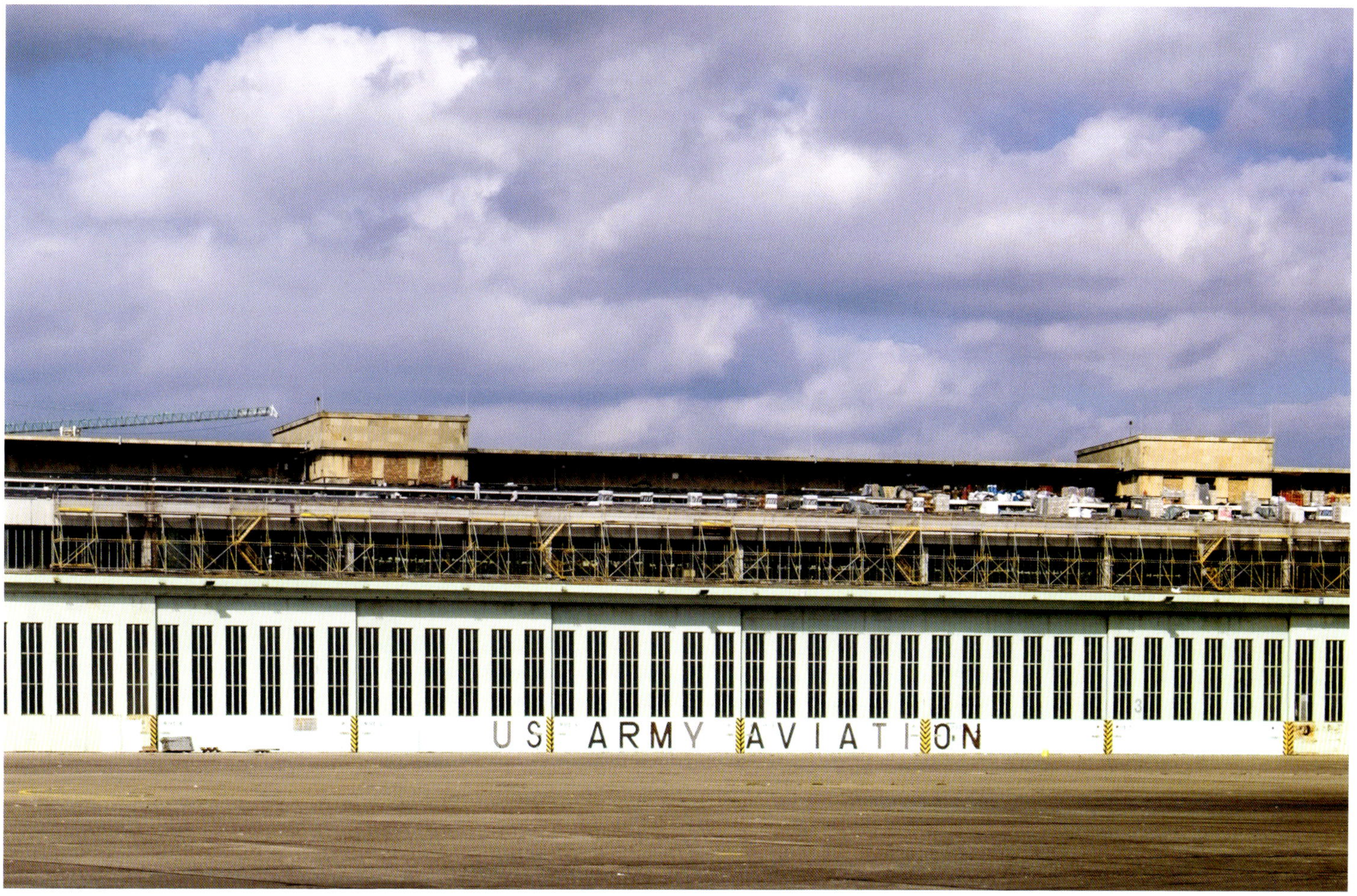

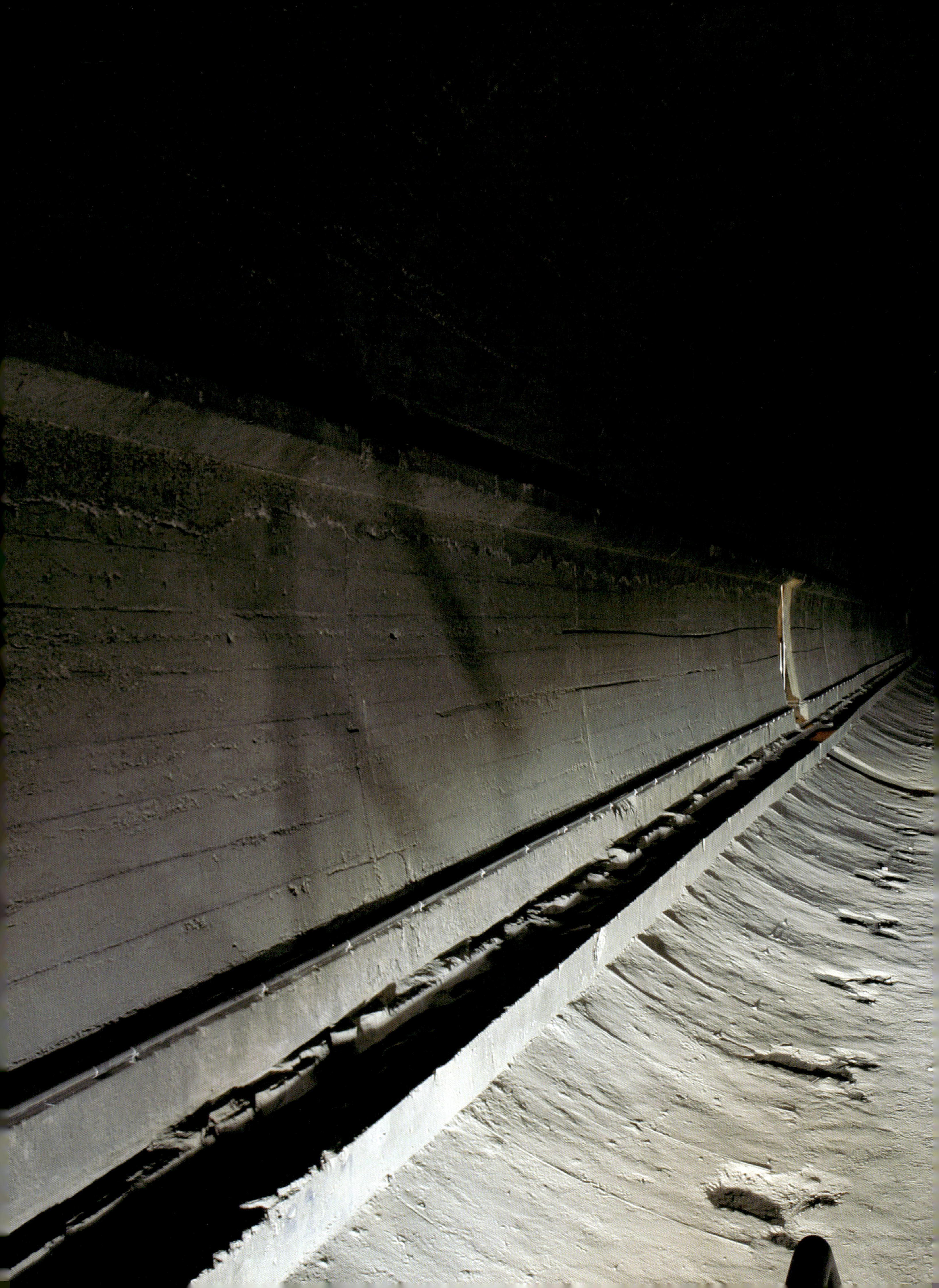

PREVIOUS PAGES:
Tunnel, Marienthal Bunker, Ahrweiler, Germany
Built in the 1960s in a former railway tunnel used by the Nazis to build V-2 rockets near the West German capital of Bonn, the 'Marienthal Office' was designed to be the seat of government in the event of a nuclear war. Up to 3000 people could be accommodated for a month.

RIGHT:
Teletype Machines, Marienthal Bunker, Ahrweiler, Germany
The bunker was well-equipped with communications systems to ensure it could remain in contact with the outside world. In the event that the various antennae on the mountain above were damaged, a retractable antenna could be used. There was even a payphone for bunker inhabitants to call their families if they and the phone system survived an attack. The bunker was finally closed in 1997.

OVERLEAF:
Former Soviet Barracks, Wünsdorf, Germany
The former communications bunker built at Zossen for the Nazi Supreme High Command (OKH – *Oberkommando des Heeres*) in WWII was converted by the Soviets into the command centre for the Group of Soviet Forces in Germany and given the codename 'Ranet'. The former Nazi barracks nearby were used by the Red Army for housing their soldiers guarding the site.

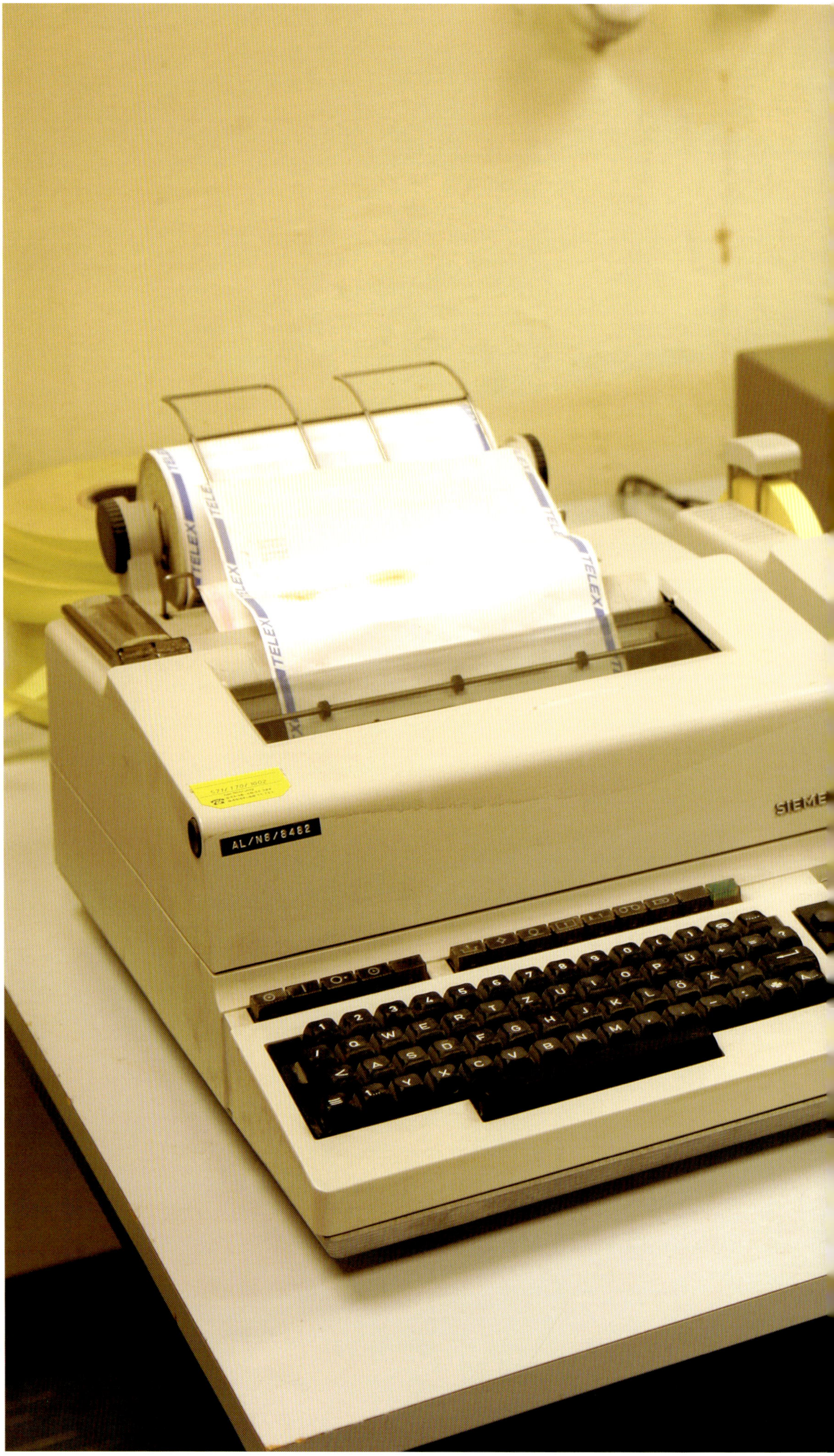

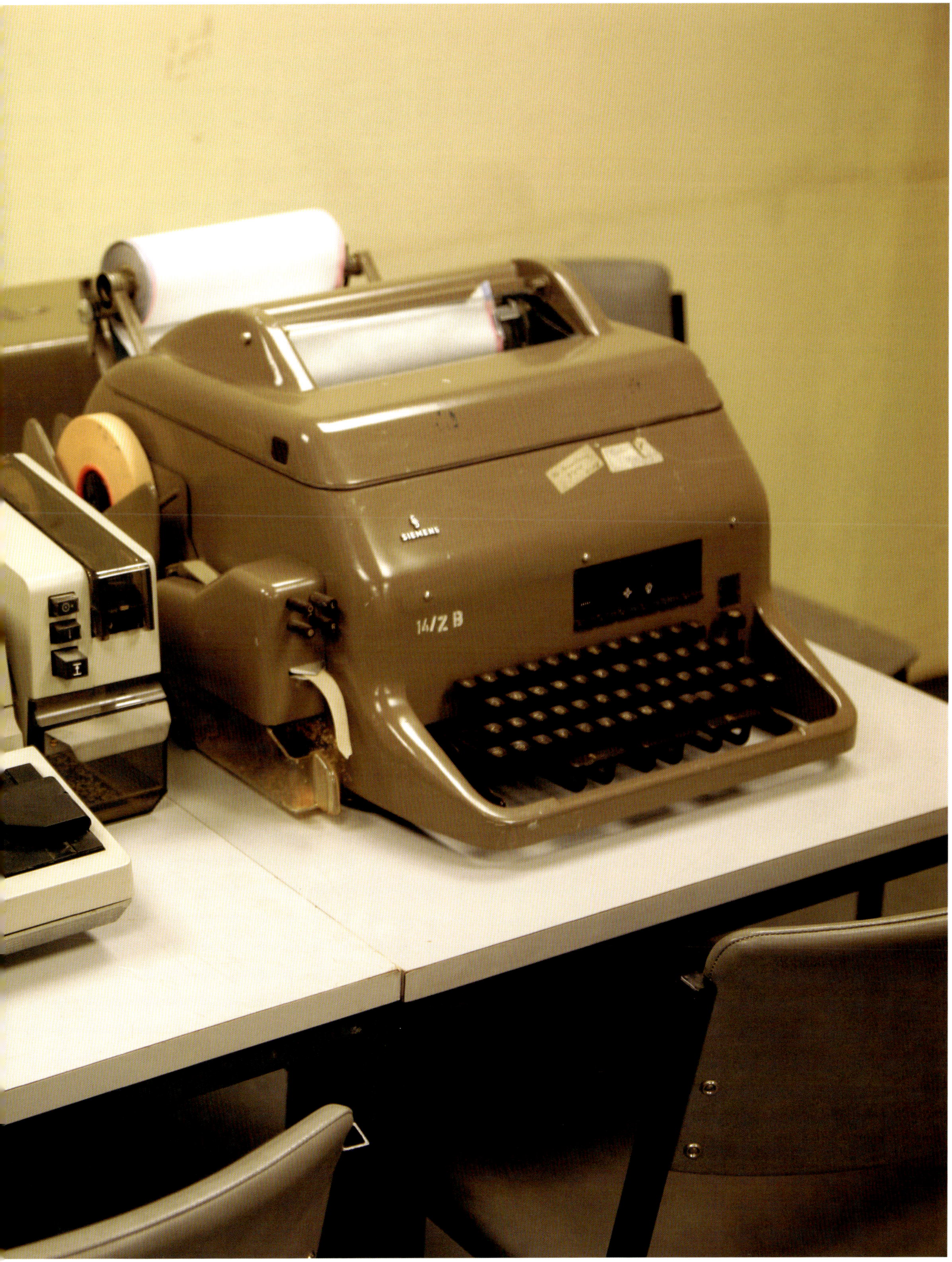
SIEMENS
14/Z B

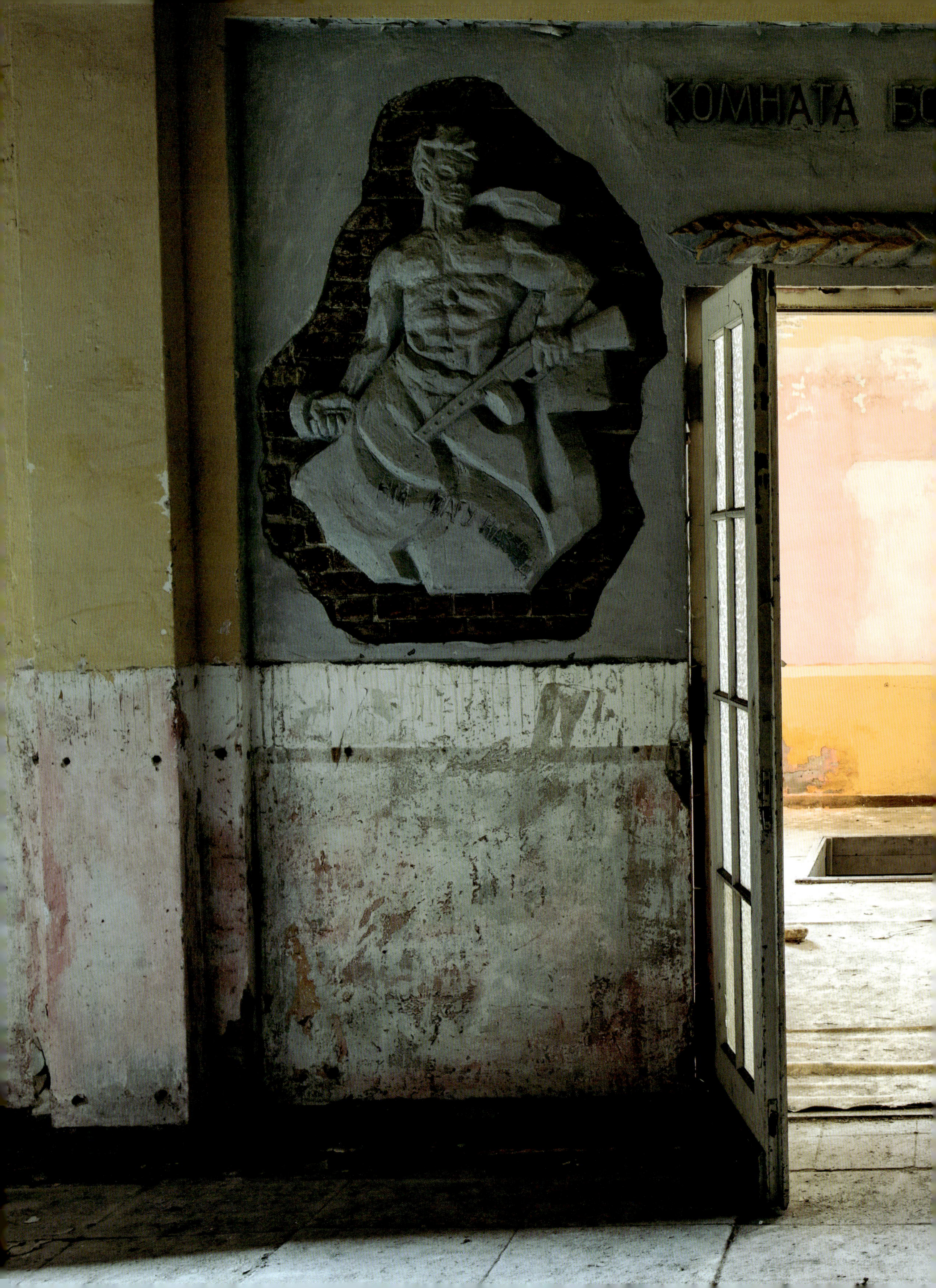
КОМНАТА

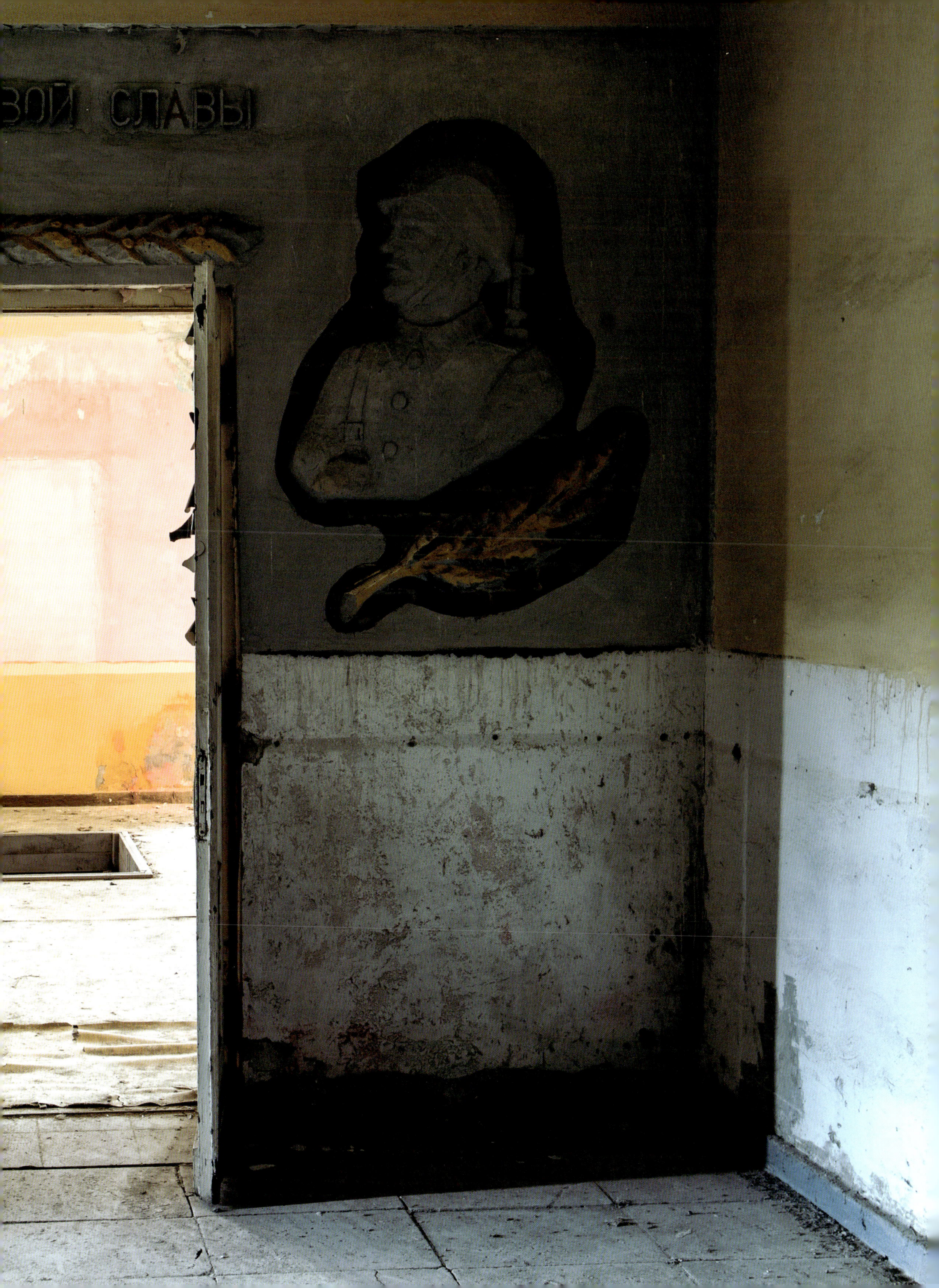
ВОЙ СЛАВЫ

RIGHT:
Former Border Watchtower, Schlagsdorf, Germany
The border between East and West Germany consisted of a series of fences and minefields, all watched over by guards either in towers like these or on patrol. Anyone living near the border in East Germany required a special stamp on their identity card as authorization to enter the restricted zone.

OVERLEAF:
Former Border Checkpoint, Marienborn, Germany
Due to its position as the closest checkpoint in West Germany to West Berlin, most vehicle traffic travelled through this crossing. It was the closing of crossing points like this and the rail routes into West Berlin by the East German government that led to the Berlin Airlift.

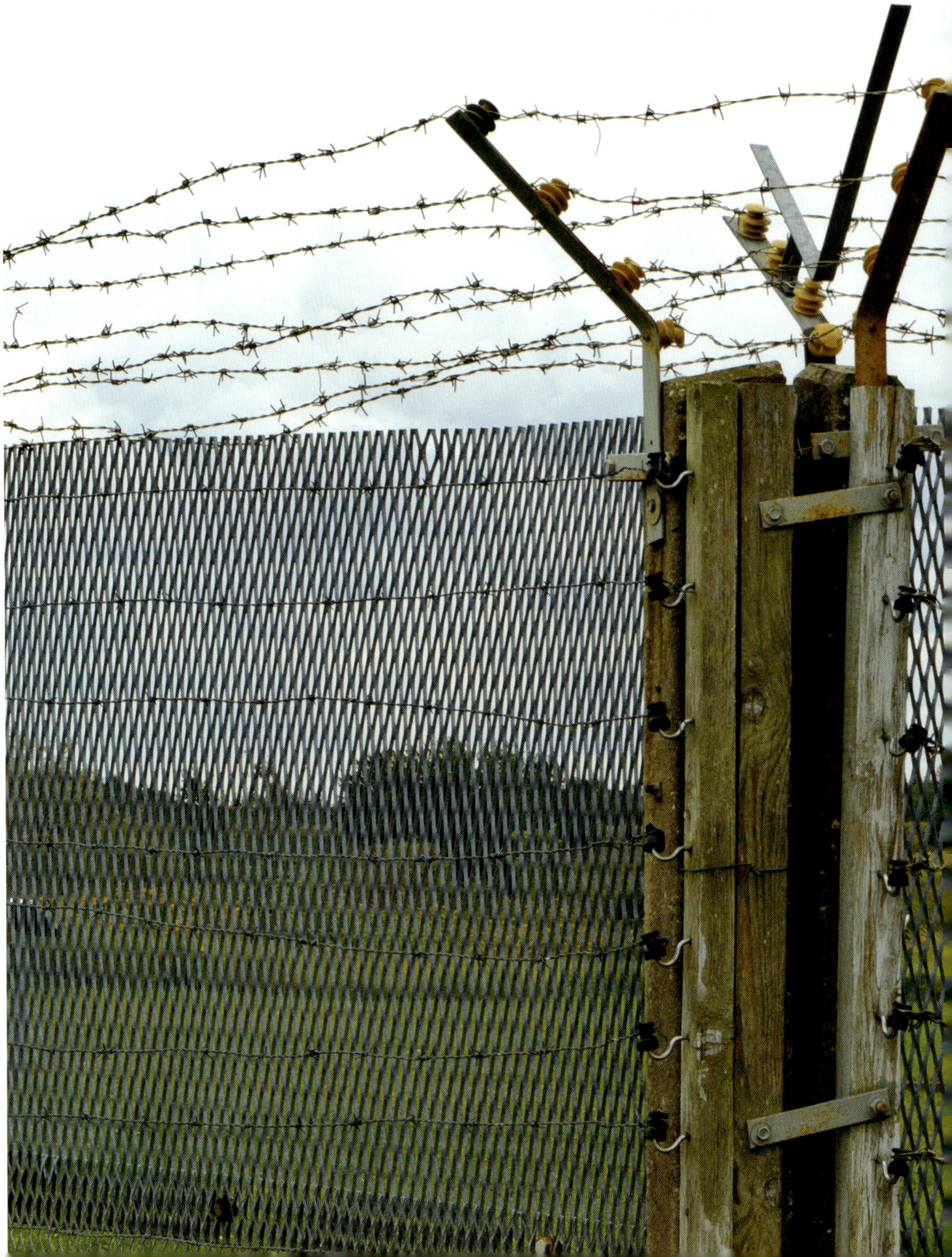

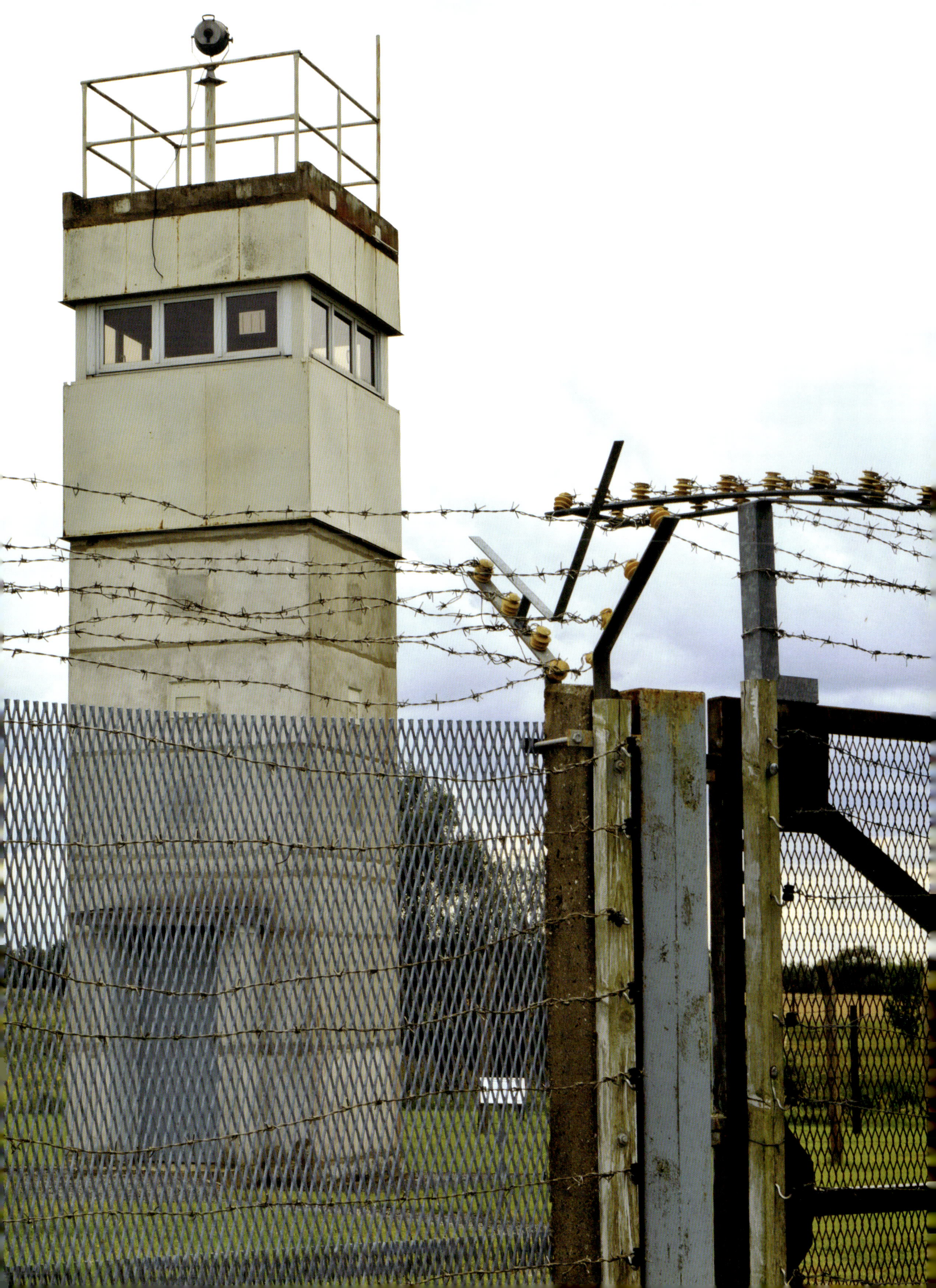

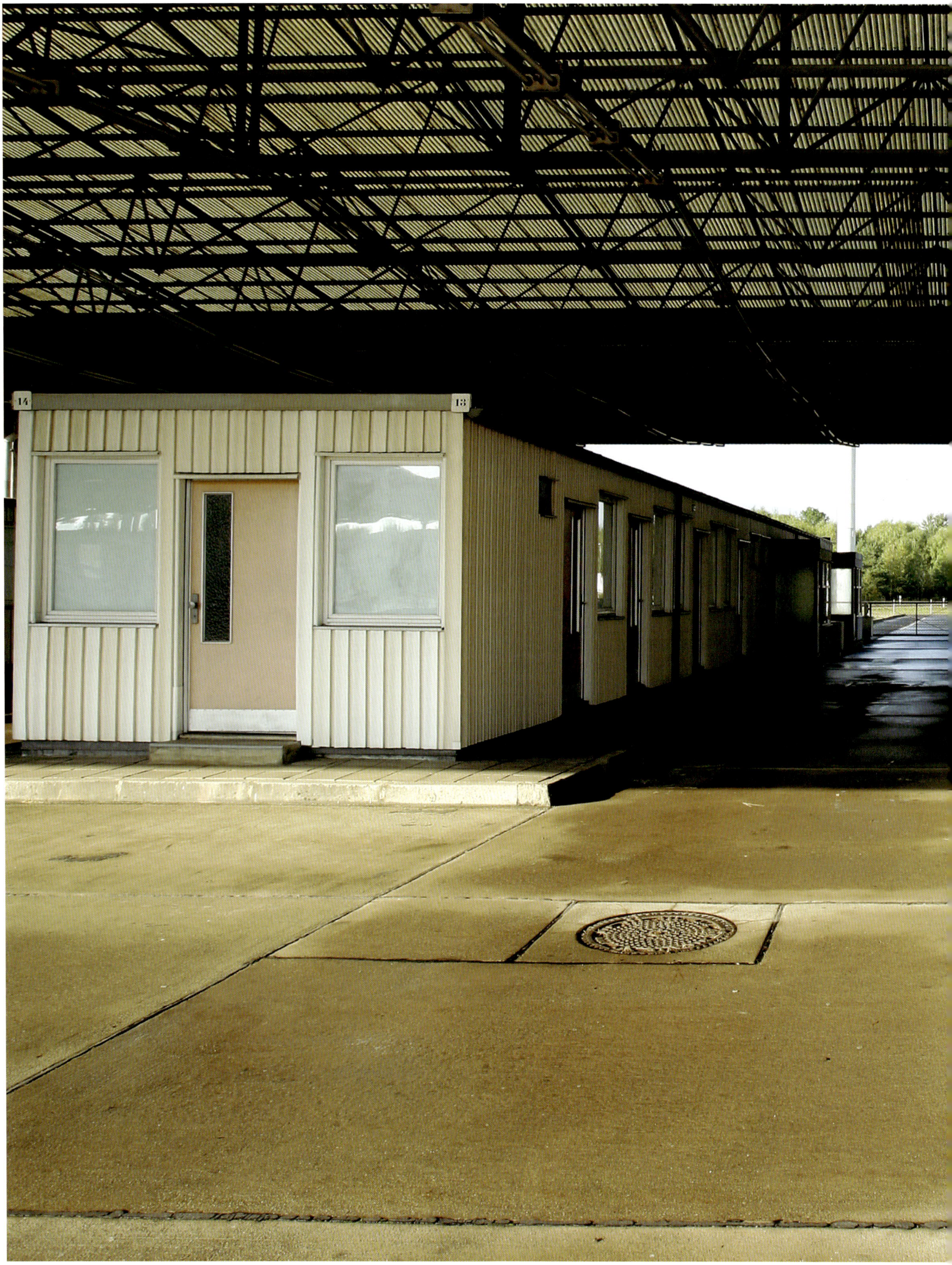
14
13

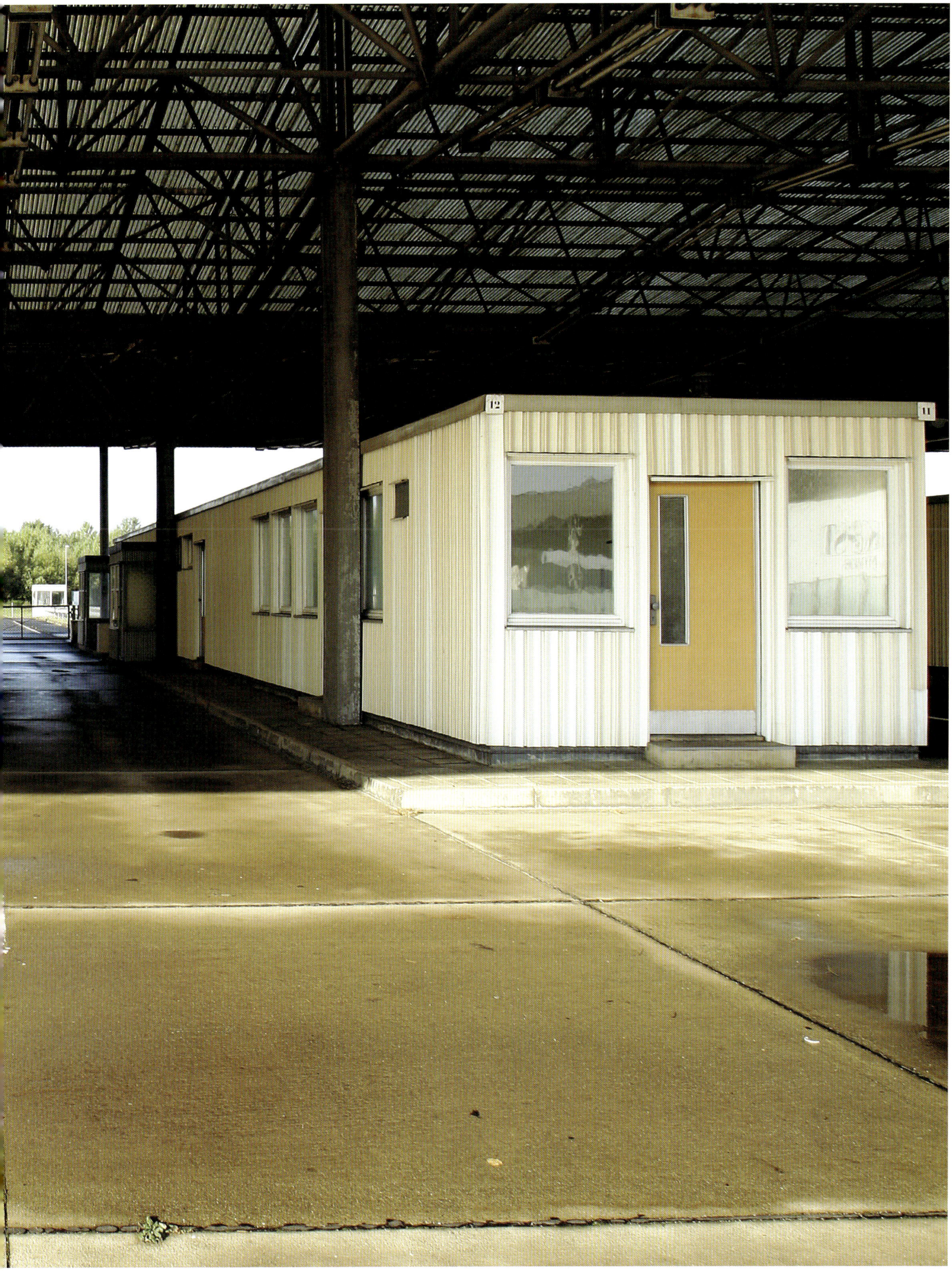

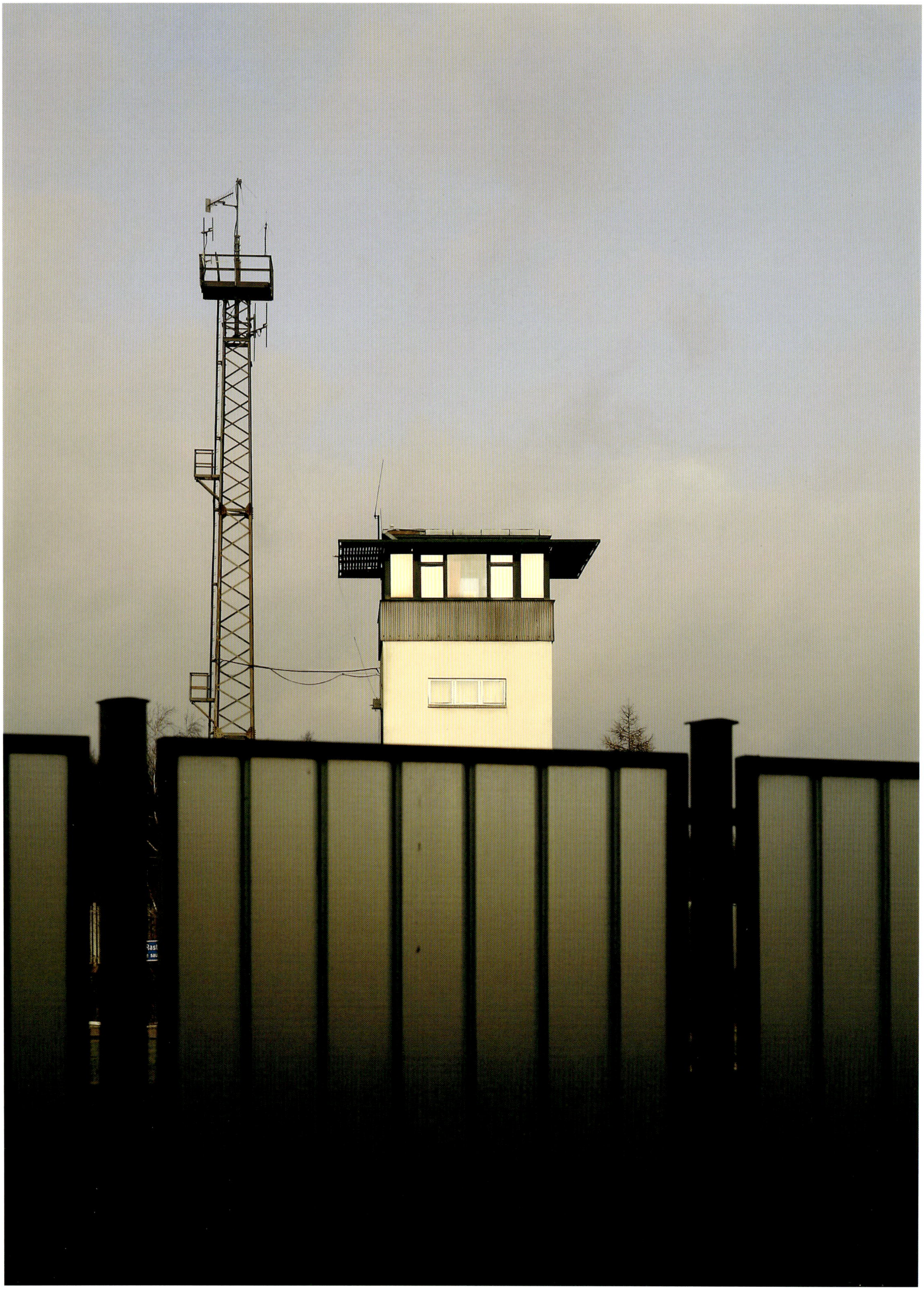

OPPOSITE:
Communications Mast and Watchtower, Marienborn, Germany
The surveillance at the border covered the East German personnel as much as the traffic transiting between Berlin and the West. Although the border staff were East German, any NATO military traffic travelling to or from West Berlin was always escorted by Soviet forces.

RIGHT:
Former Border Checkpoint, Marienborn, Germany
High-powered lights ensured that the crossing was brightly lit at all times. The facility at Marienborn was known to the West as Checkpoint Alpha. At the other end of the autobahn the crossing point into West Berlin at Drewitz-Dreilinden was known as Checkpoint Bravo.

BELOW:
Vehicle Inspection Room, Marienborn, Germany
If a vehicle aroused the suspicions of the border guards, it would be brought into this room for closer inspection. With steel doors at either end to prevent any escape, the inspection pit allowed the officials to examine the vehicle for anything suspicious.

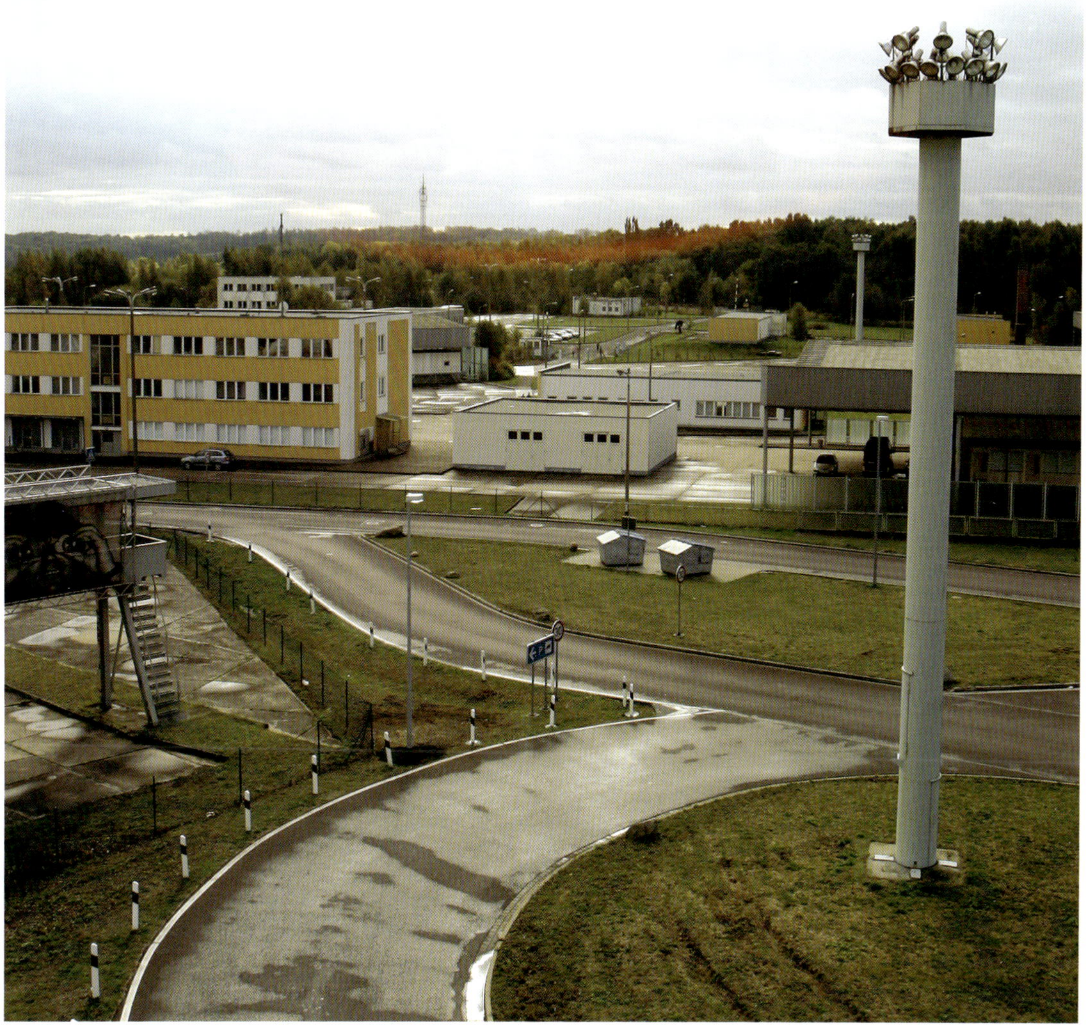

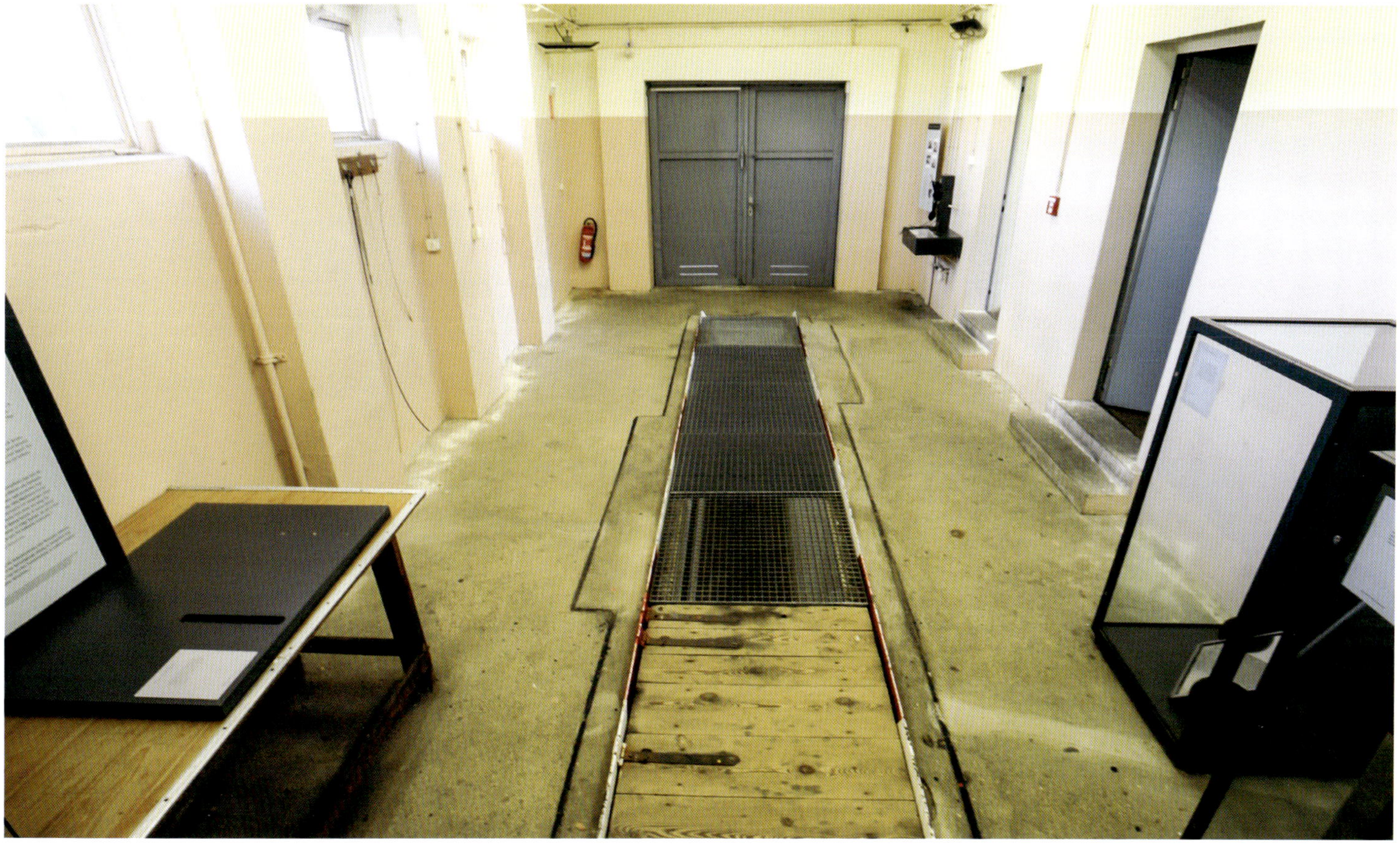

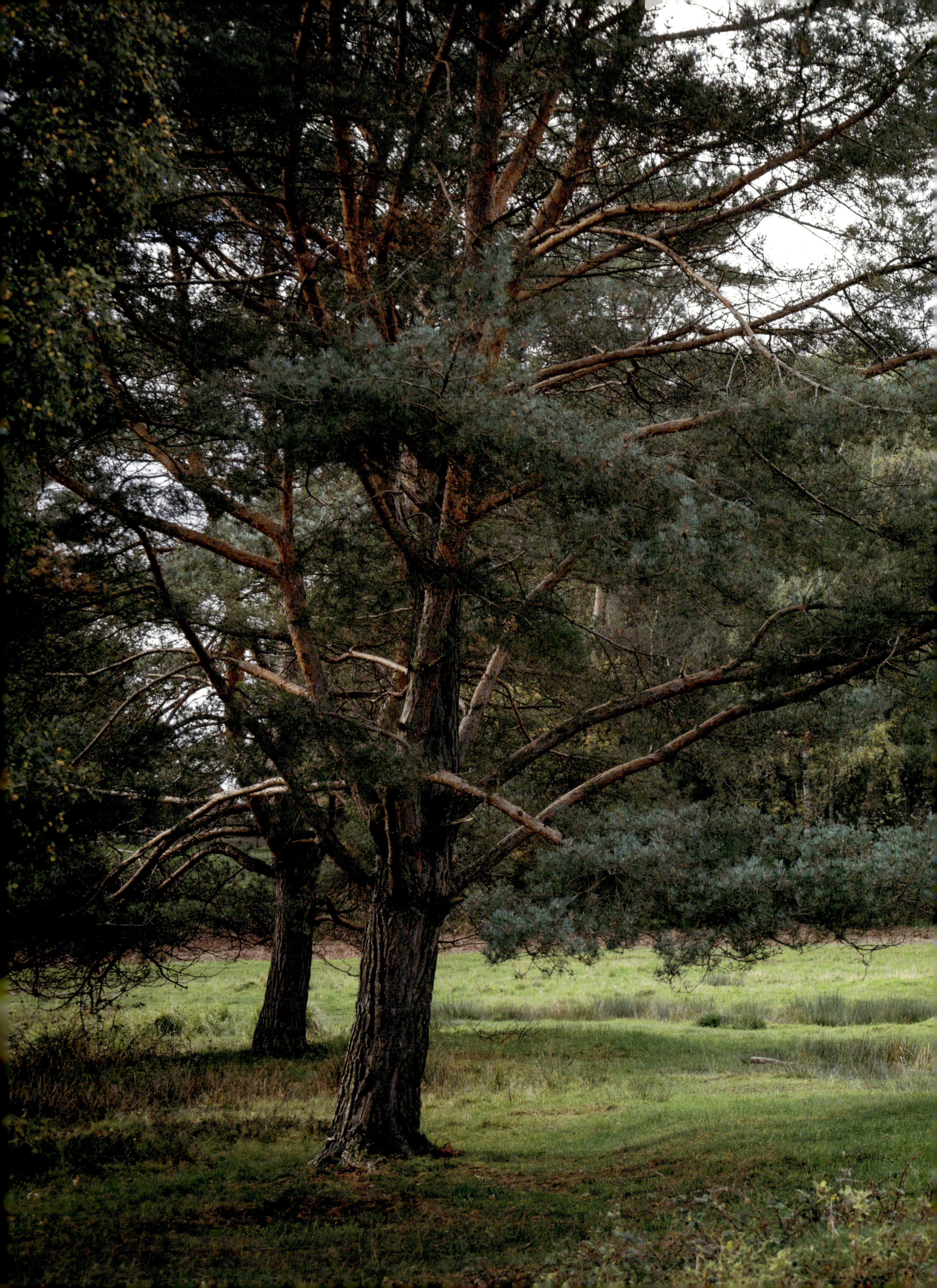

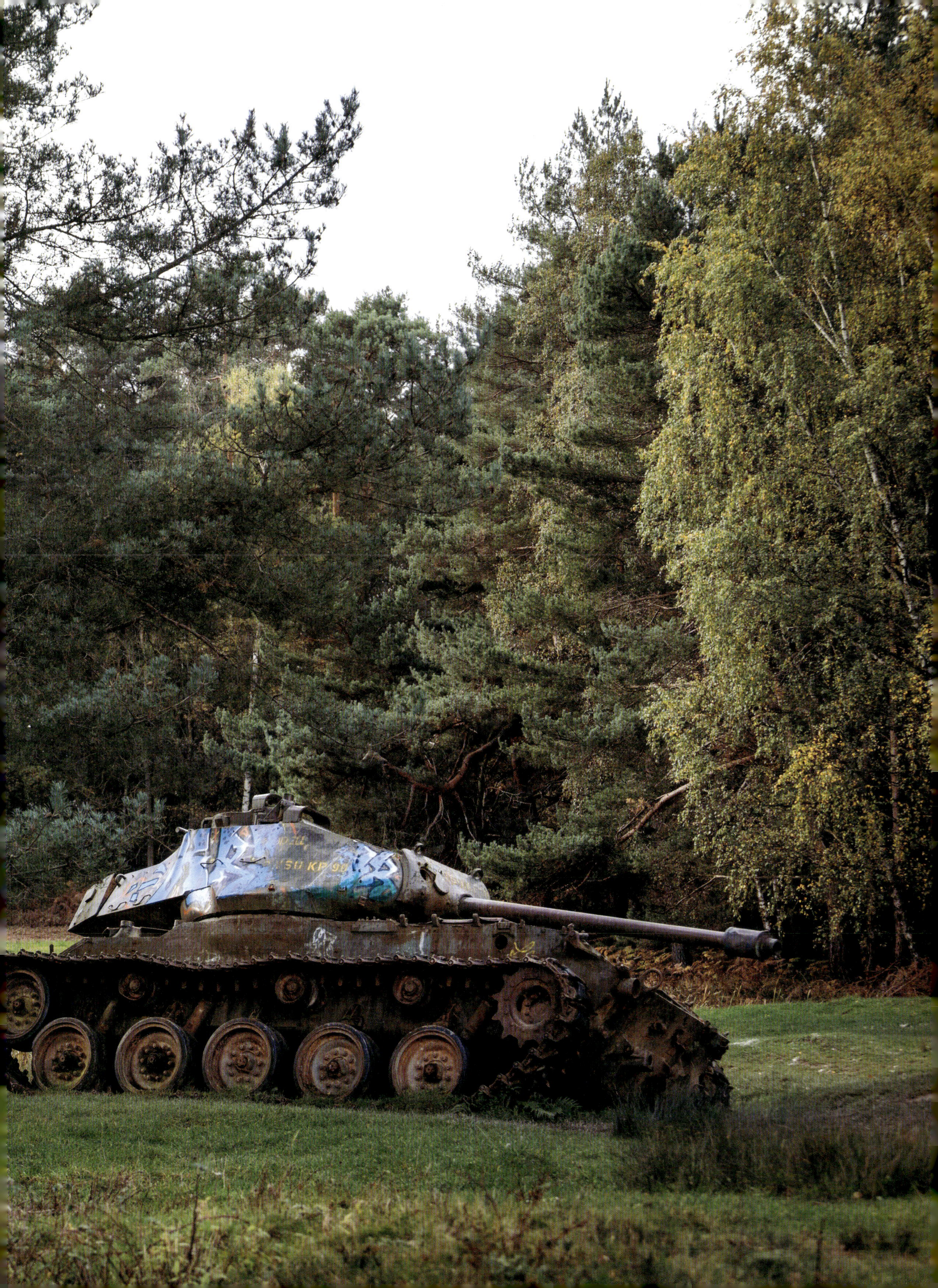

PREVIOUS PAGES:
M47 Patton Tank, Aachen, Germany
The M47 Patton was produced in 1951 in response to the failings of the M46 tank and the outbreak of the Korean War. It had a troubled development but eventually over 8500 were produced, and it became the standard tank of the US forces in Germany and the West German Army. However, it was quickly replaced in US service by the superior M48 tank.

RIGHT:
Civil Defence Shelter, Katrineholm, Sweden
The entrance to the shelter built under the city hall in Katrineholm, Sweden. The Swedish government made extensive preparations during the Cold War to protect its citizens against nuclear war, including the establishment of civil defence schools in key municipalities such as Katrineholm.

OVERLEAF:
Civil Defence Shelter, Interior, Katrineholm, Sweden
Sweden was officially neutral throughout the Cold War, but maintained significant armed forces to protect her neutrality. The Royal Swedish Air Force built underground hangars for its aircraft that included workshops, an air operations centre, canteen and accommodation protected by large blast doors.

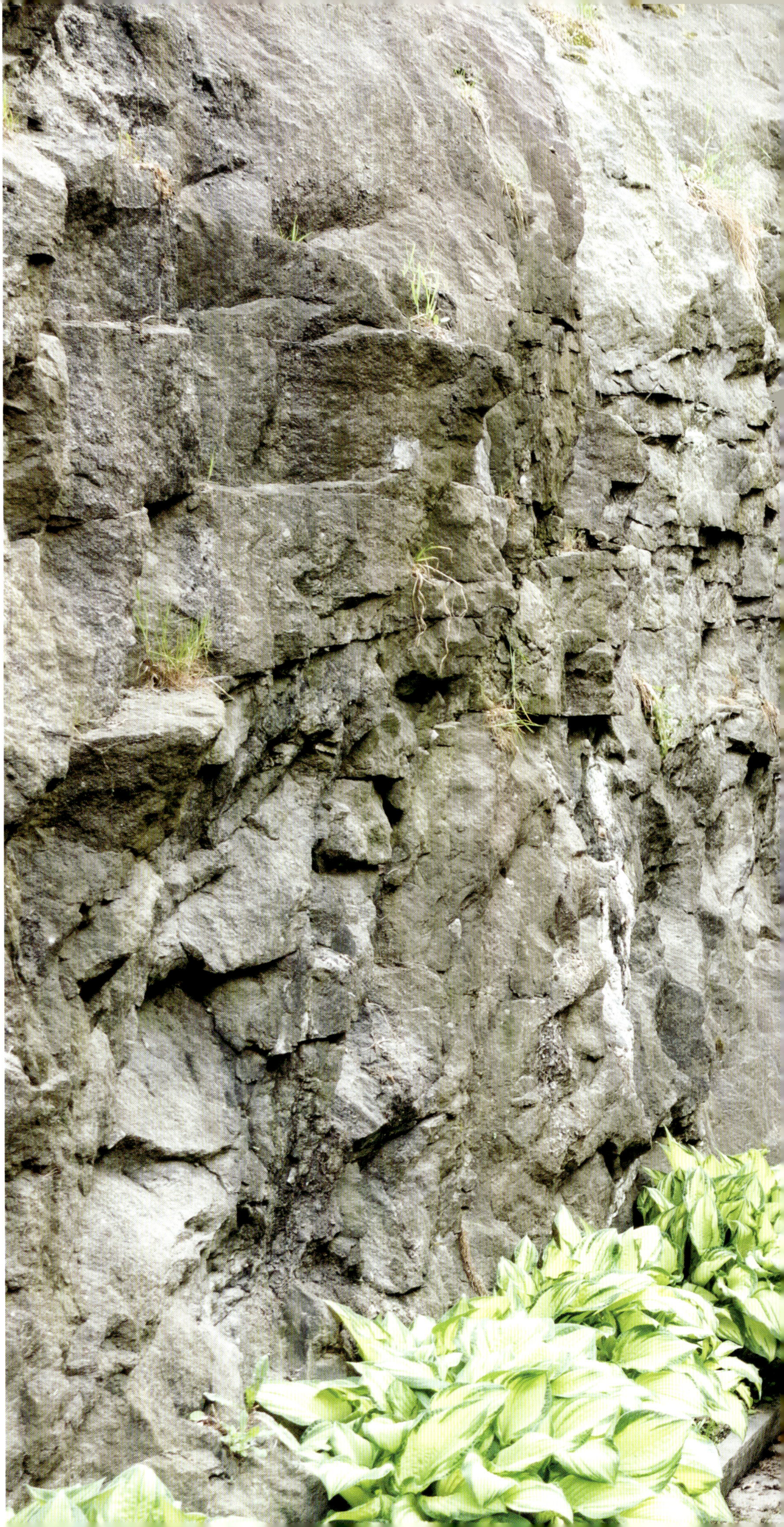

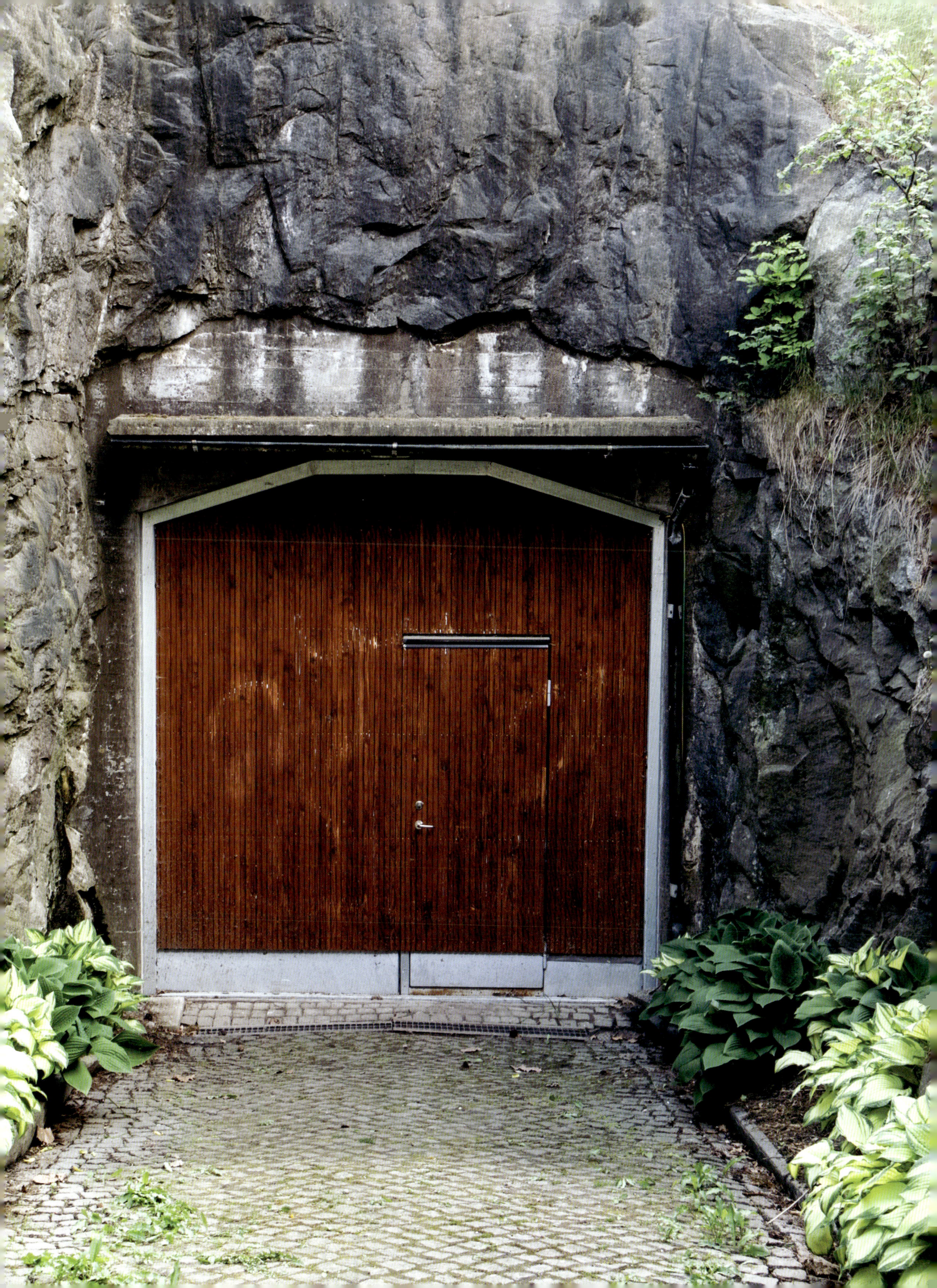

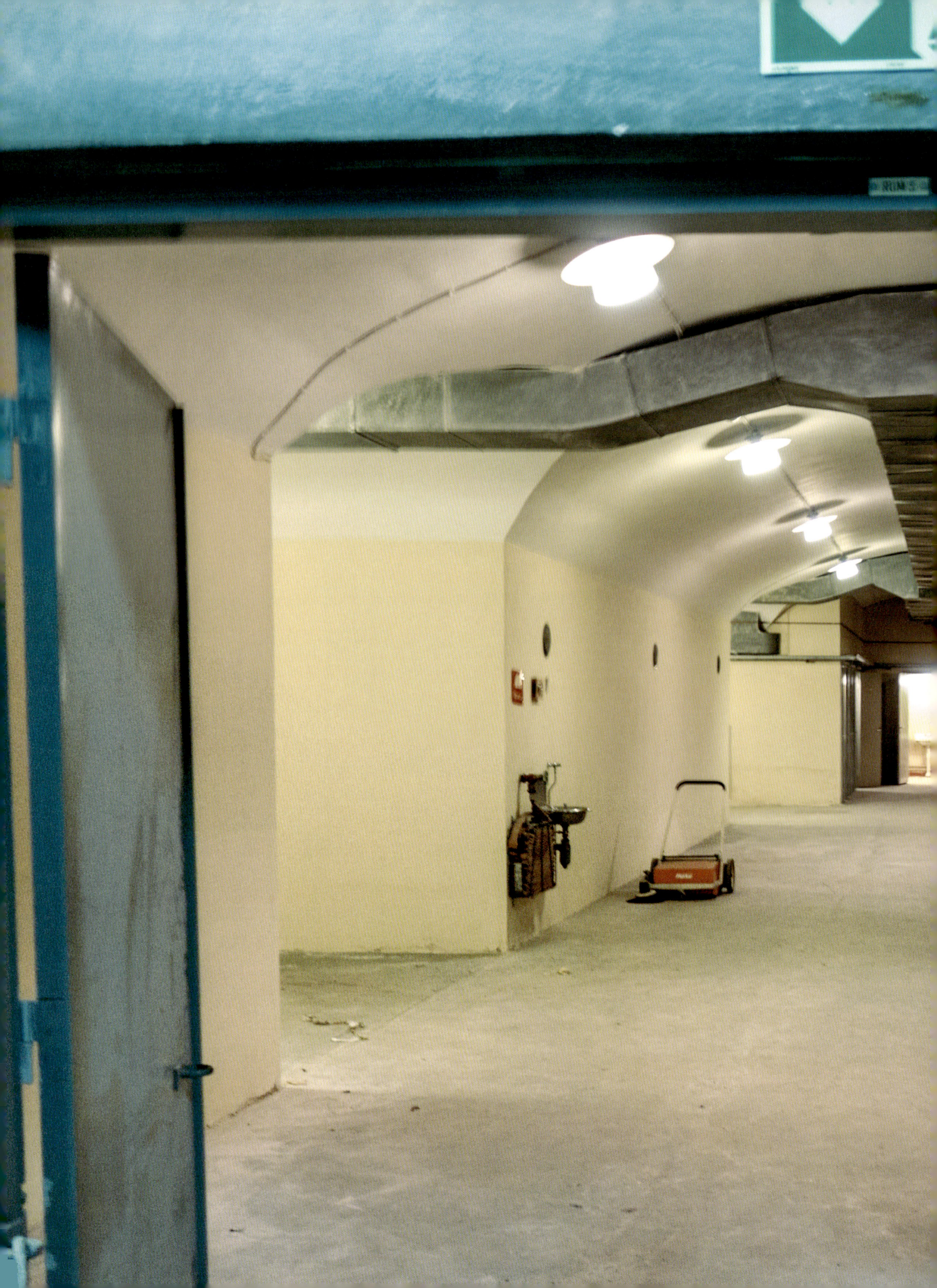

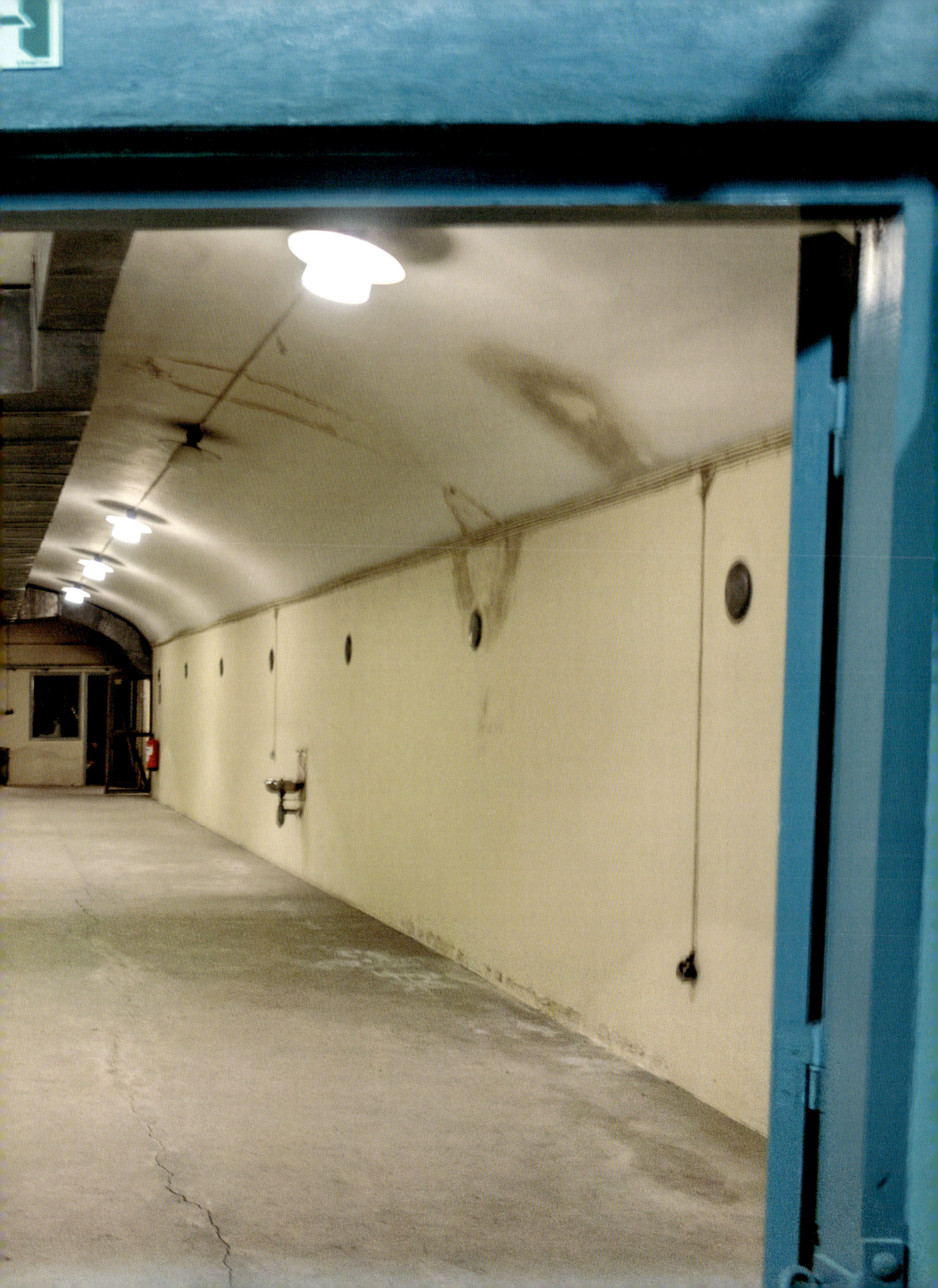

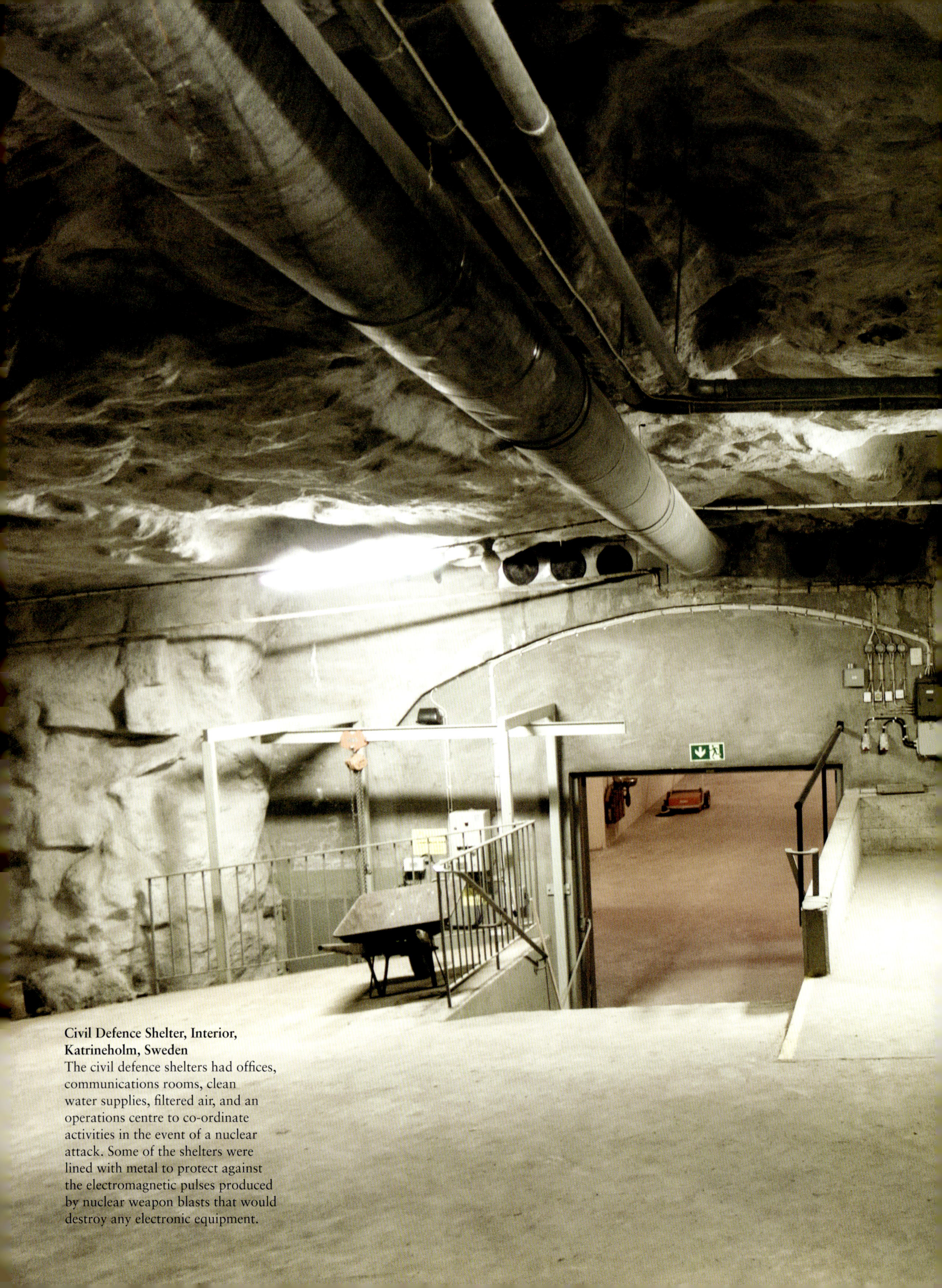

Civil Defence Shelter, Interior, Katrineholm, Sweden
The civil defence shelters had offices, communications rooms, clean water supplies, filtered air, and an operations centre to co-ordinate activities in the event of a nuclear attack. Some of the shelters were lined with metal to protect against the electromagnetic pulses produced by nuclear weapon blasts that would destroy any electronic equipment.

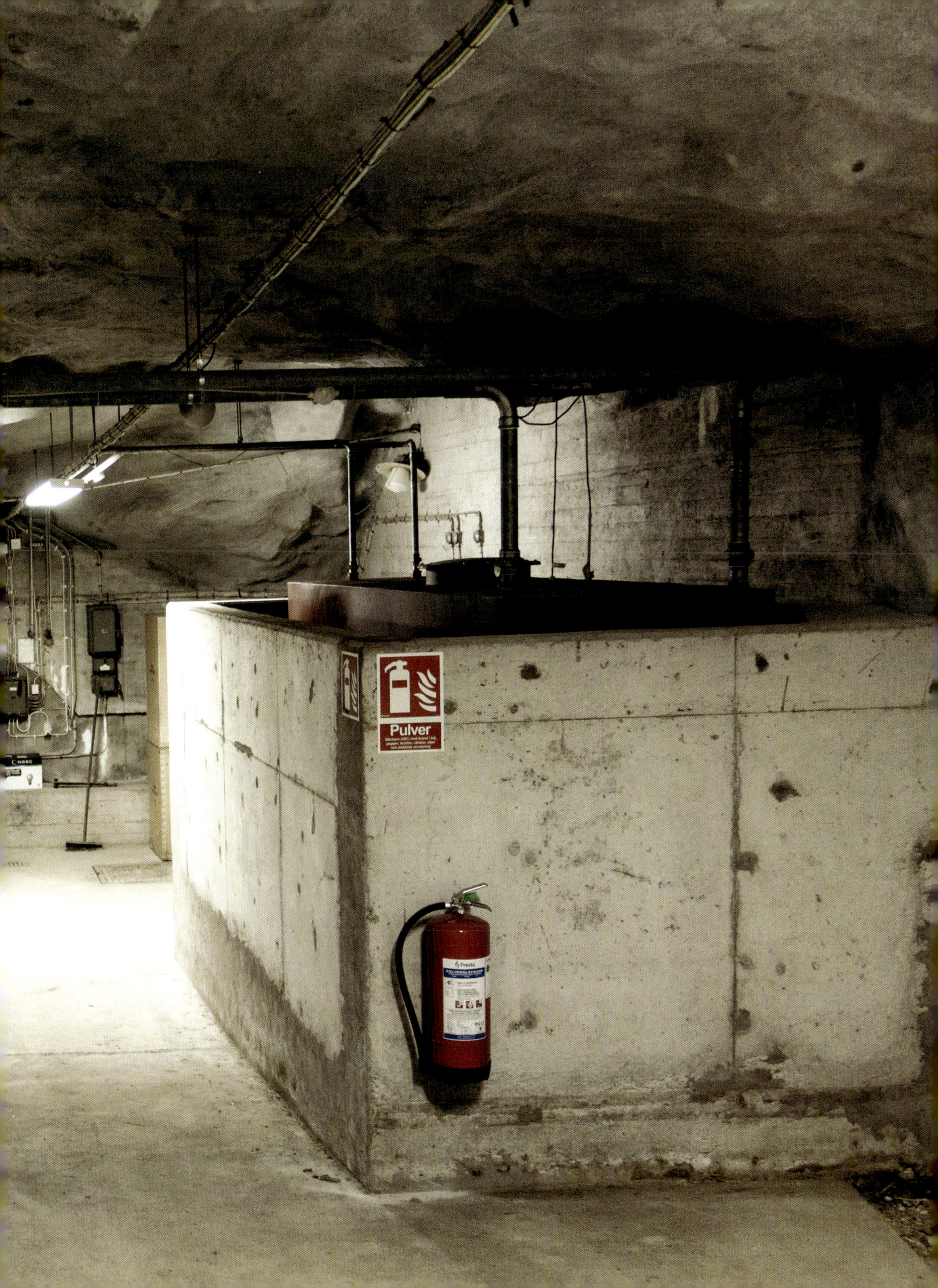
Pulver

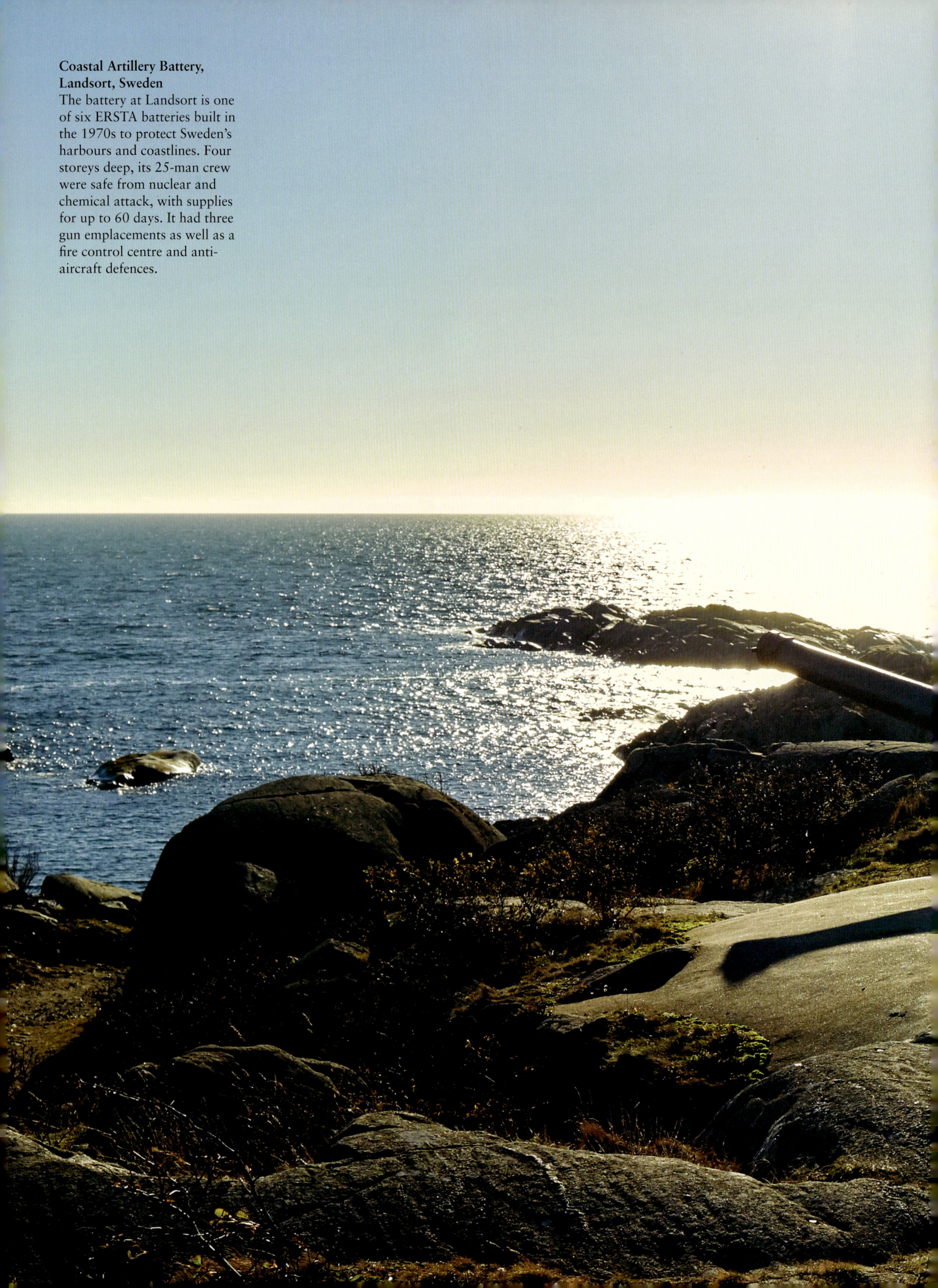

Coastal Artillery Battery, Landsort, Sweden
The battery at Landsort is one of six ERSTA batteries built in the 1970s to protect Sweden's harbours and coastlines. Four storeys deep, its 25-man crew were safe from nuclear and chemical attack, with supplies for up to 60 days. It had three gun emplacements as well as a fire control centre and anti-aircraft defences.

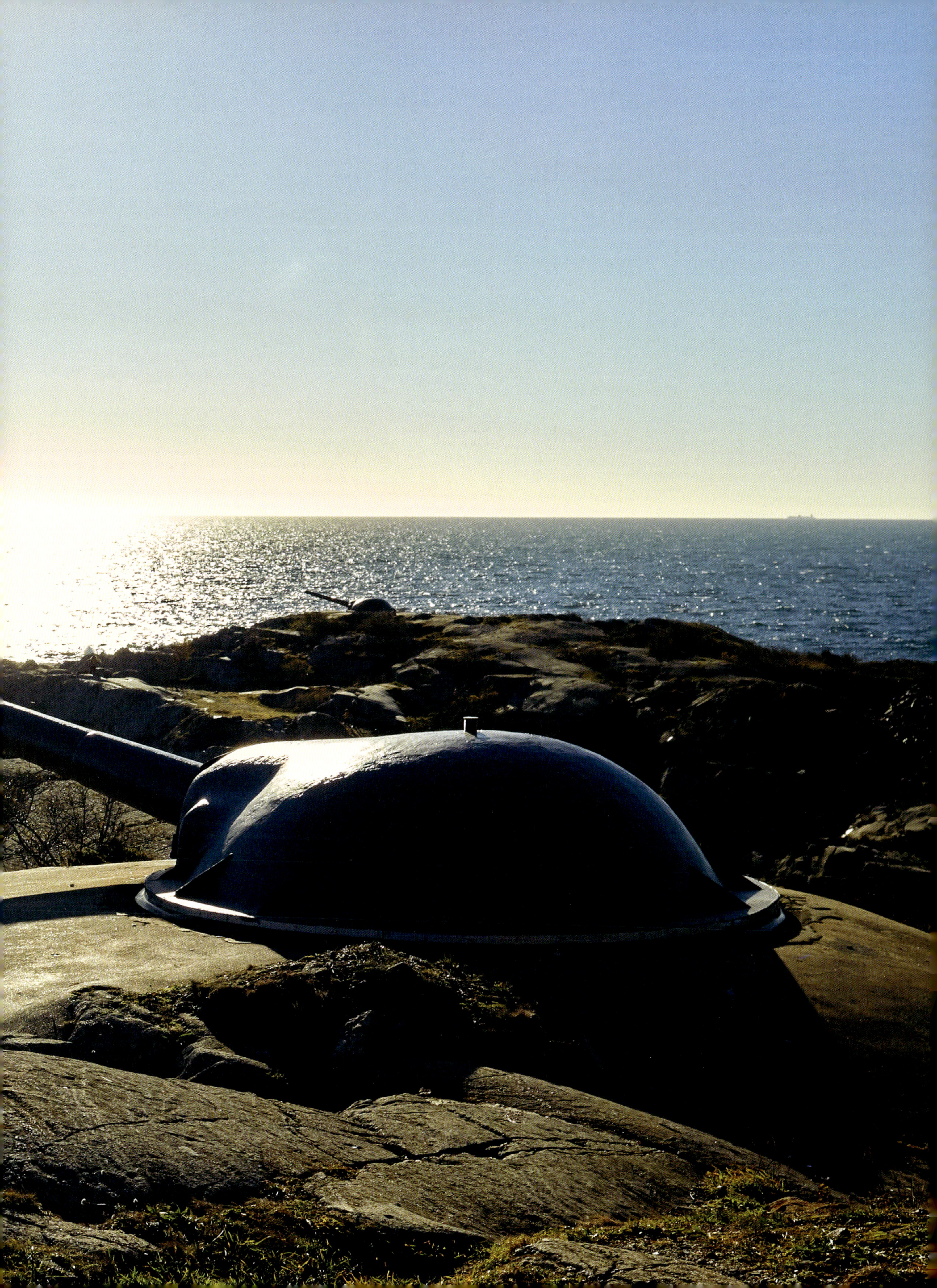

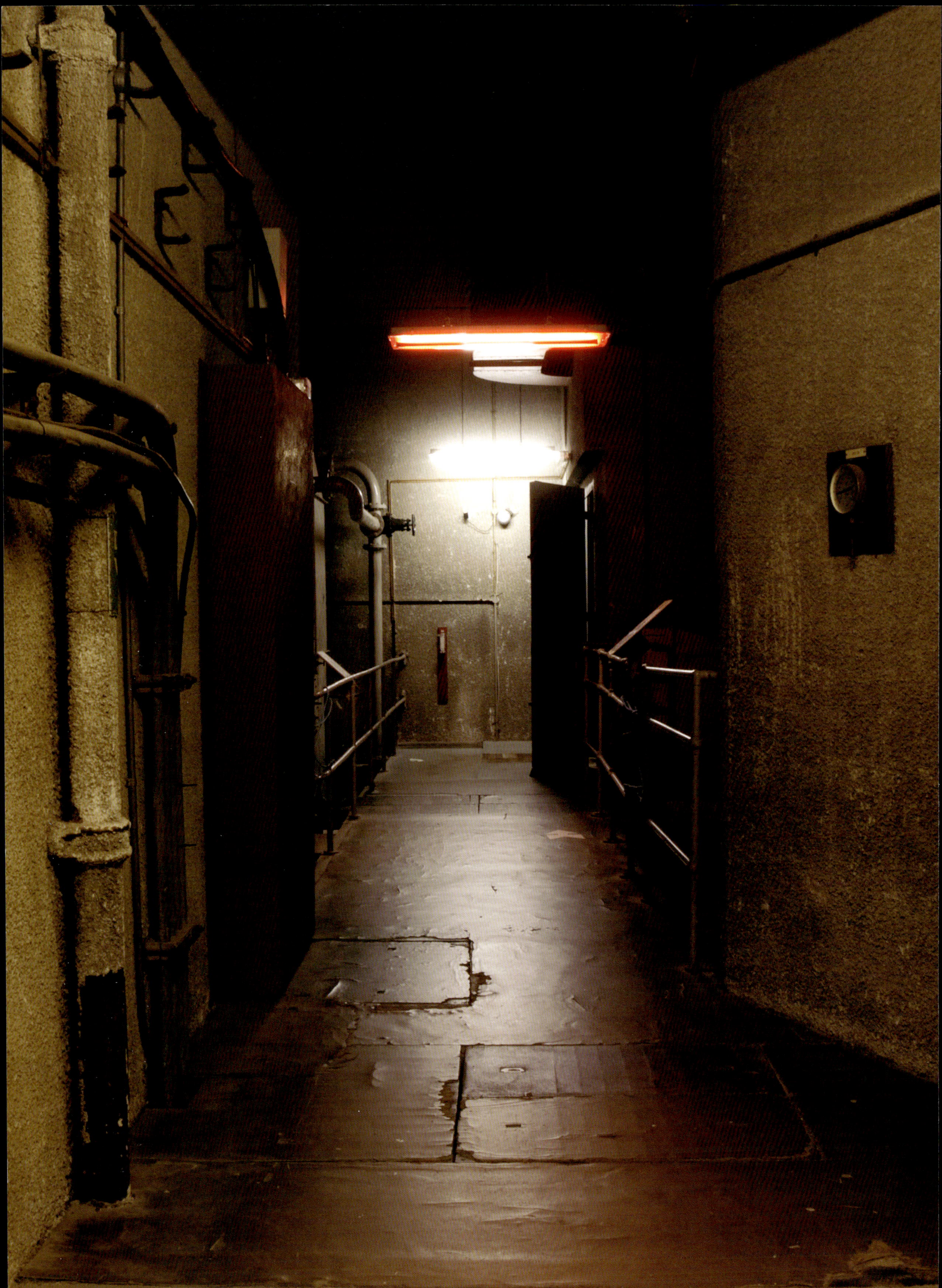

United Kingdom

Once referred to as an 'unsinkable aircraft carrier', its position on the edge of Europe made the UK a vital staging post for US forces as it had in World War II, as well as forming part of the GIUK (Greenland – Iceland – UK) gap defences designed to intercept Soviet submarines attempting to enter the Atlantic. In fact, due to the relative small geographic size of the UK and its high population density, only a small proportion of the Soviet nuclear arsenal would have been required to cause such extensive damage and casualties that the UK would have been unable to continue to function as an active combatant. Initial post-war UK defences against any Soviet attack relied on plans that differed very little from those that had been successful against the Nazis, but by the 1950s a new awareness of the likely nature of any nuclear conflict spurred a re-evaluation of their preparations.

As the Cold War entered the 1980s, the UK controversially became host to two US cruise missile bases. After the collapse of the Soviet Union, the subsequent withdrawal of the majority of US service personnel and rapid reduction in the size of UK forces left many former Cold War bases shuttered or sold off to developers.

OPPOSITE:

Air Filtration Room, Kelvedon Hatch Bunker, Essex

The British radar network had helped win the Battle of Britain in WWII. In the early 1950s a replacement network named ROTOR was built to detect potential Soviet bomber attacks as early as possible. Six Sector Operations Centre (SOC) stations were built to co-ordinate information from the radar stations, including one at Kelvedon Hatch, and they reported to RAF Fighter Command at Bentley Priory.

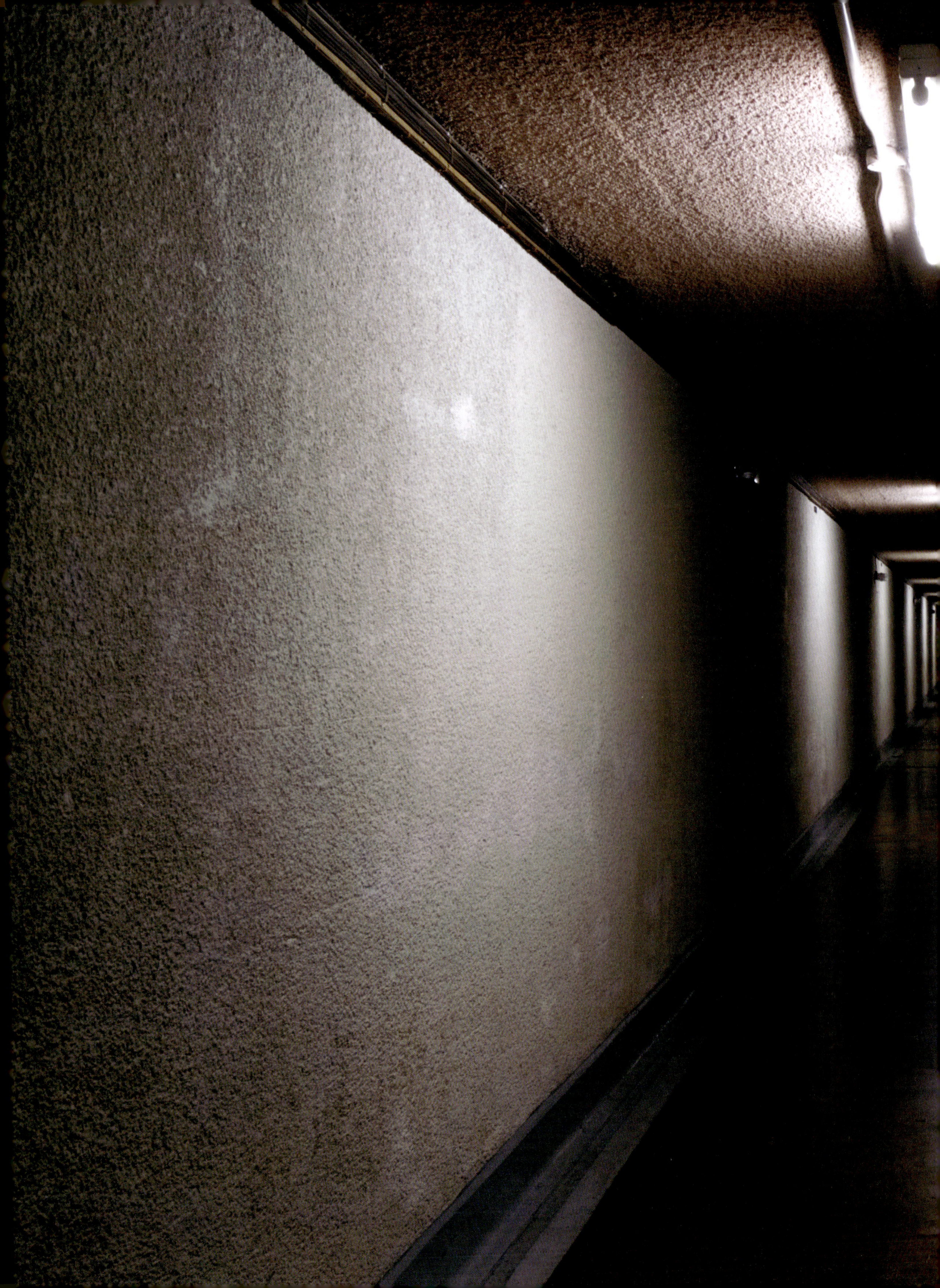

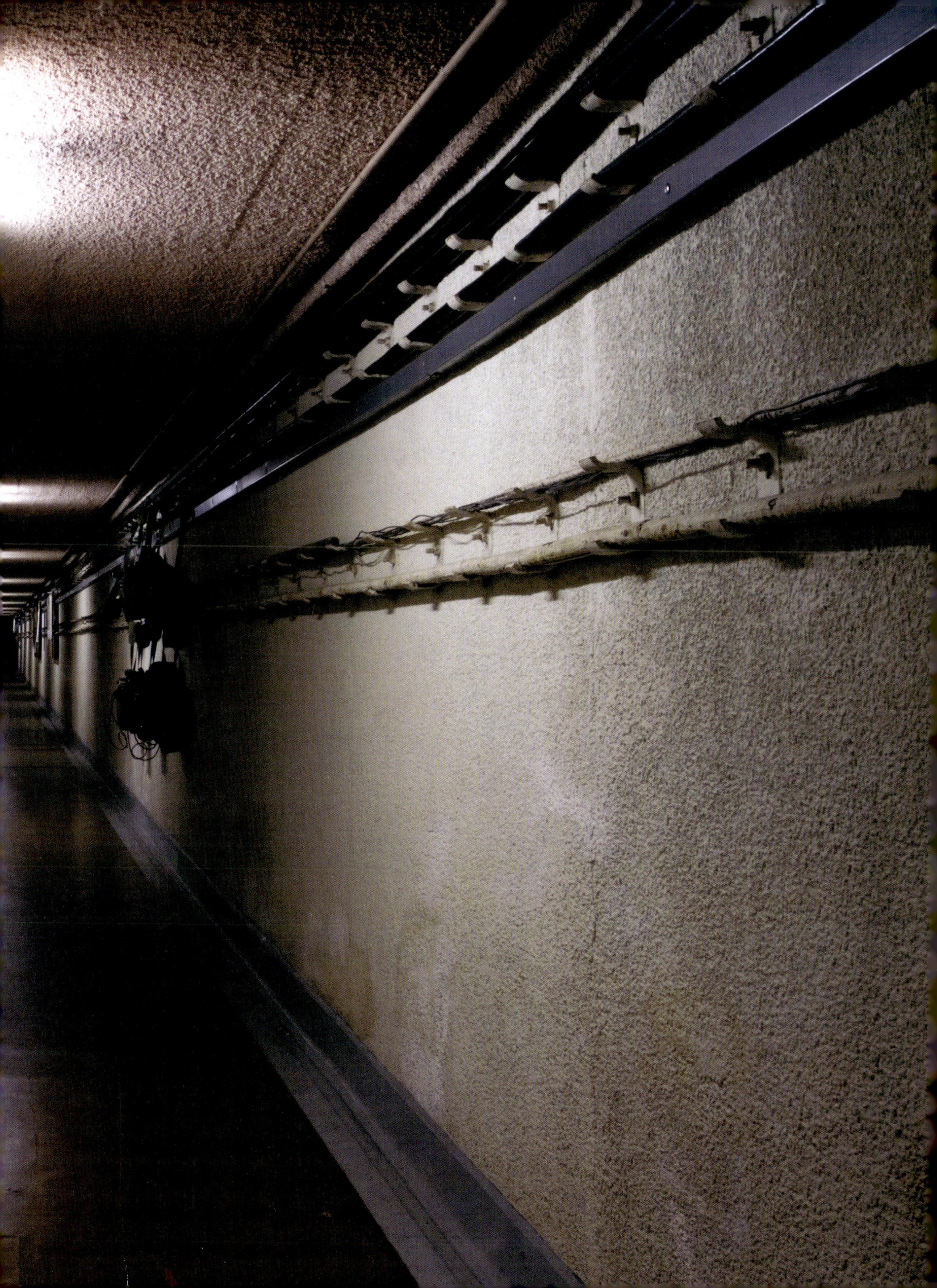

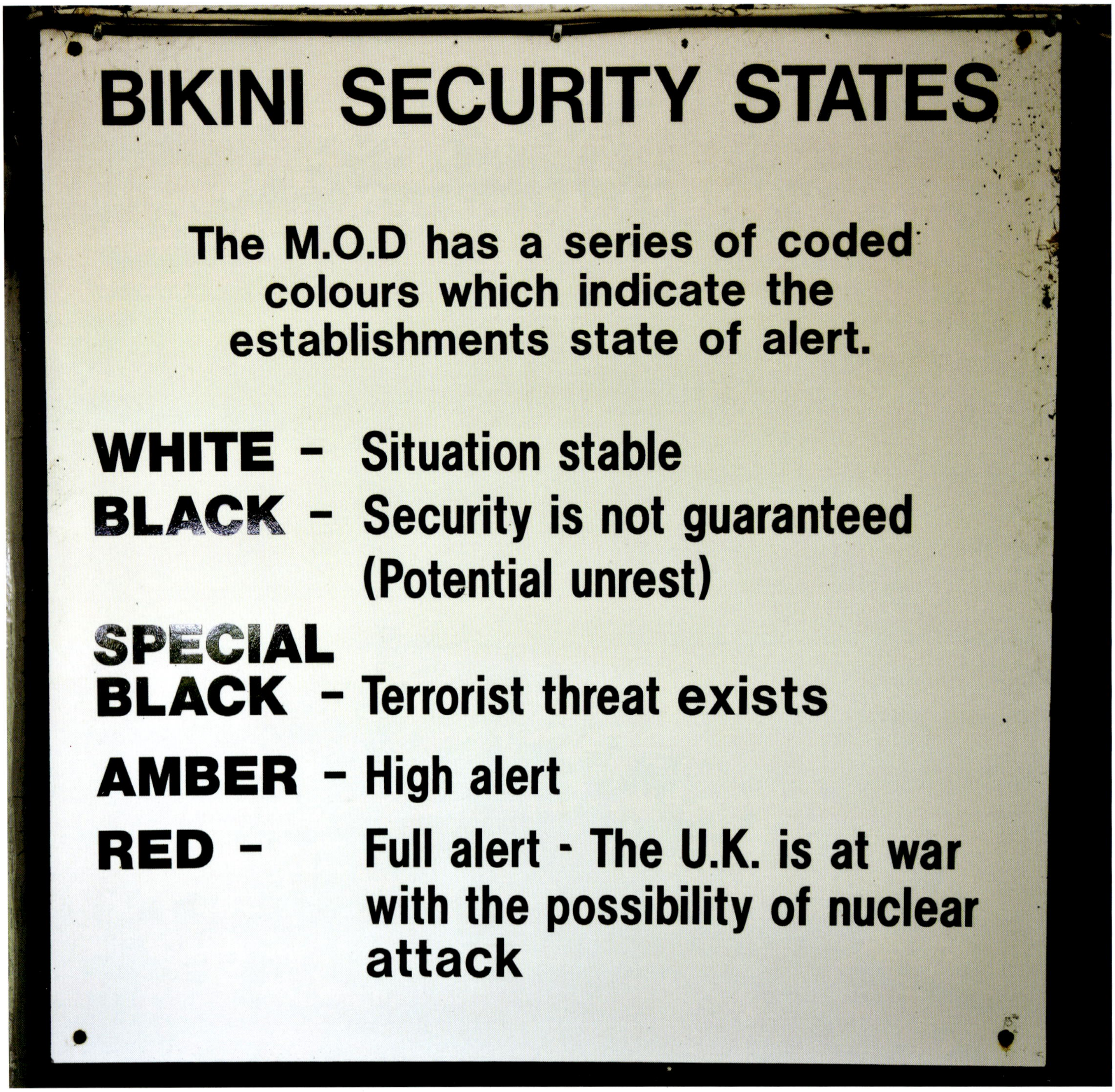

PREVIOUS PAGES:
Corridor, Kelvedon Hatch Bunker, Essex
The entrance to the three-storey bunker was disguised as a bungalow to allay suspicion. Equipped with its own power plant, air filtration, and thick blast doors, the bunker had a plotting table to show the movement of Soviet bombers and the fighters sent to intercept them.

ABOVE:
Alert Level Notice, Kelvedon Hatch Bunker, Essex
The Bikini State system was introduced in 1970 to give an easily-understandable level of alert, similar to the US DEFCON system, except the Bikini States were for individual units or establishments rather than applied on a national level. The system has now been replaced.

OPPOSITE:
Hardened Aircraft Shelter, RAF Woodbridge, Suffolk
Reinforced concrete shelters were used by the US Air Force to protect individual aircraft from attack. With each shelter having direct access to the runway, they ensured aircraft could quickly get into the air in an emergency. RAF Woodbridge operated as a USAF air base from 1952 until 1993.

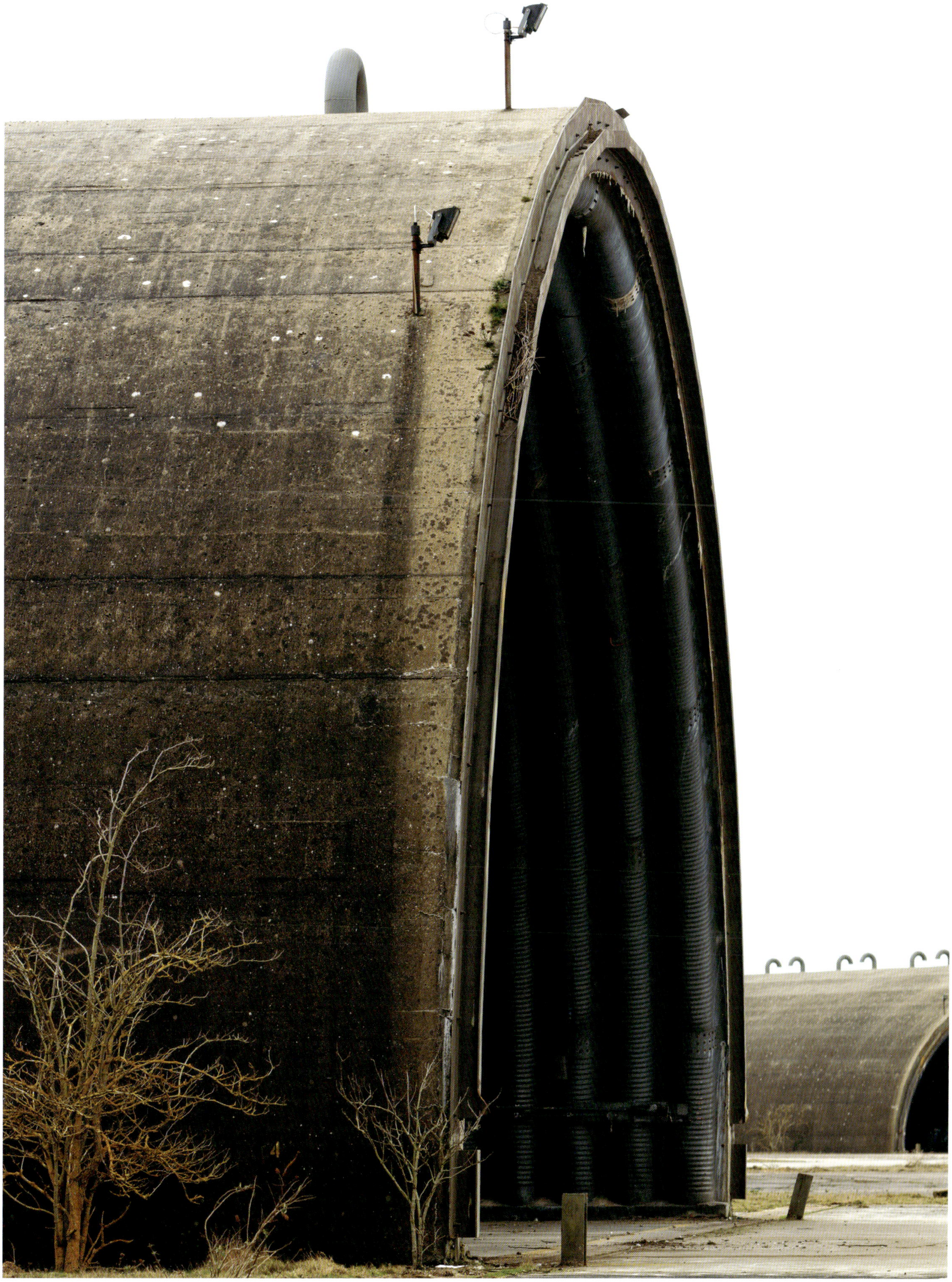

Radar, RAF Neatishead, Norfolk
Part of the original ROTOR network, RAF Neatishead was a key component of the UK air defence network during the Cold War. In February 1966 one of the staff deliberately started a fire in the bunker that burned for nine days before it could be extinguished and cost the lives of three civilian firemen. He was sentenced to seven years' imprisonment.

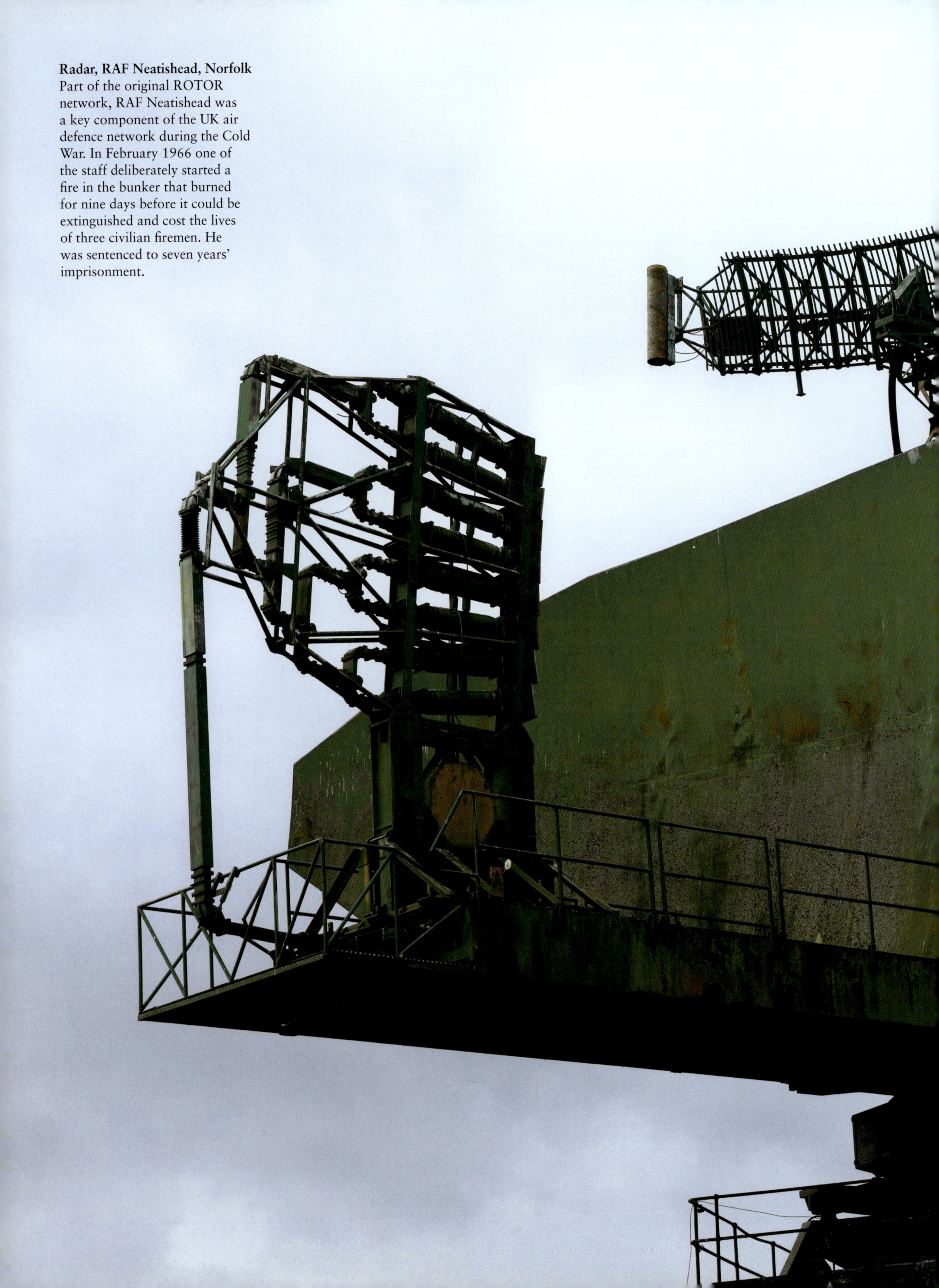

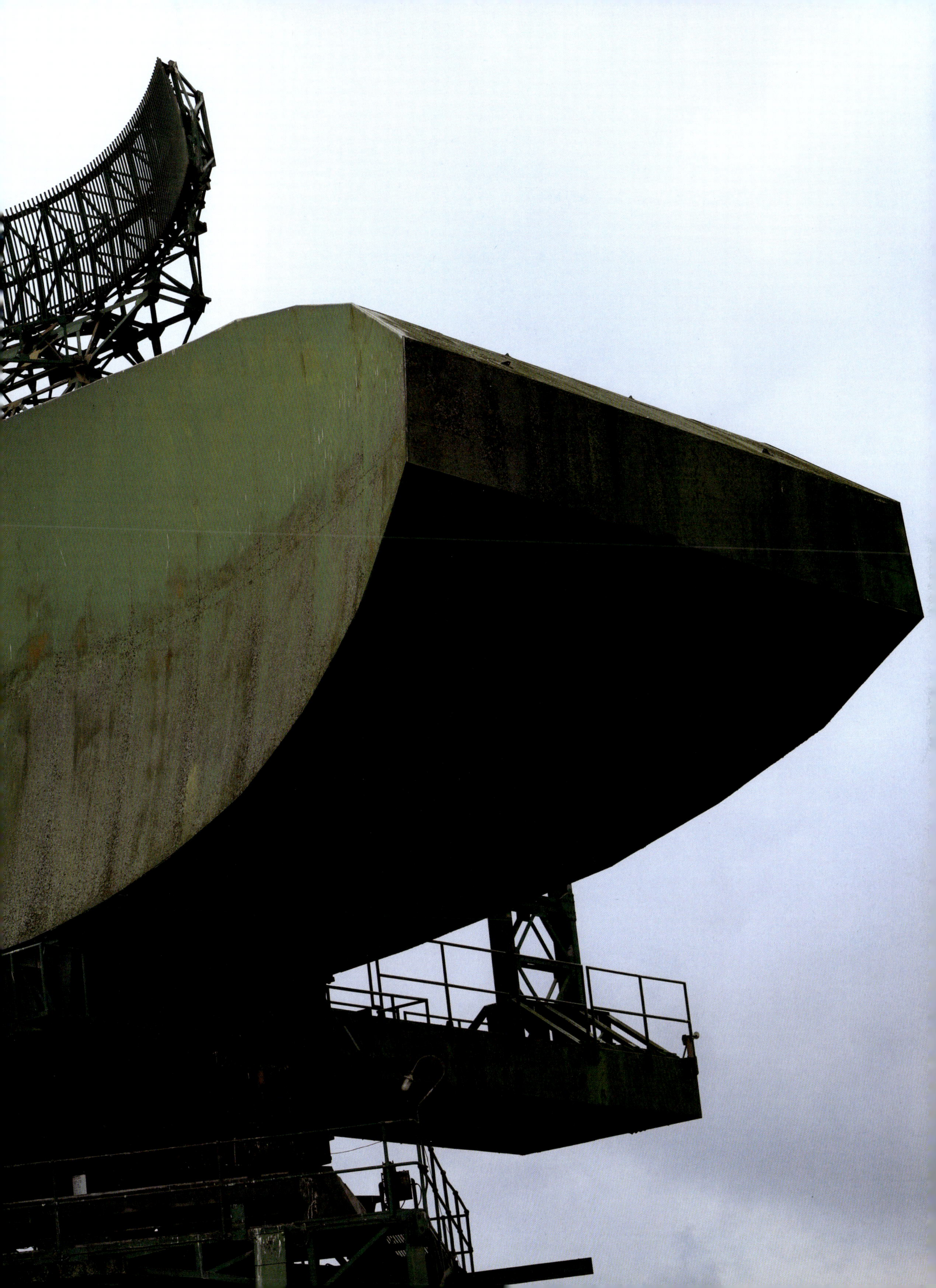

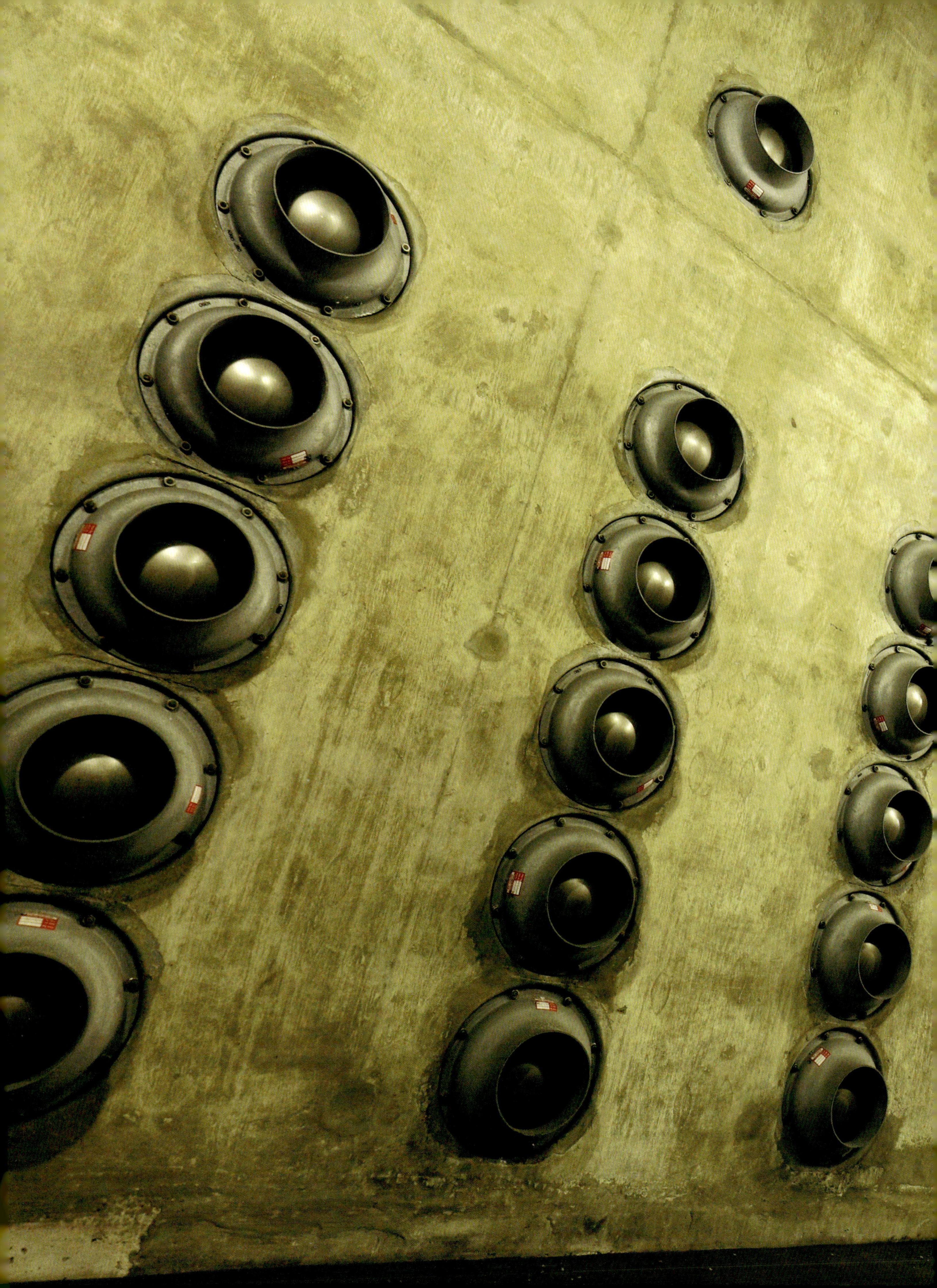

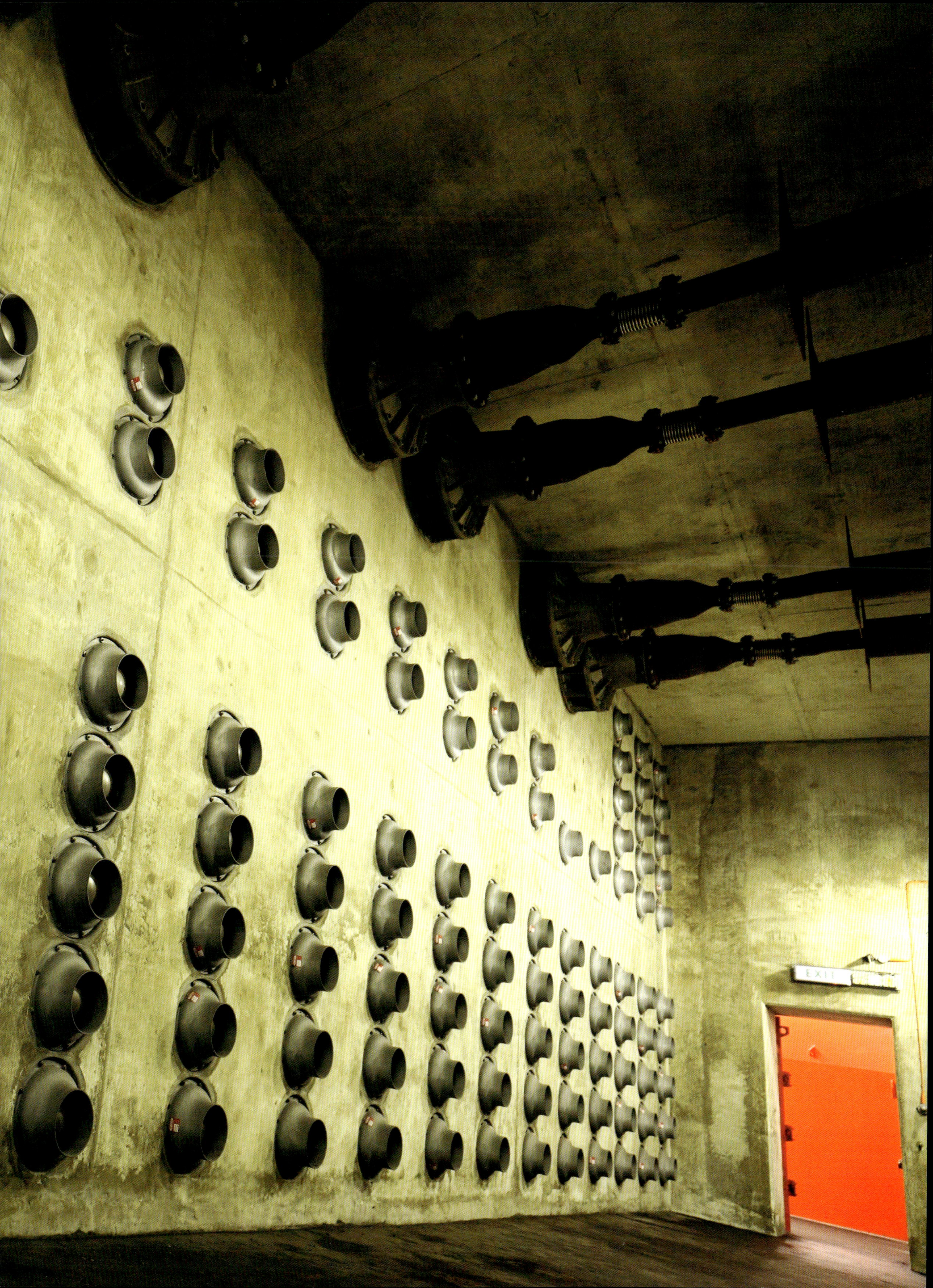

PREVIOUS PAGES:
Vents, RAF Neatishead, Norfolk
The station was extensively rebuilt after the fire, and not brought back into action until 1974, with new radars and communications equipment that acted as part of the NATO early warning system. The base closed in 2006.

TOP RIGHT:
Bunkers, GAMA, RAF Greenham Common, Berkshire
Entering service in the early 1980s, the US BGM-109G Gryphon Ground-Launched Cruise Missile (GLCM) was launched from a mobile transporter that made it hard for Soviet planners to be sure of destroying all nuclear weapons in a pre-emptive strike. Each of these bunkers contained 16 GLCMs and their launchers.

BOTTOM RIGHT:
Bunker Door, GAMA, RAF Greenham Common, Berkshire
In the event of a high alert, the launchers would have left the GAMA (Ground Launched Cruise Missile Alert and Maintenance Area) and dispersed throughout the countryside ready to launch a nuclear response to any strike. However the bunkers were protected against a nuclear and chemical attack should the Soviet forces achieve surprise.

OPPOSITE:
Building 704, GAMA, RAF Greenham Common, Berkshire
The GAMA stored 64 GLCMs ready for 16 convoys to deploy in groups of four launcher vehicles each (plus associated command and protection vehicles), with another 32 GLCMs in reserve. There was a similar GLCM base at RAF Molesworth in Cambridgeshire.

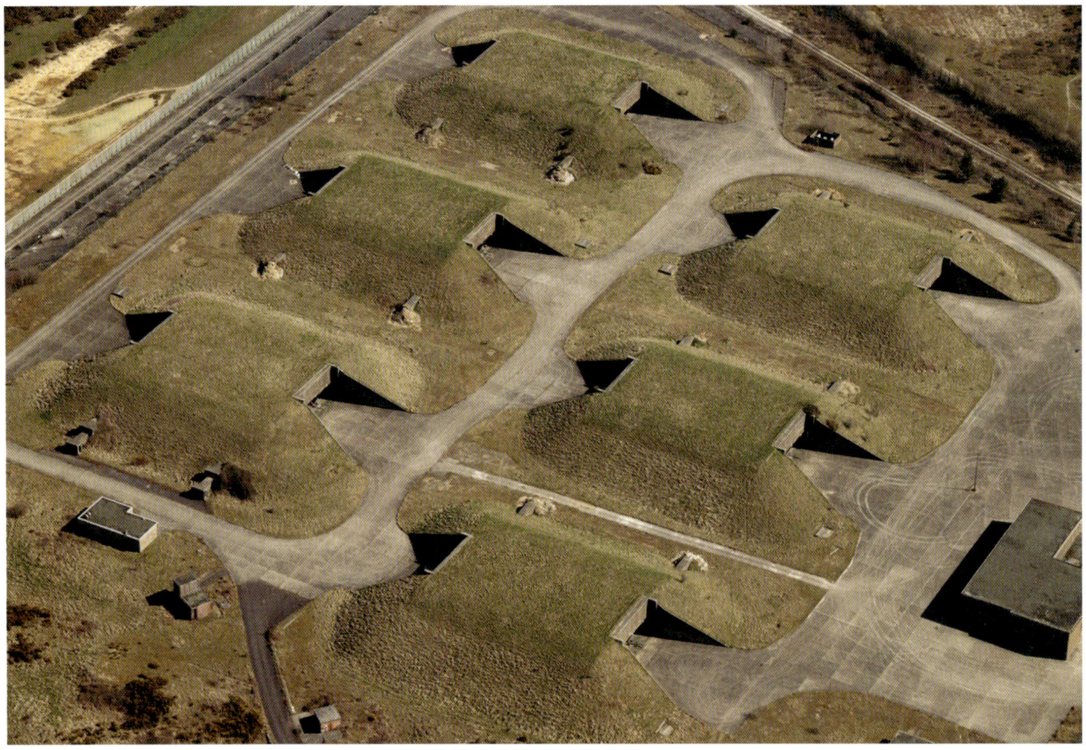

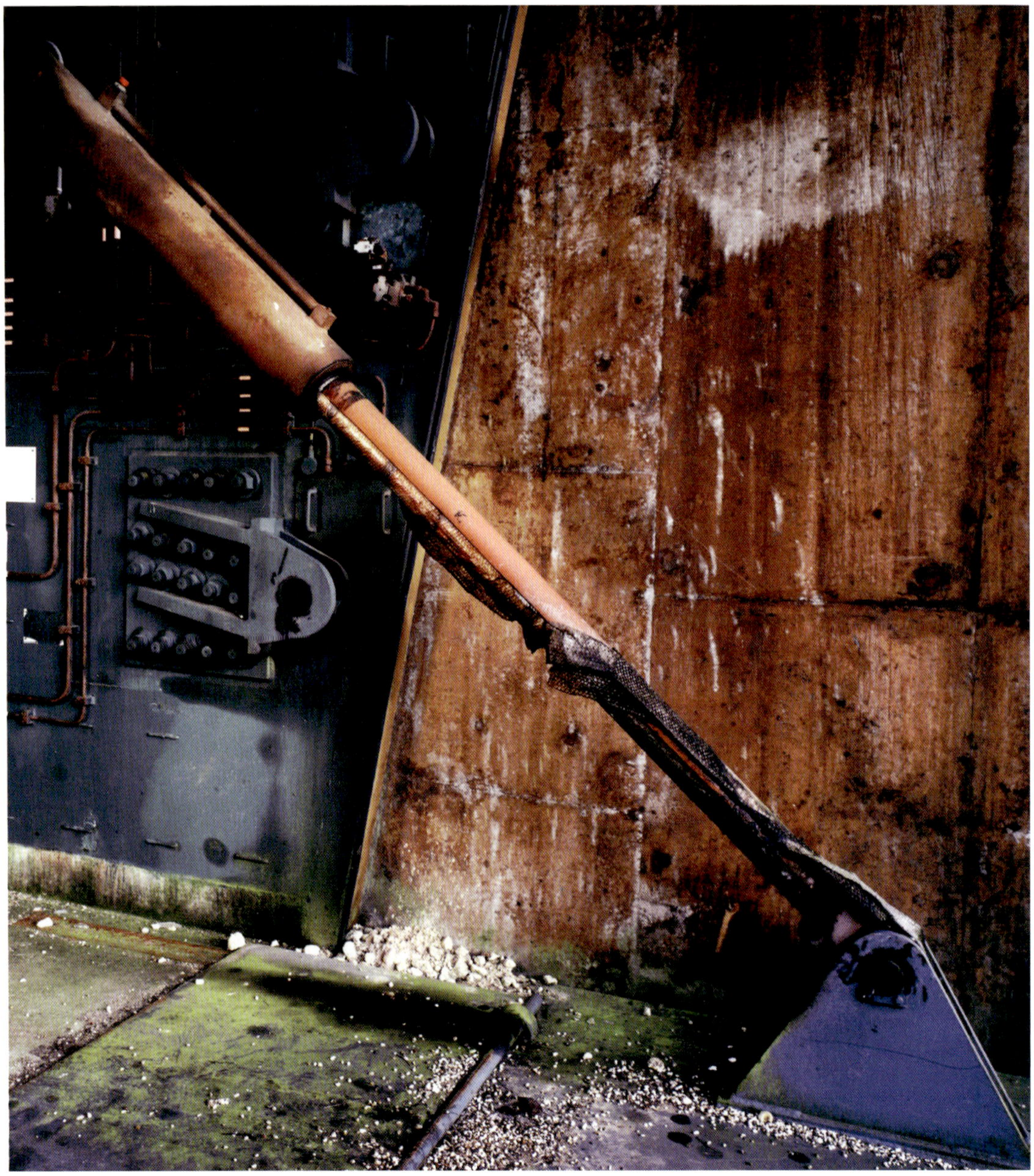

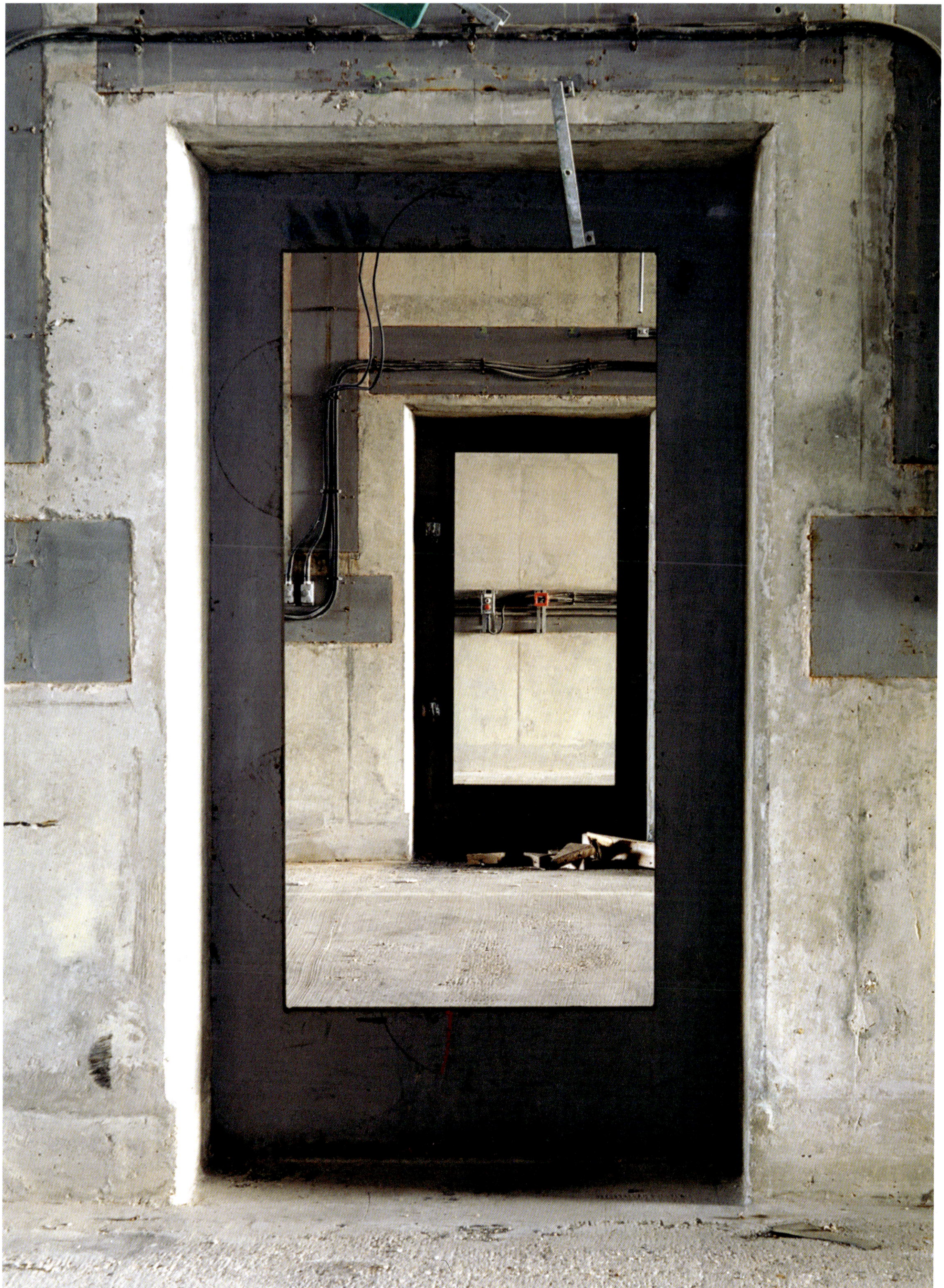

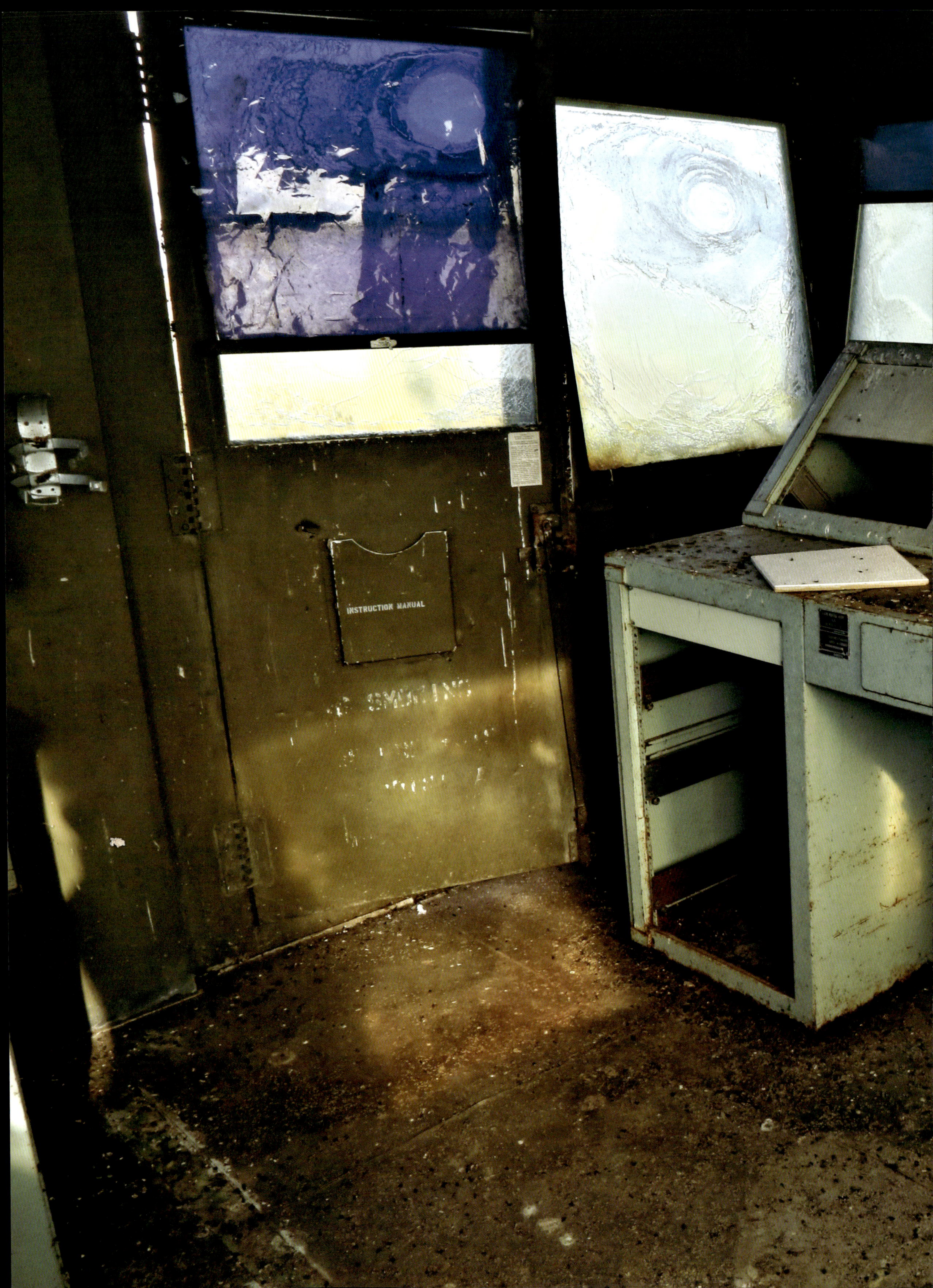
INSTRUCTION MANUAL

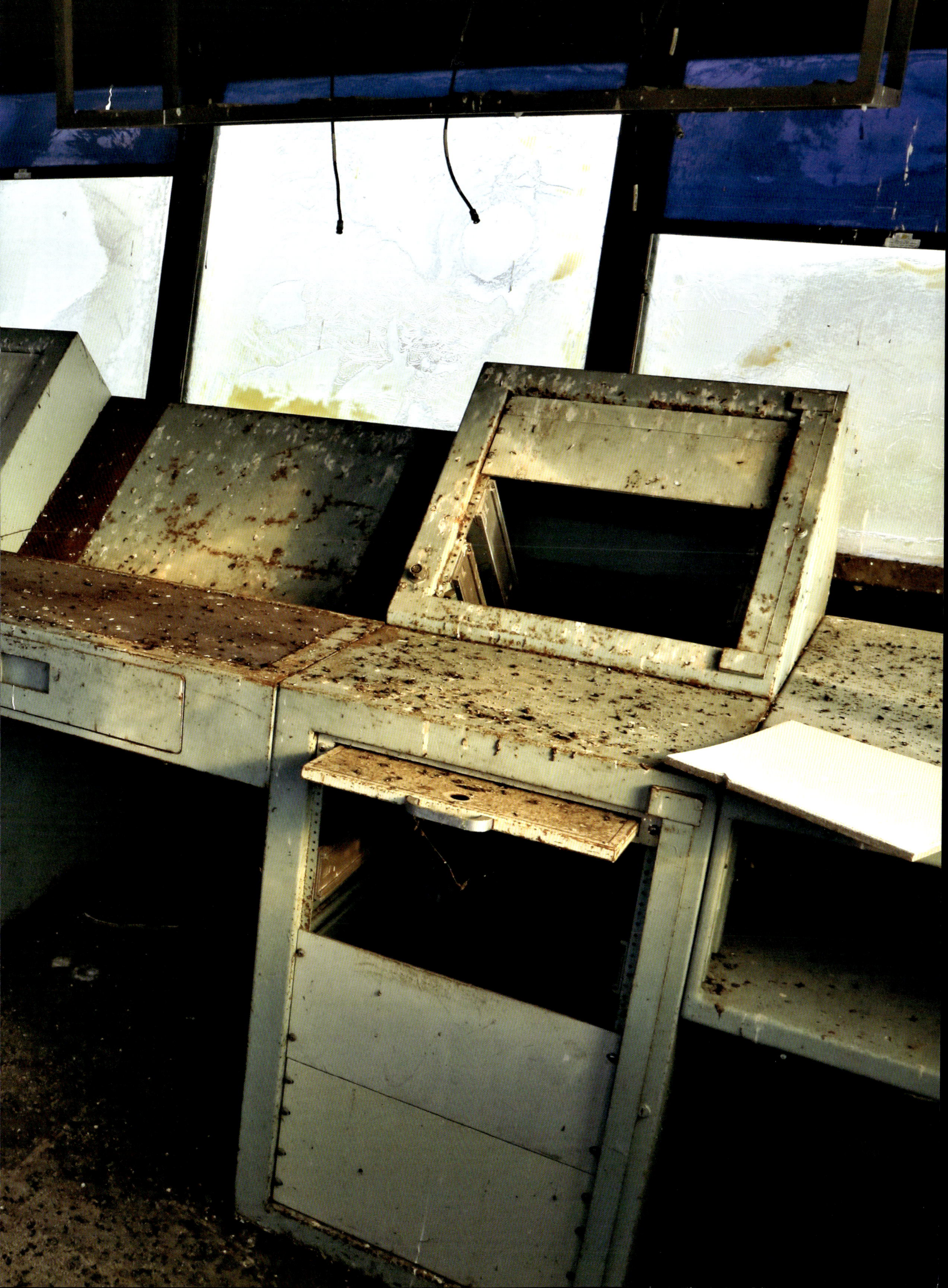

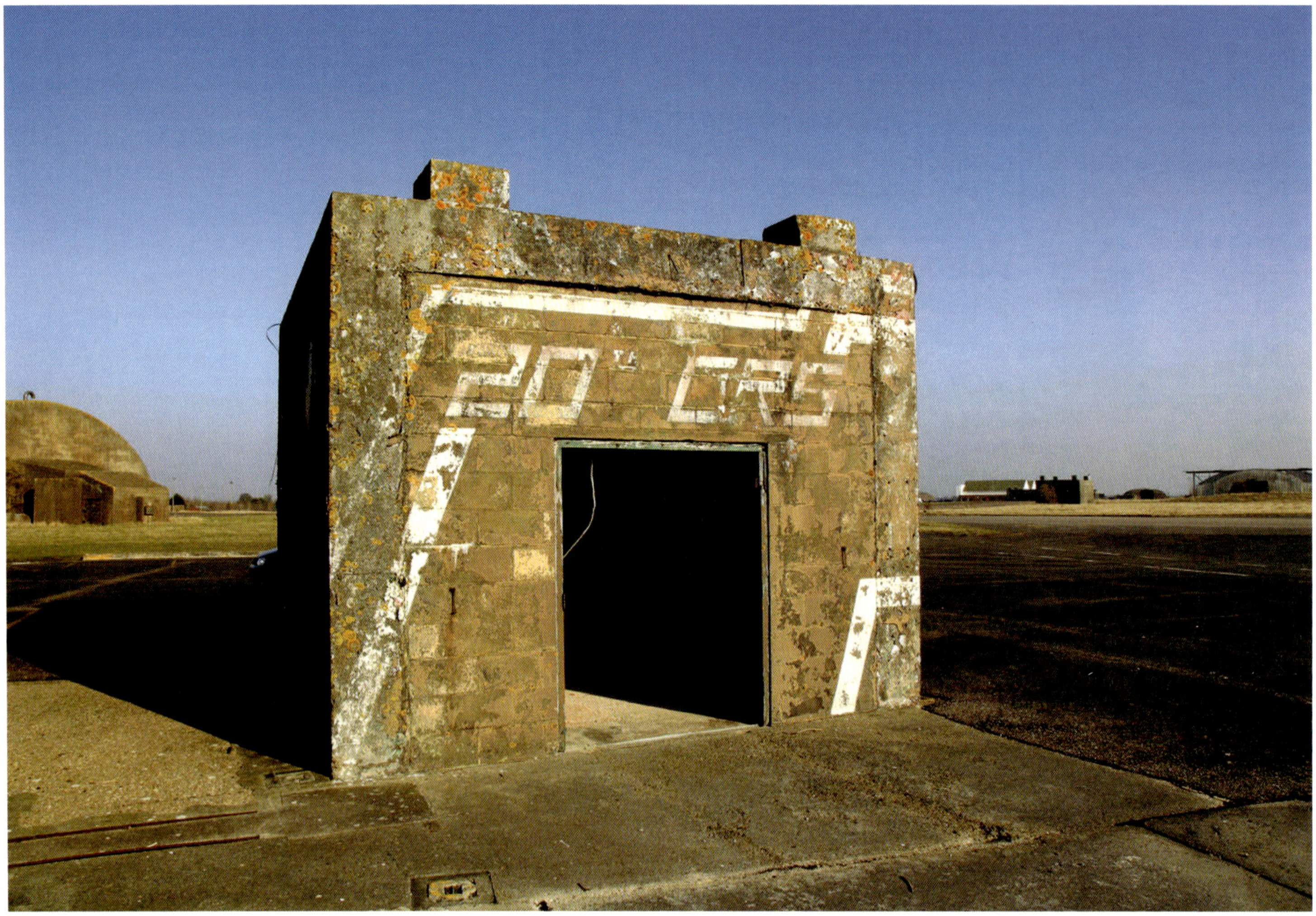
20 CRS

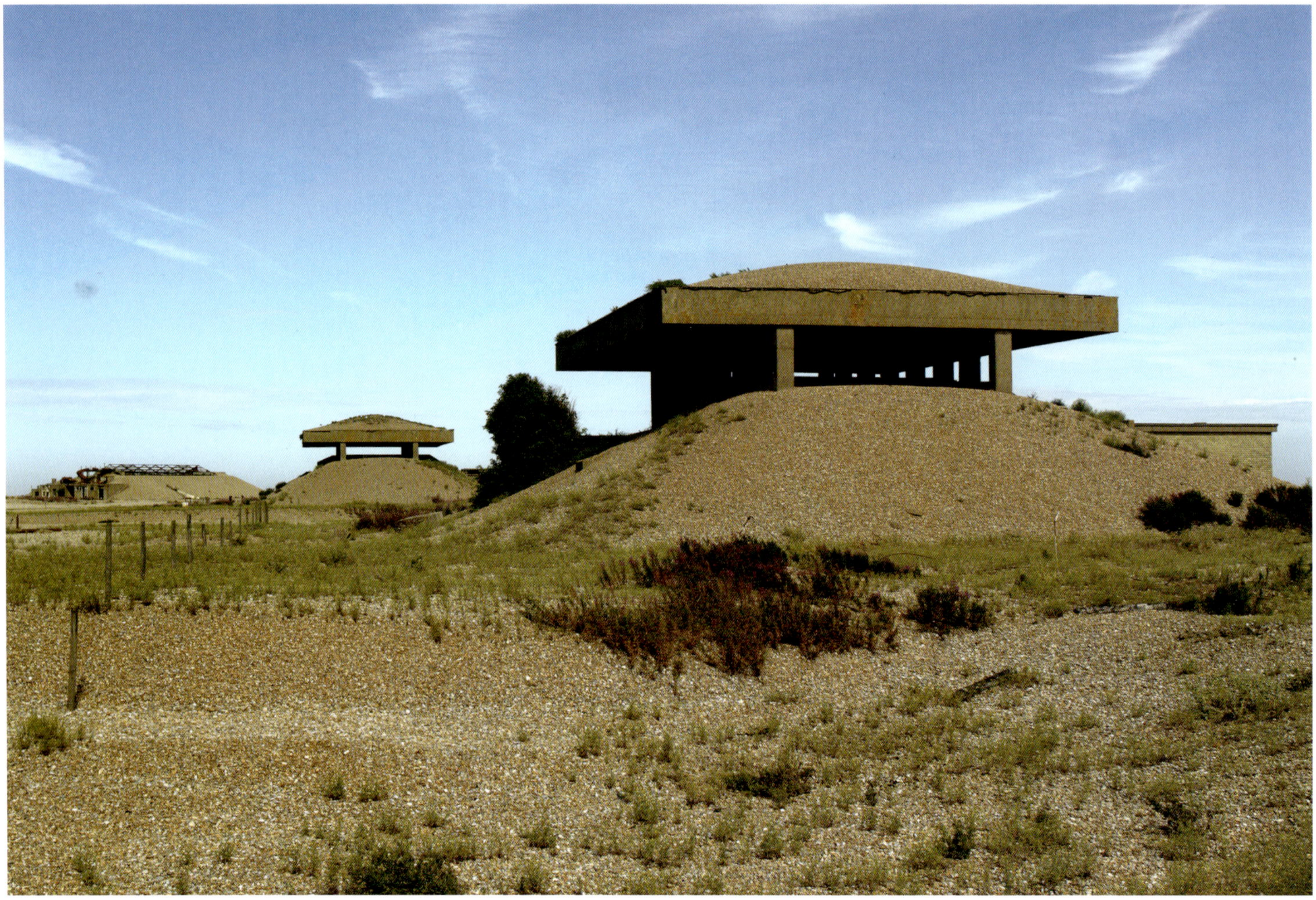

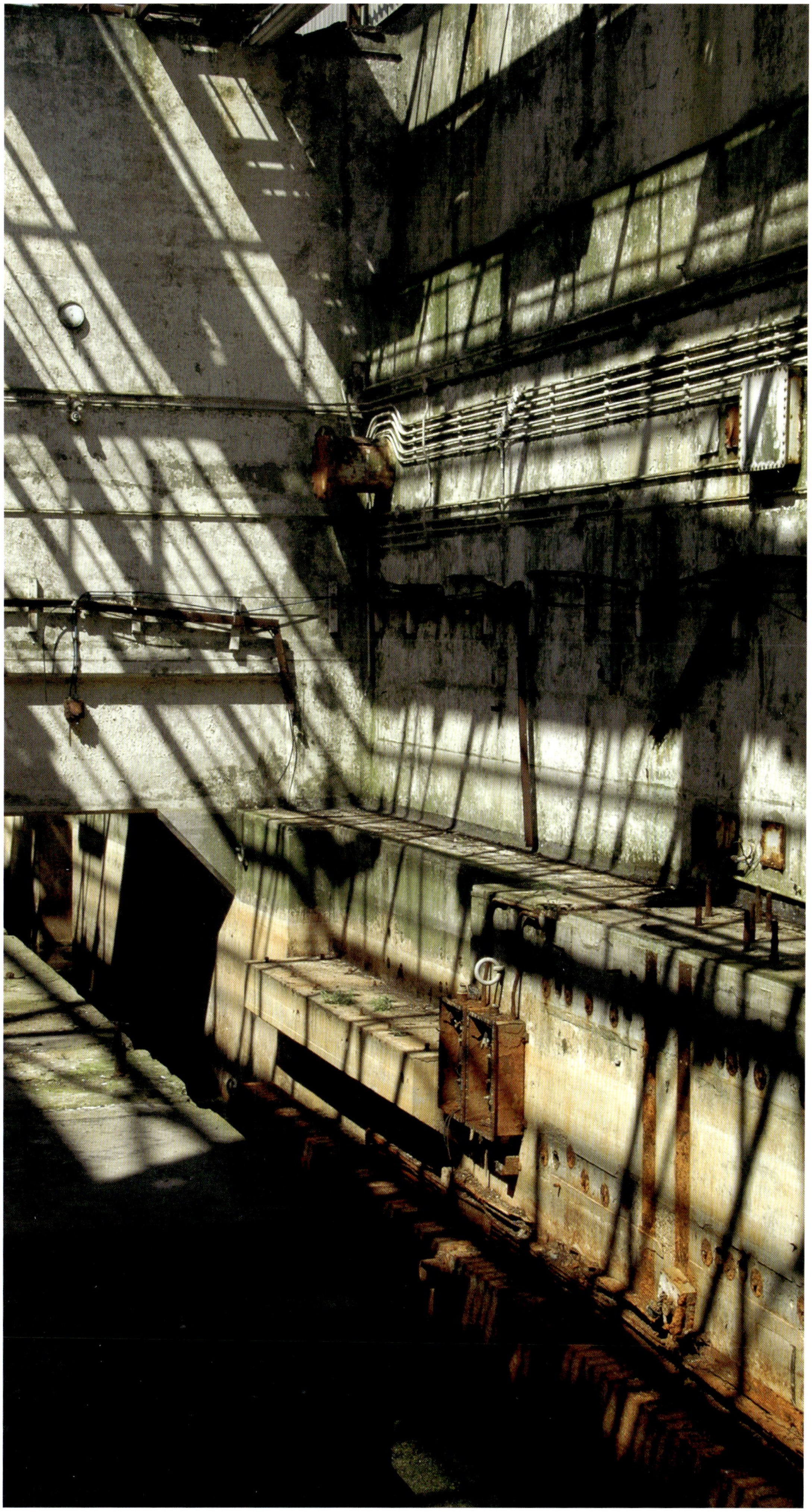

PREVIOUS PAGES:
Former Control Tower, RAF Upper Heyford, Oxfordshire
During the 1960s, the USAF's strategic bombers were being replaced by ICBMs as the United States' primary nuclear option. The large bombers returned to the US to be replaced by smaller tactical aircraft, and Upper Heyford became a key base for the new F-111 tactical bomber in the 1970s. The end of the Cold War saw the rapid removal of the USAF personnel and aircraft, and the base was returned to the RAF in 1993.

OPPOSITE TOP:
Former Store, RAF Upper Heyford, Oxfordshire
First developed as an airbase in 1916, the first US personnel arrived in 1950. The runways were subsequently lengthened and concrete ammunition, fuel and other store depots constructed for the strategic bombers that were based here, such as the B-36 Peacemaker, B-47 Stratojet, B-50 Superfortress and B-52 Stratofortress.

OPPOSITE BOTTOM:
Concrete Pagodas, AWRE, Orford Ness, Suffolk
Orford Ness on the east coast of England was used as a military test site for much of the twentieth century. The distinctive concrete pagodas were built by the Atomic Weapons Research Establishment (AWRE) for testing in real-world conditions. The design allowed explosive gases to vent safely, but a larger-than-expected explosion would cause the roof to fall and safely seal the testing pit.

LEFT:
Lab 1, AWRE, Orford Ness, Suffolk
Nuclear bomb cases and detonators for the UK's first atomic bomb, the Blue Danube, were once tested in this 1950s-era lab that was built to reproduce the conditions found in an aircraft's bomb bay in order to minimise the risk of an accidental explosion during transit to the bomb's intended target.

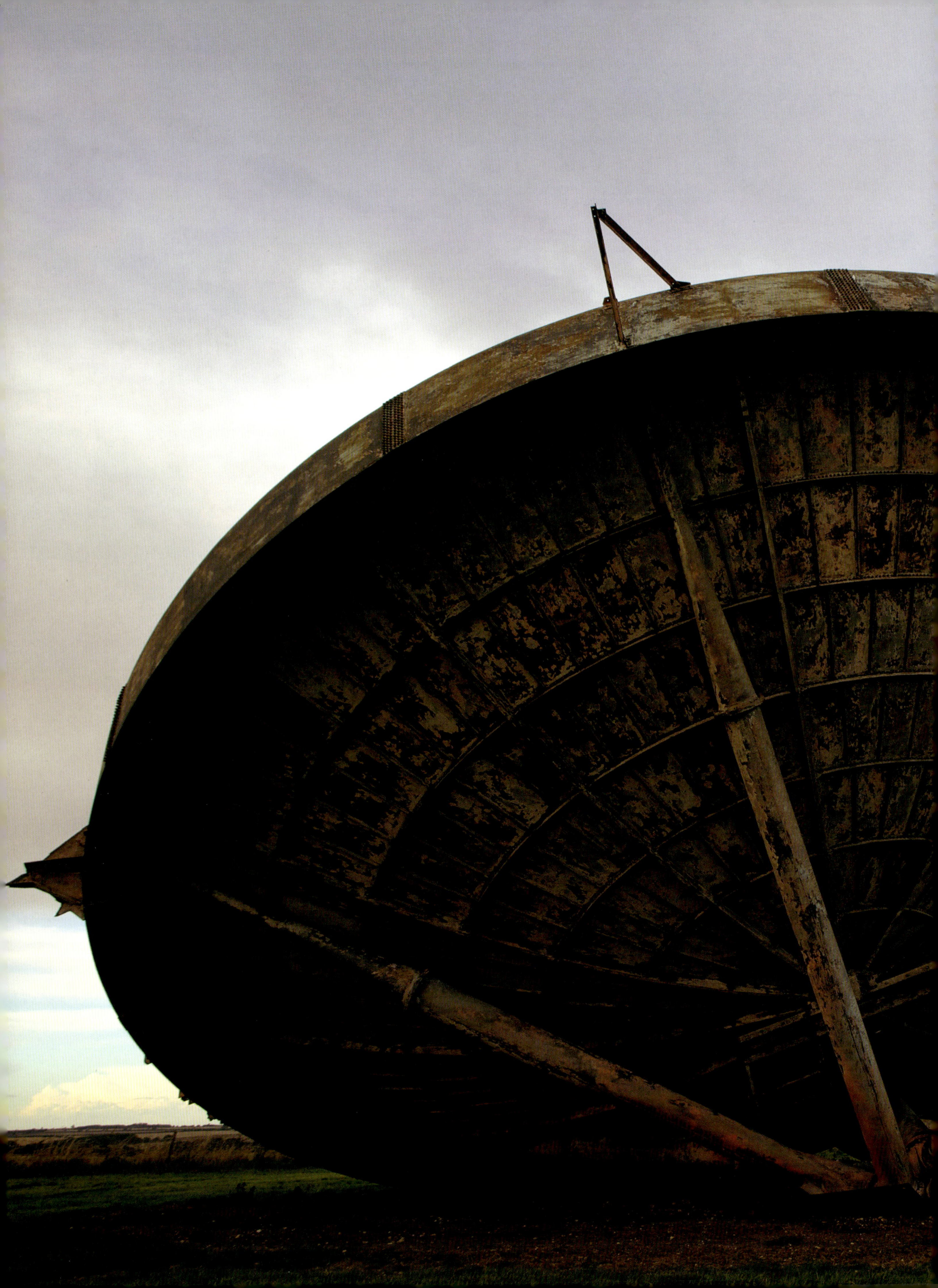

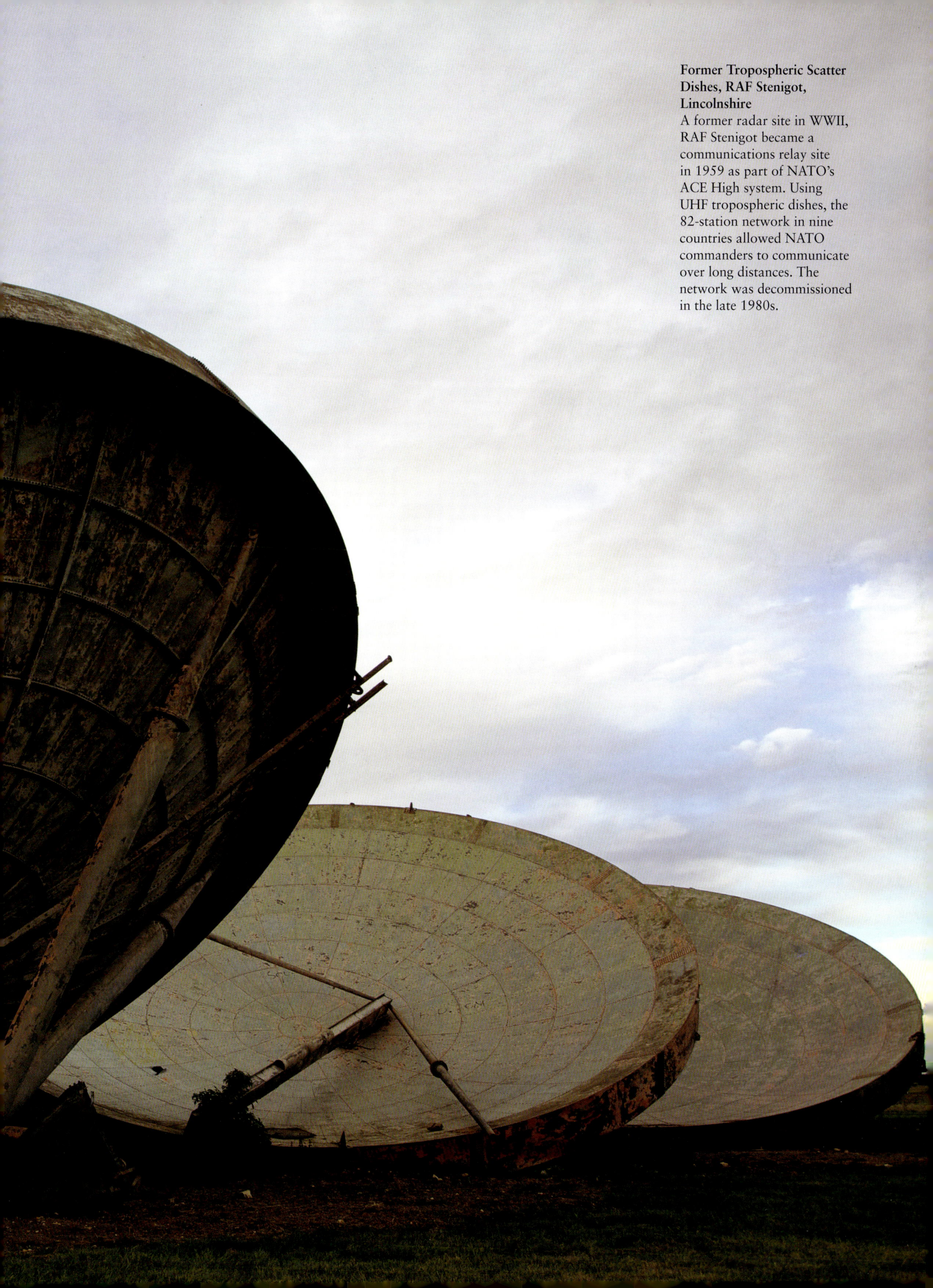

Former Tropospheric Scatter Dishes, RAF Stenigot, Lincolnshire
A former radar site in WWII, RAF Stenigot became a communications relay site in 1959 as part of NATO's ACE High system. Using UHF tropospheric dishes, the 82-station network in nine countries allowed NATO commanders to communicate over long distances. The network was decommissioned in the late 1980s.

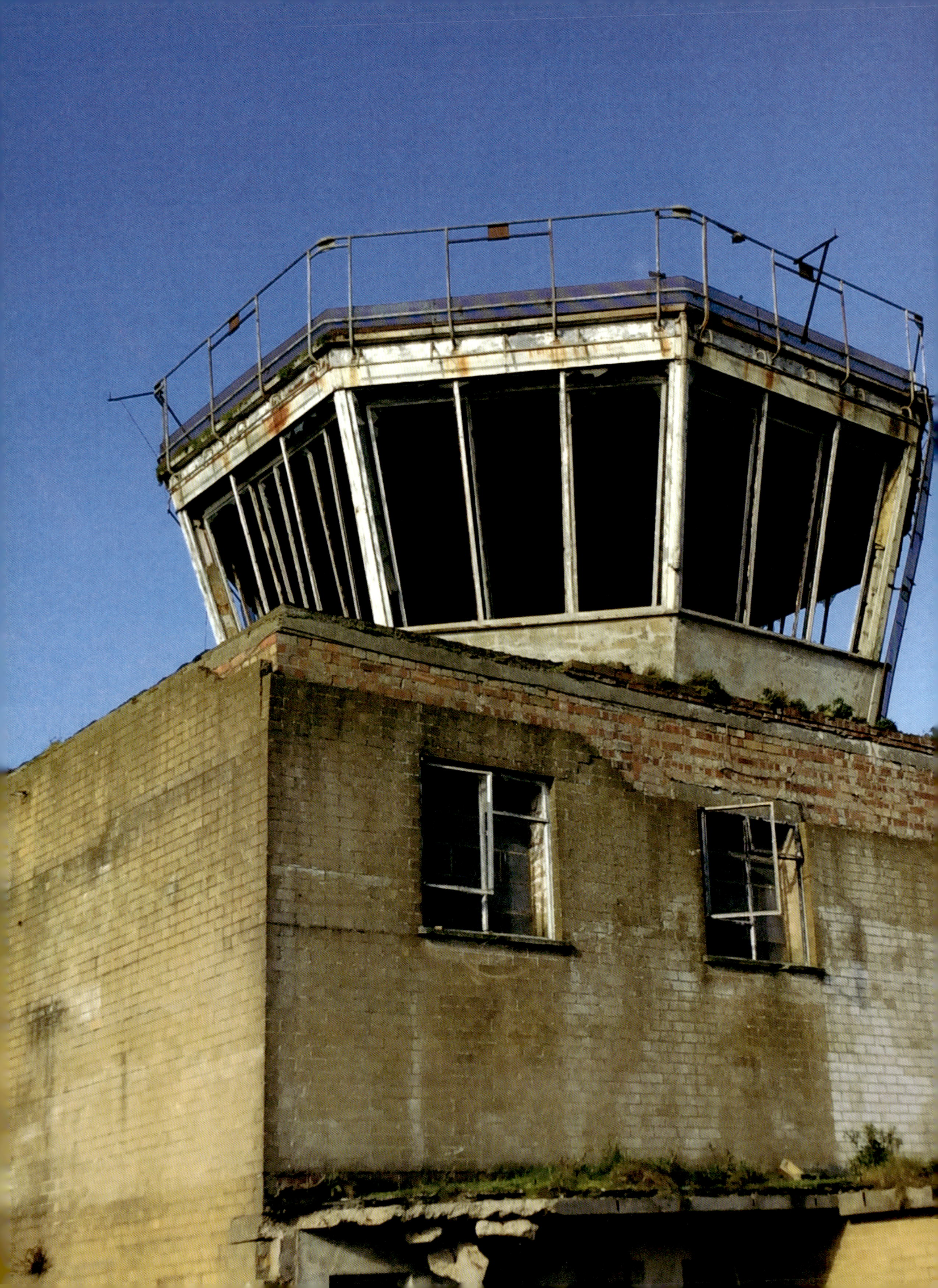

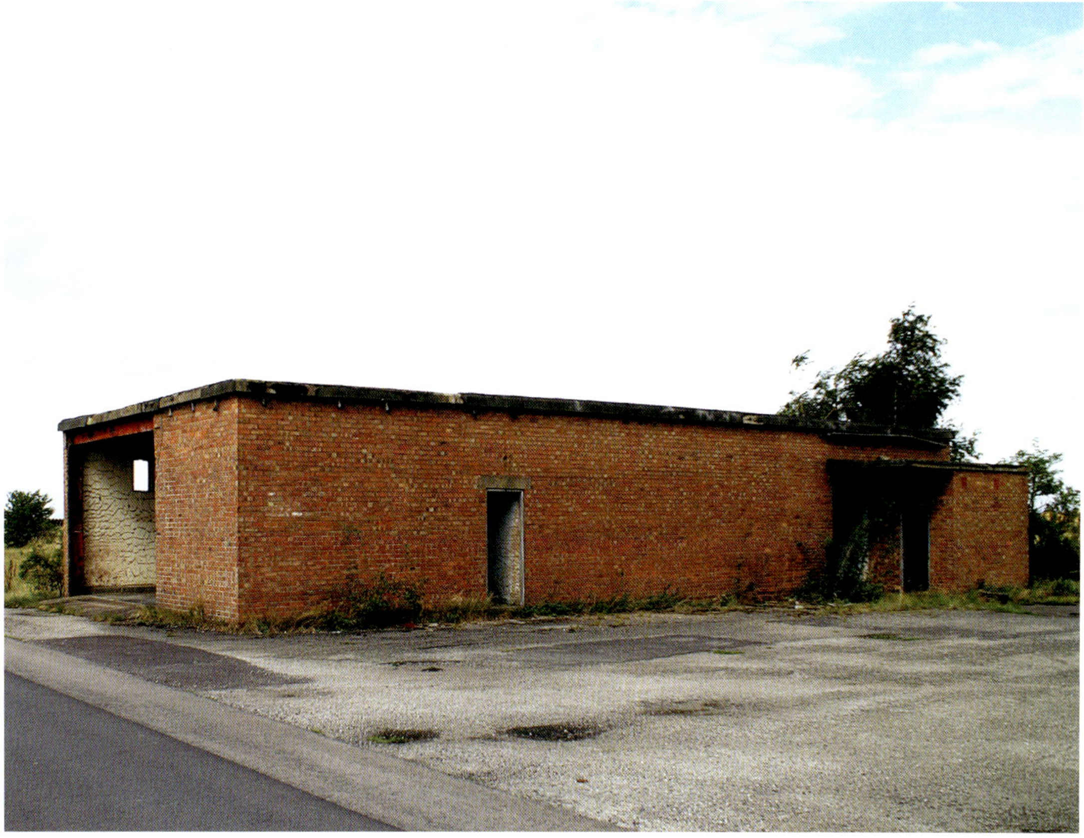

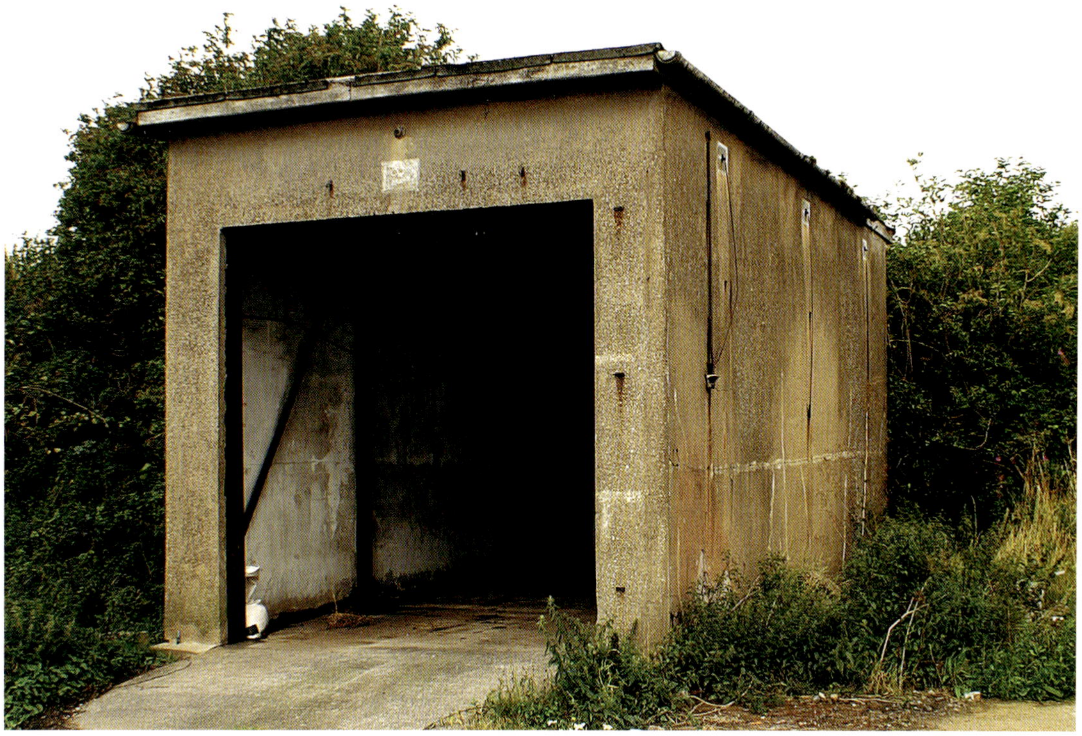

OPPOSITE:

Former Control Tower, RAF Manby, Lincolnshire

Built just before WWII, in July 1949 RAF Manby became home to the RAF Flying College and the RAF Handling Squadron, the latter used to evaluate new aircraft types before they entered service until it disbanded in 1954. The base closed in 1974.

TOP AND ABOVE:

Former Stores, RAF Breighton, Yorkshire

Project Emily was a joint US/UK initiative to station intermediate range nuclear missiles (IRBMs) on UK soil beginning in 1958. RAF Breighton was one of the sites chosen for the 60 Thor missiles in the UK, each of which carried a 1.44 megaton warhead. The missile launch keys were operated by one British and one American officer.

The Thor missile was always seen as a temporary measure until the first US ICBMs in continental America were ready. The three missiles operated by 240 Squadron at Breighton were only active between 1959 and 1963, and the base closed the following year.

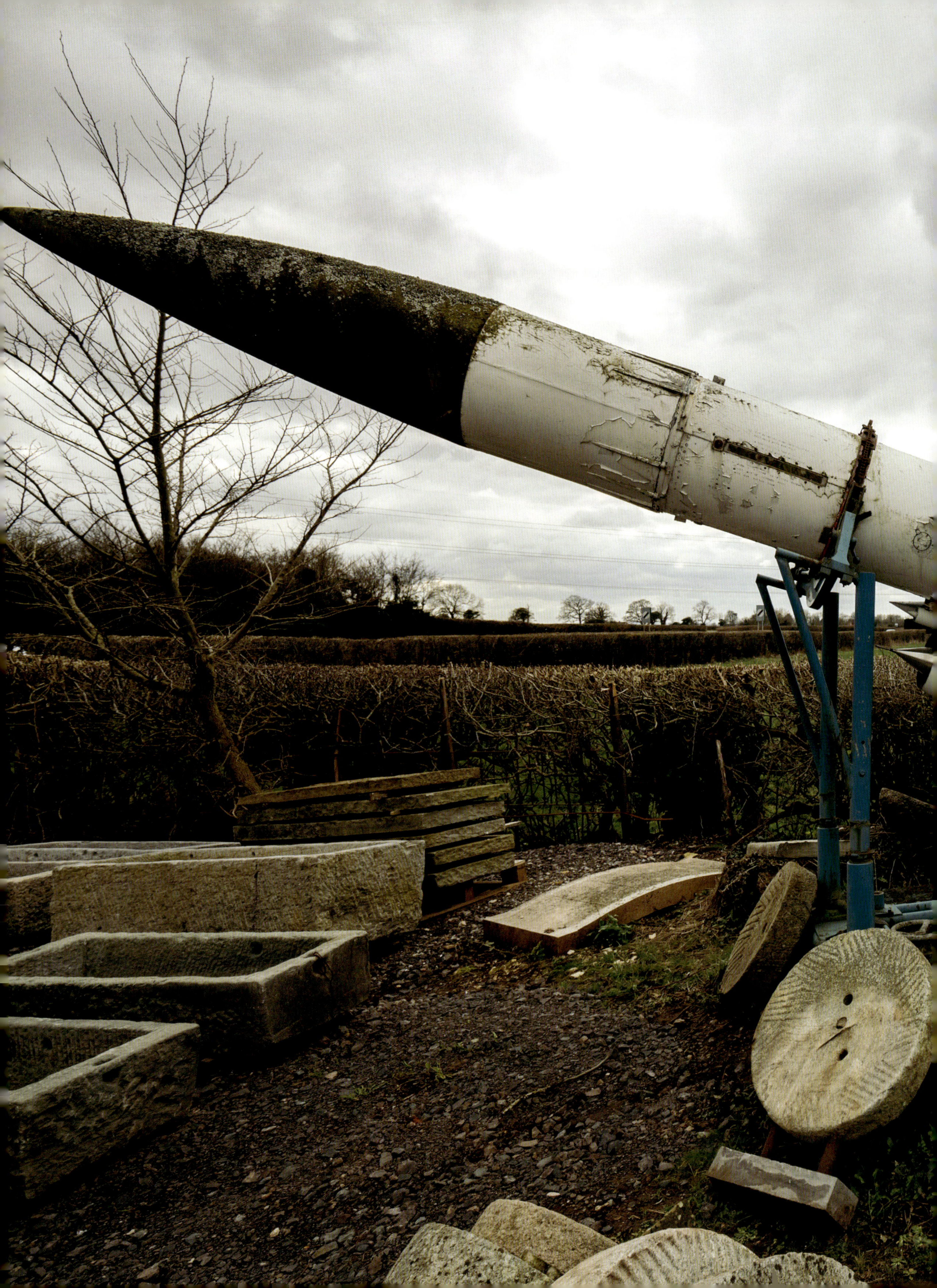

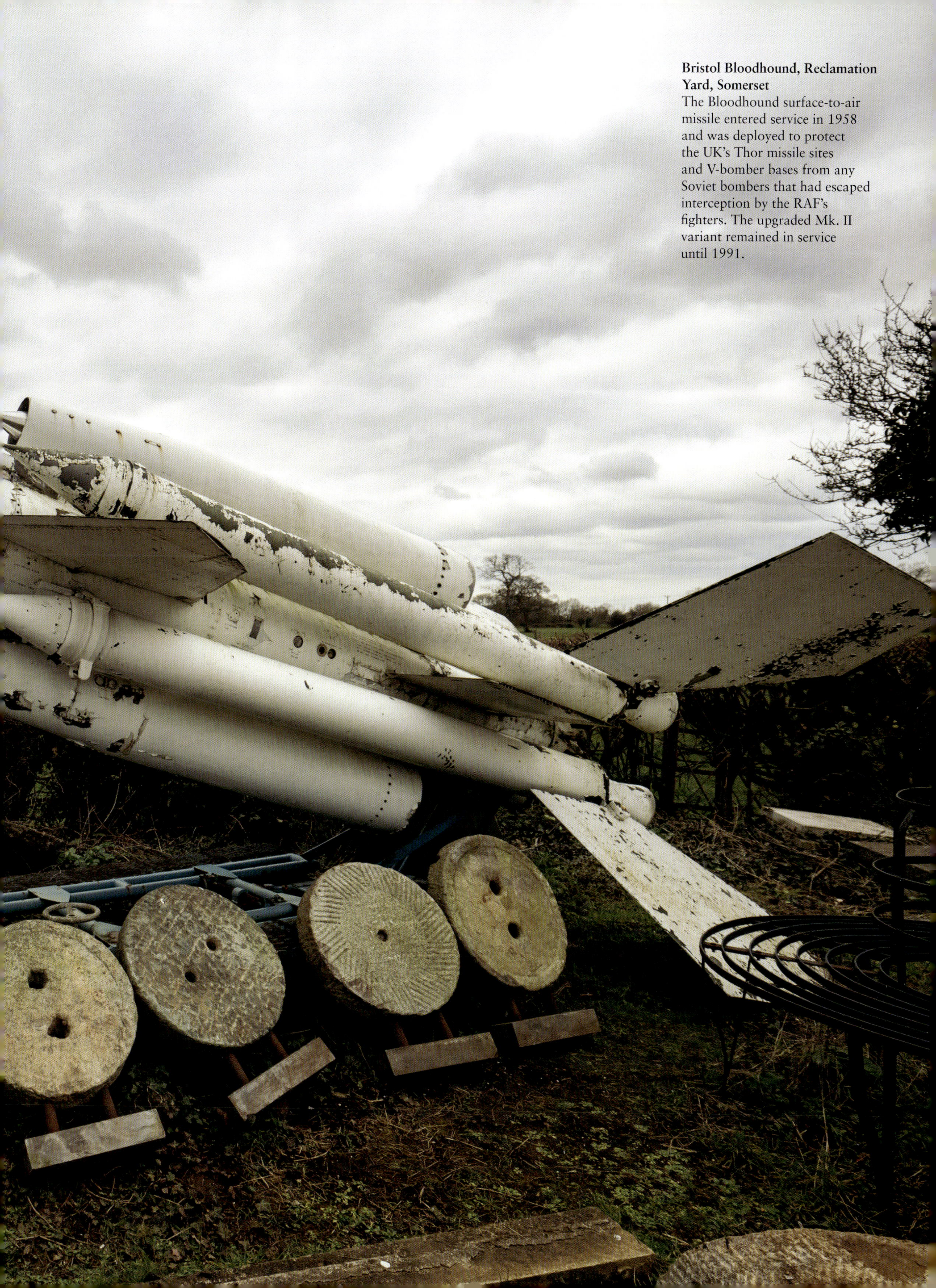

Bristol Bloodhound, Reclamation Yard, Somerset
The Bloodhound surface-to-air missile entered service in 1958 and was deployed to protect the UK's Thor missile sites and V-bomber bases from any Soviet bombers that had escaped interception by the RAF's fighters. The upgraded Mk. II variant remained in service until 1991.

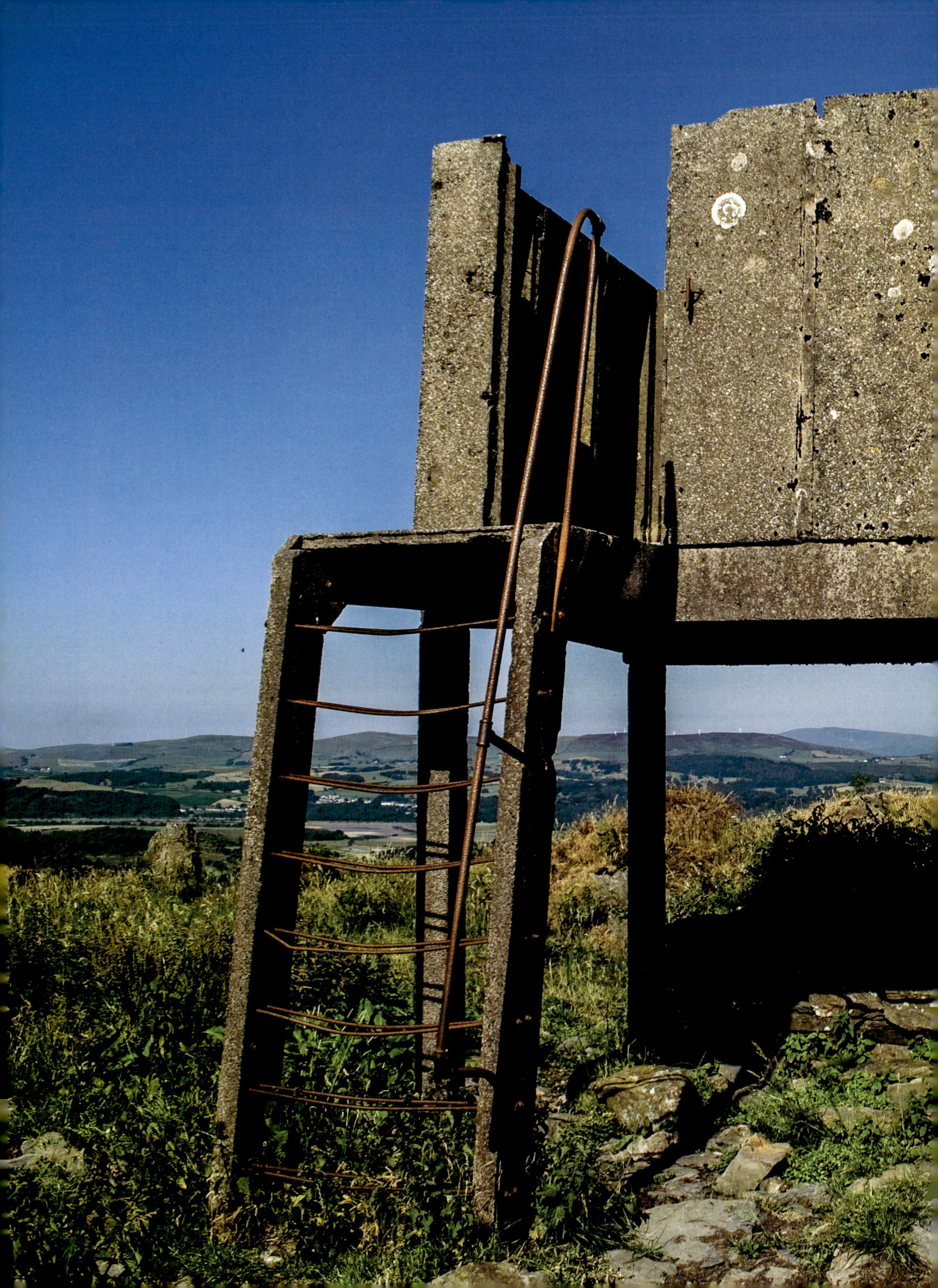

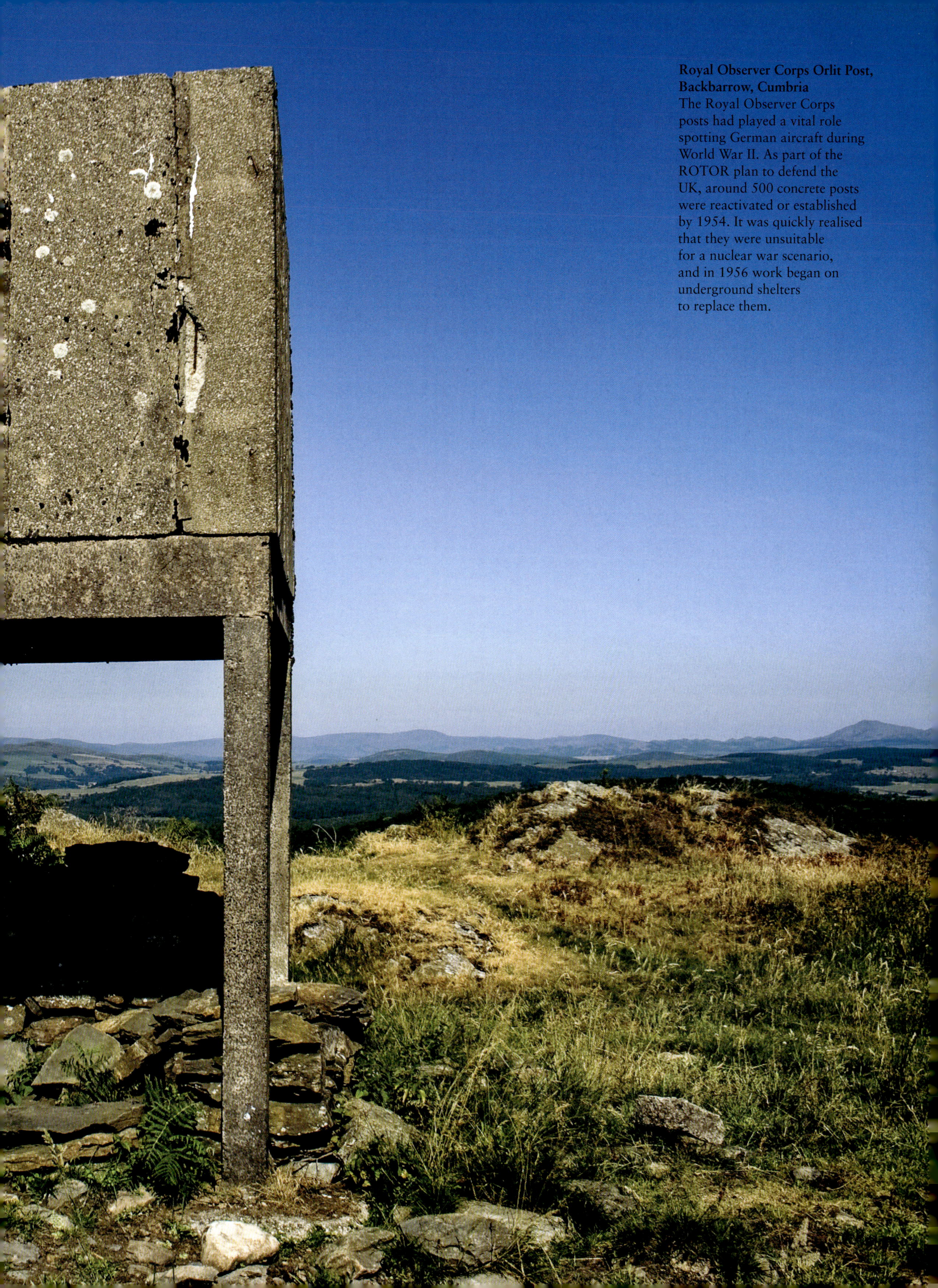

Royal Observer Corps Orlit Post, Backbarrow, Cumbria
The Royal Observer Corps posts had played a vital role spotting German aircraft during World War II. As part of the ROTOR plan to defend the UK, around 500 concrete posts were reactivated or established by 1954. It was quickly realised that they were unsuitable for a nuclear war scenario, and in 1956 work began on underground shelters to replace them.

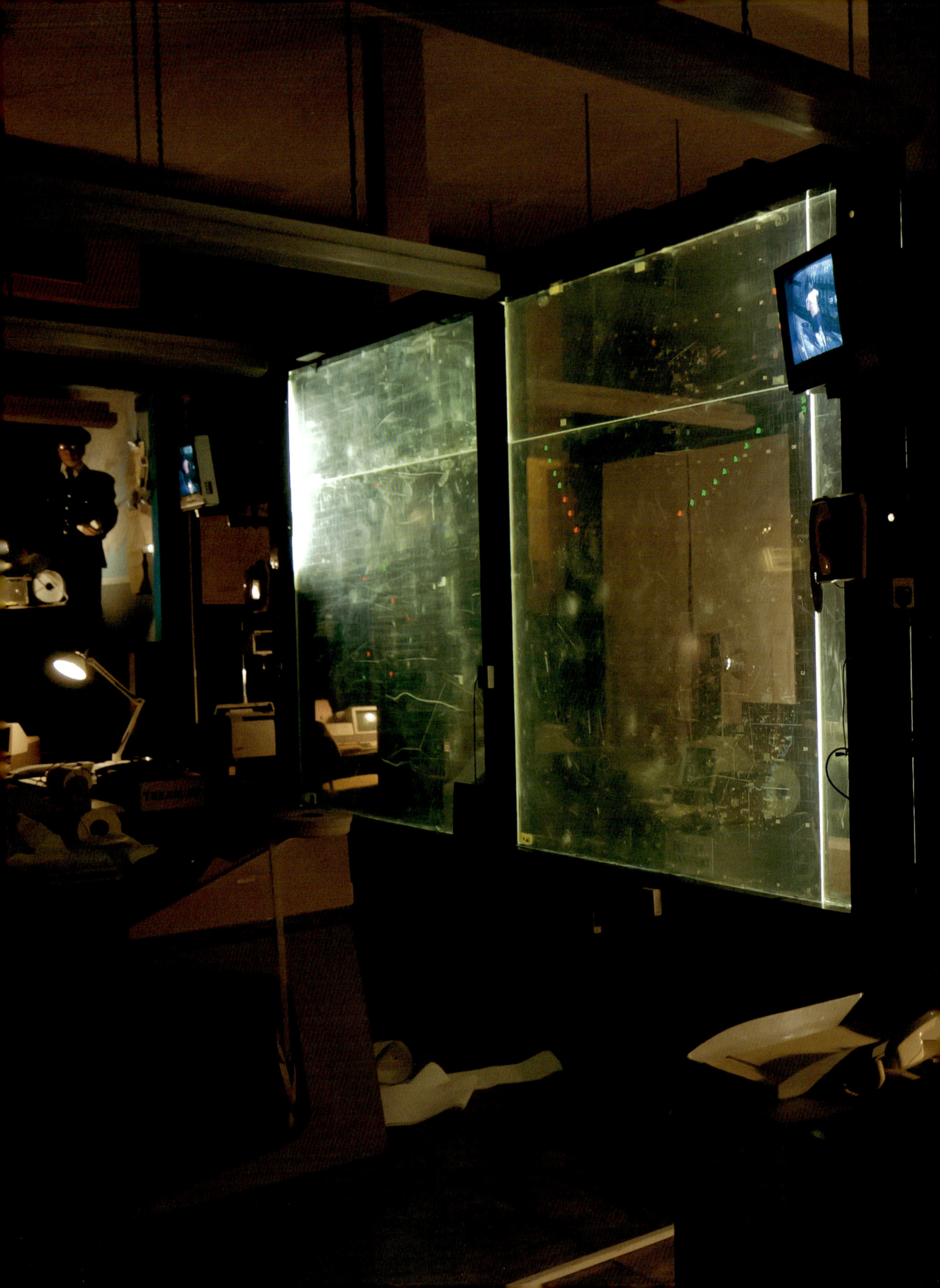

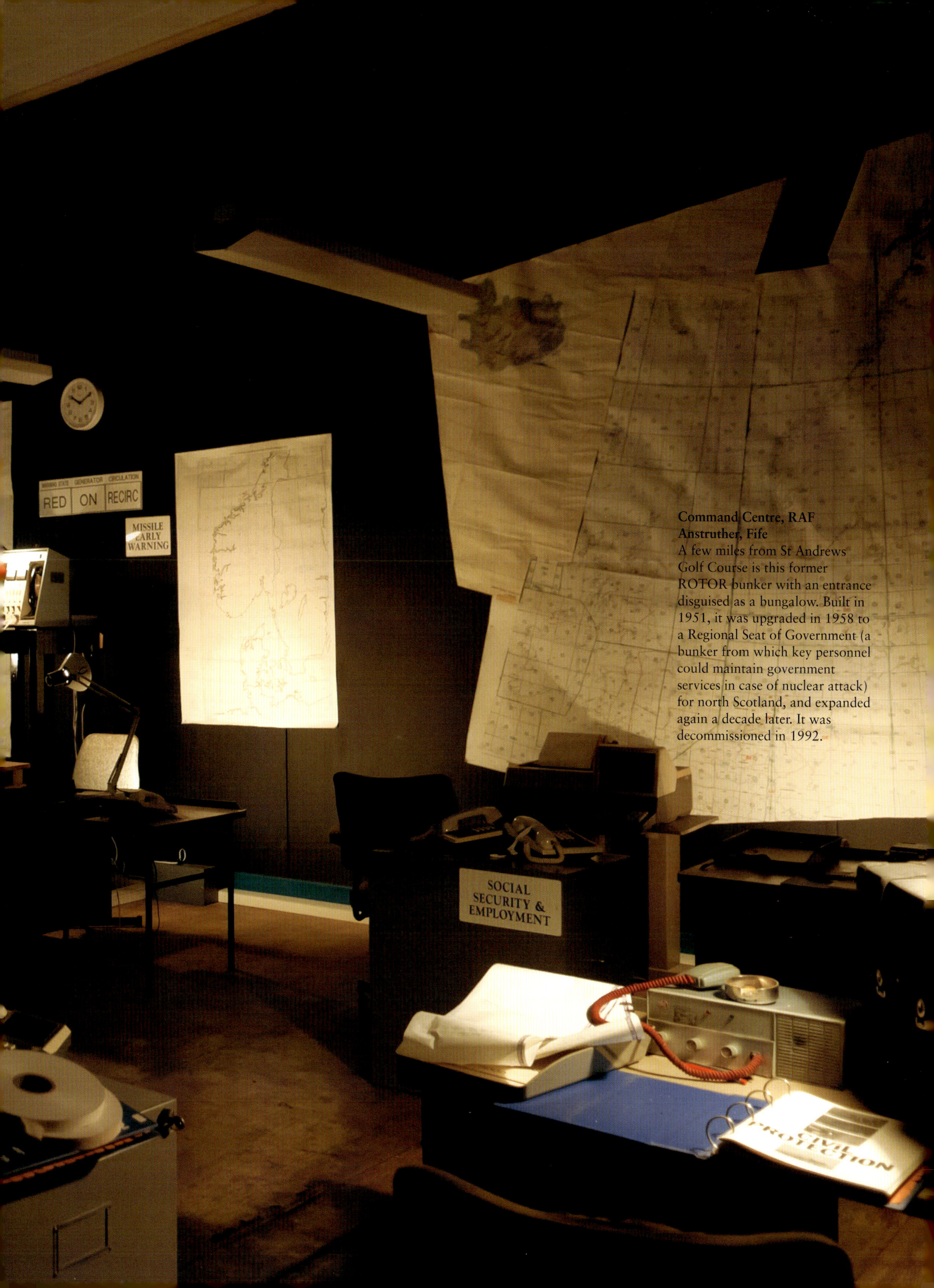

Command Centre, RAF Anstruther, Fife
A few miles from St Andrews Golf Course is this former ROTOR bunker with an entrance disguised as a bungalow. Built in 1951, it was upgraded in 1958 to a Regional Seat of Government (a bunker from which key personnel could maintain government services in case of nuclear attack) for north Scotland, and expanded again a decade later. It was decommissioned in 1992.

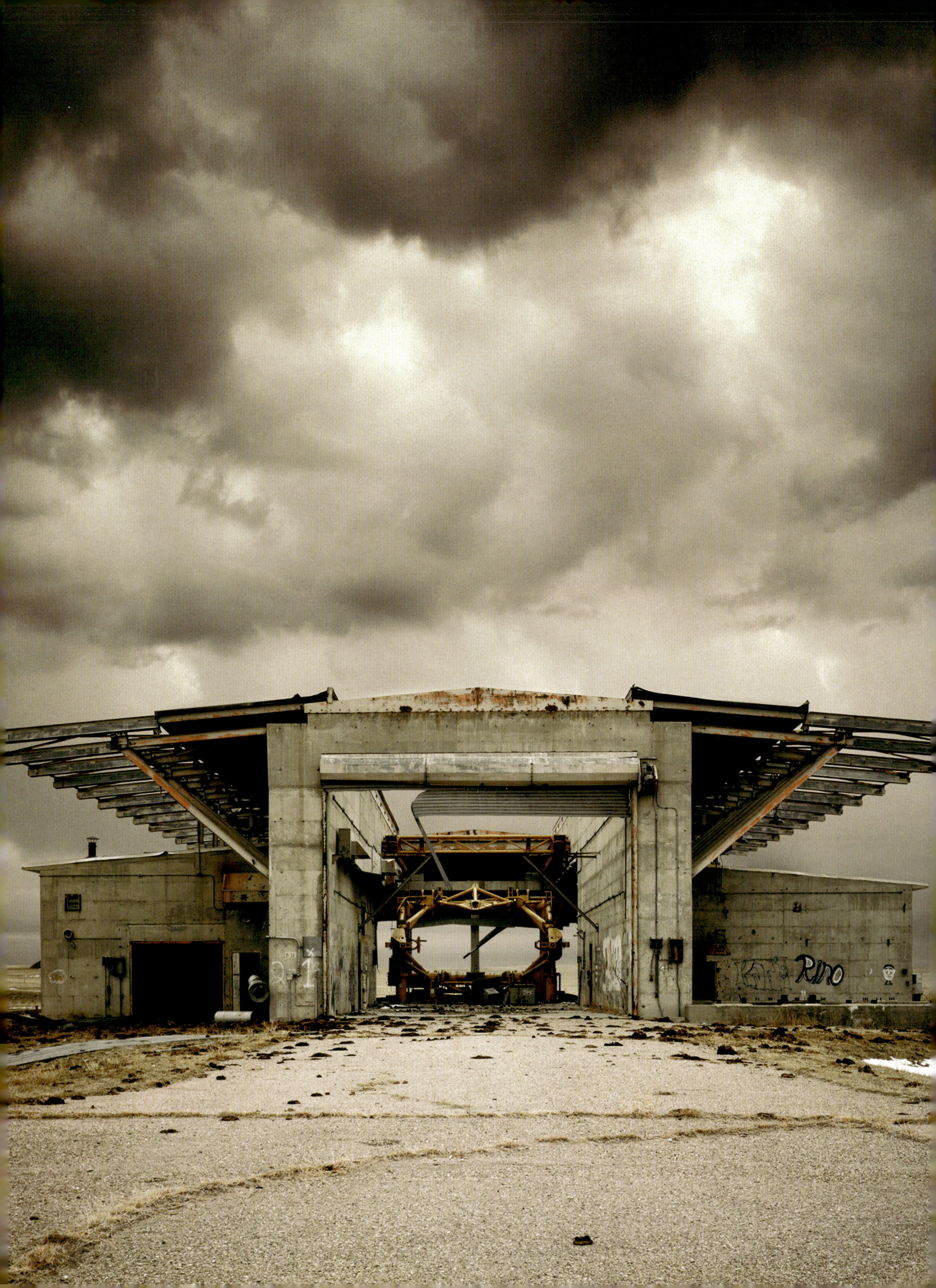
Rino

The Americas

From the frozen north of Canada to the Caribbean warmth of Cuba, preparation for war – or in the latter's case, invasion – stretched across the northern half of the Americas. The nature of the threat to the US and its Canadian neighbour changed as the fear of waves of Soviet bombers flying across the Arctic Circle was replaced by the much more immediate threat of the new ICBM missiles, and the lurking missile submarines of the Soviet Navy.

One of the closest moments to war between the superpowers came with the Cuban Missile Crisis in 1962, when the Soviet Union built R-12 and R-14 missile facilities on the island of Cuba. Although the US had already based similar Jupiter missiles in Turkey, close to the Soviet Union, President Kennedy demanded the removal of offensive weapons from Cuba. The Soviet Premier, Khrushchev, backed down, and as a consequence, the hotline between the Kremlin and the White House was established to avoid accidental escalation of a future crisis. The continued rivalry between the US and the Soviet Union led to the escalation of the space race, and eventually President Reagan's 'Star Wars' space-based anti-ballistic missile programme that threatened to nullify the Soviet nuclear threat.

OPPOSITE:
Former Atlas D ICBM Launch Facility, Wyoming, USA
The first operational ICBM, the Atlas D, entered service in 1959 and had a range of up to 14,500km (9,000 miles). The missile was guided by radio signal to its target. After it became obsolete in the mid-1960s, former Atlas ICBMs were used as satellite launchers until 2004.

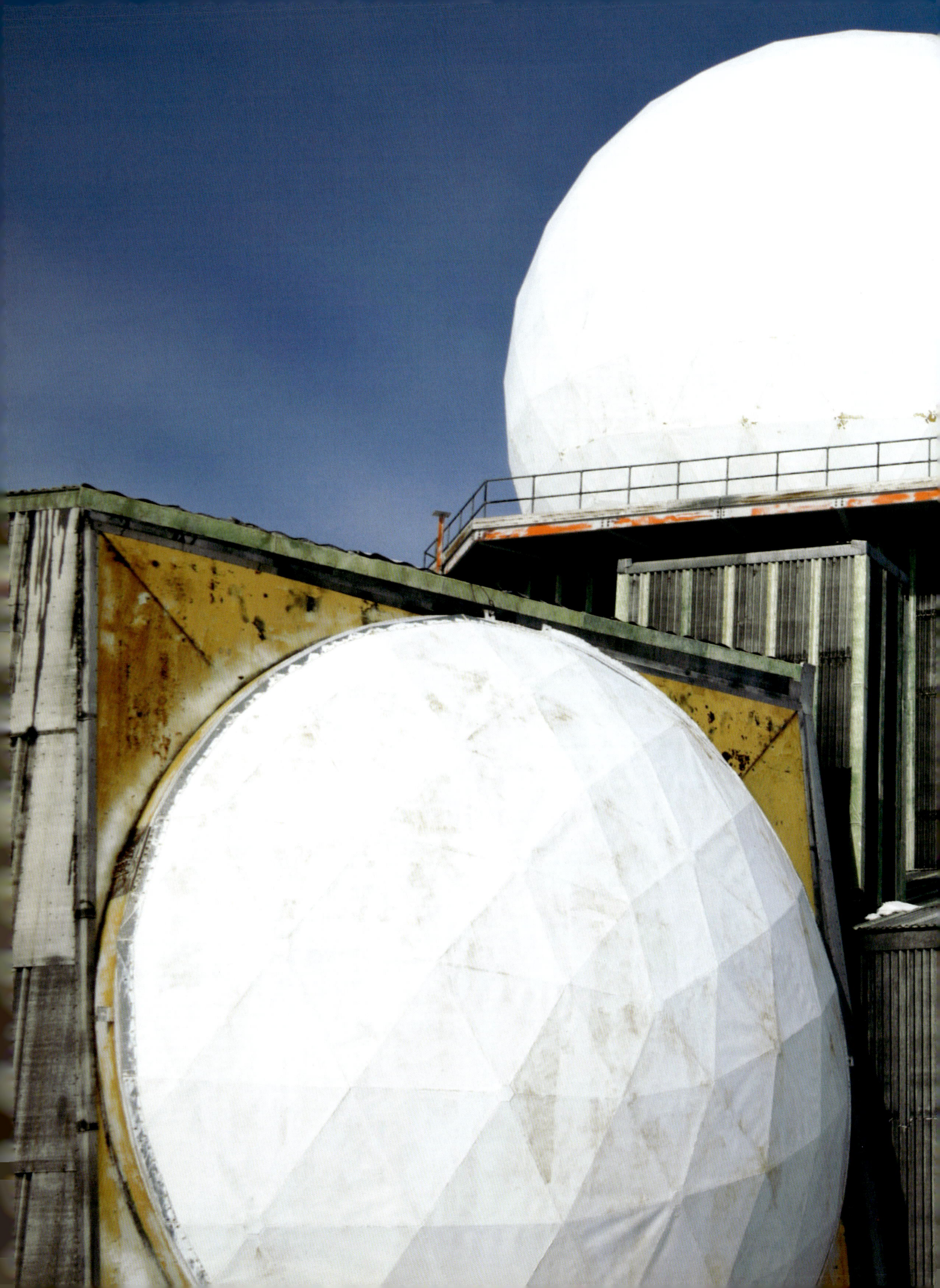

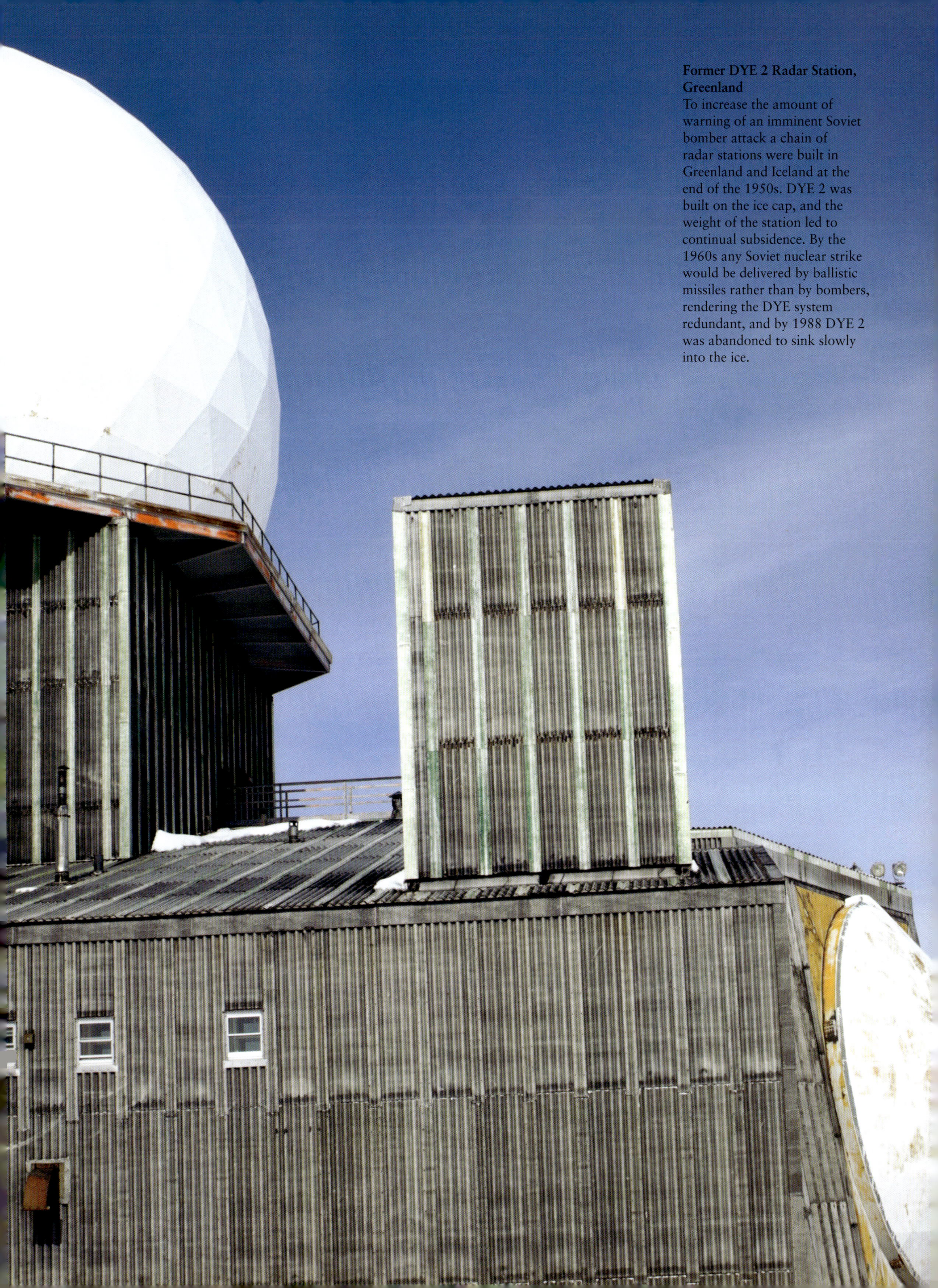

Former DYE 2 Radar Station, Greenland
To increase the amount of warning of an imminent Soviet bomber attack a chain of radar stations were built in Greenland and Iceland at the end of the 1950s. DYE 2 was built on the ice cap, and the weight of the station led to continual subsidence. By the 1960s any Soviet nuclear strike would be delivered by ballistic missiles rather than by bombers, rendering the DYE system redundant, and by 1988 DYE 2 was abandoned to sink slowly into the ice.

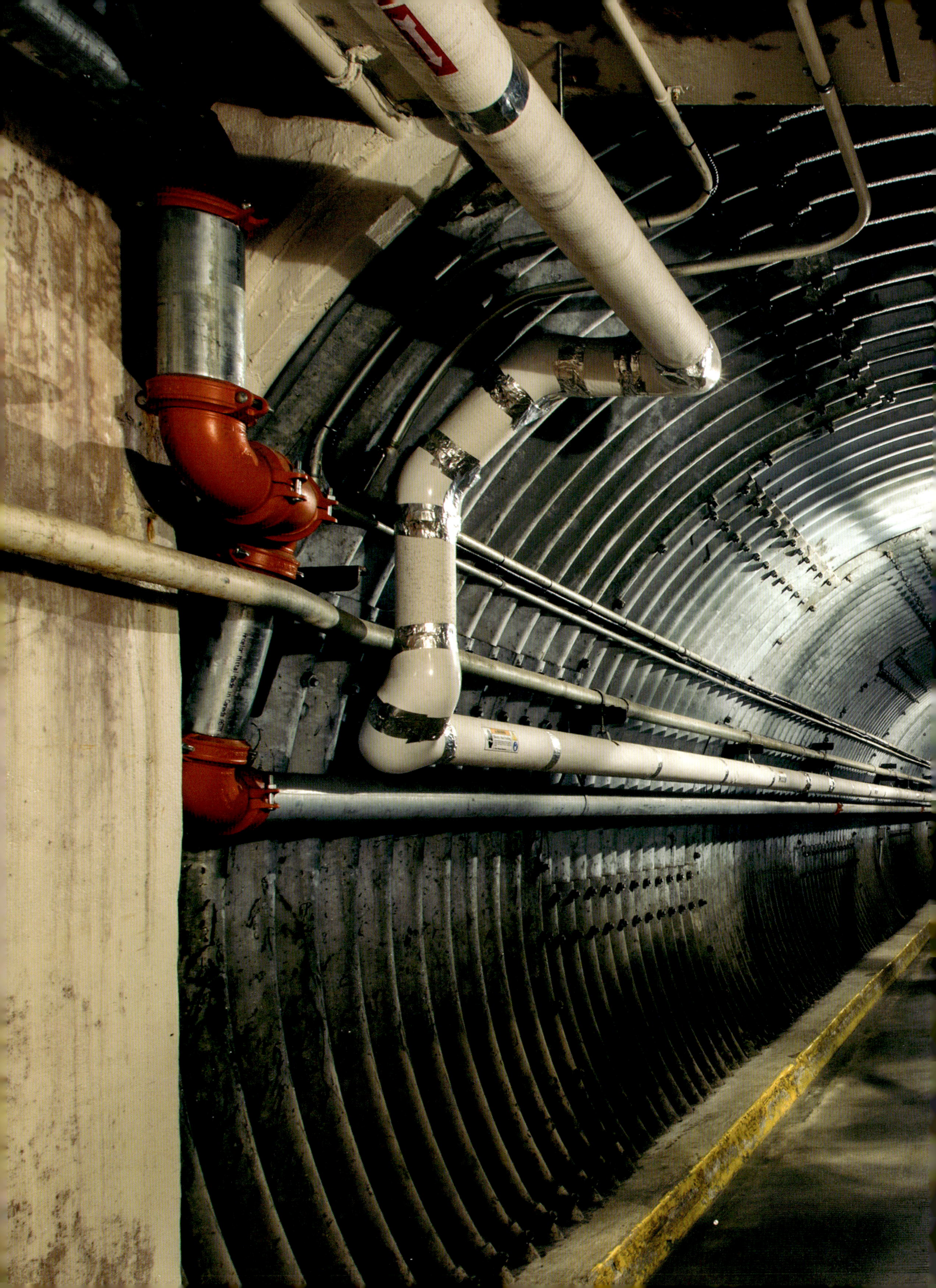

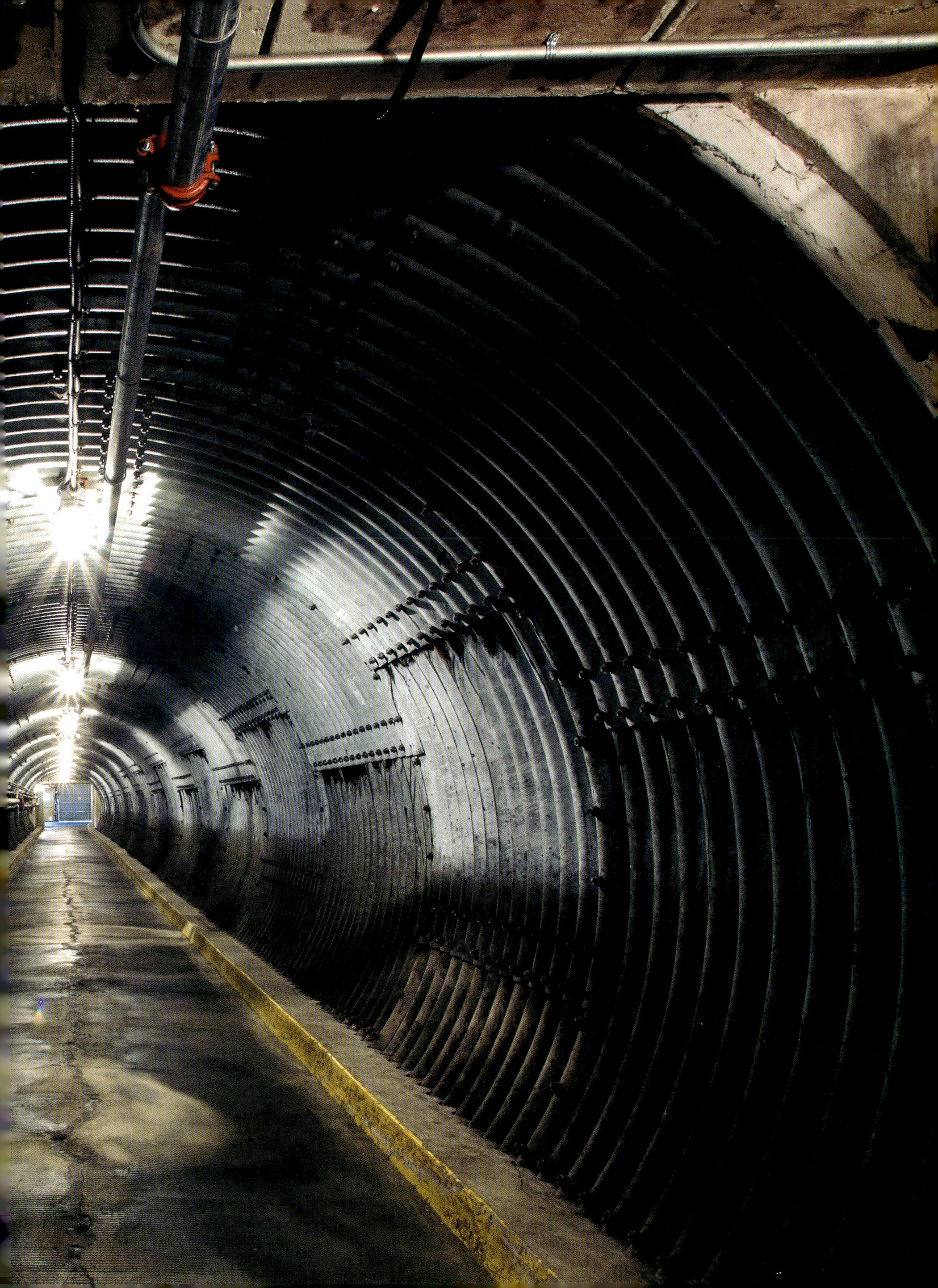

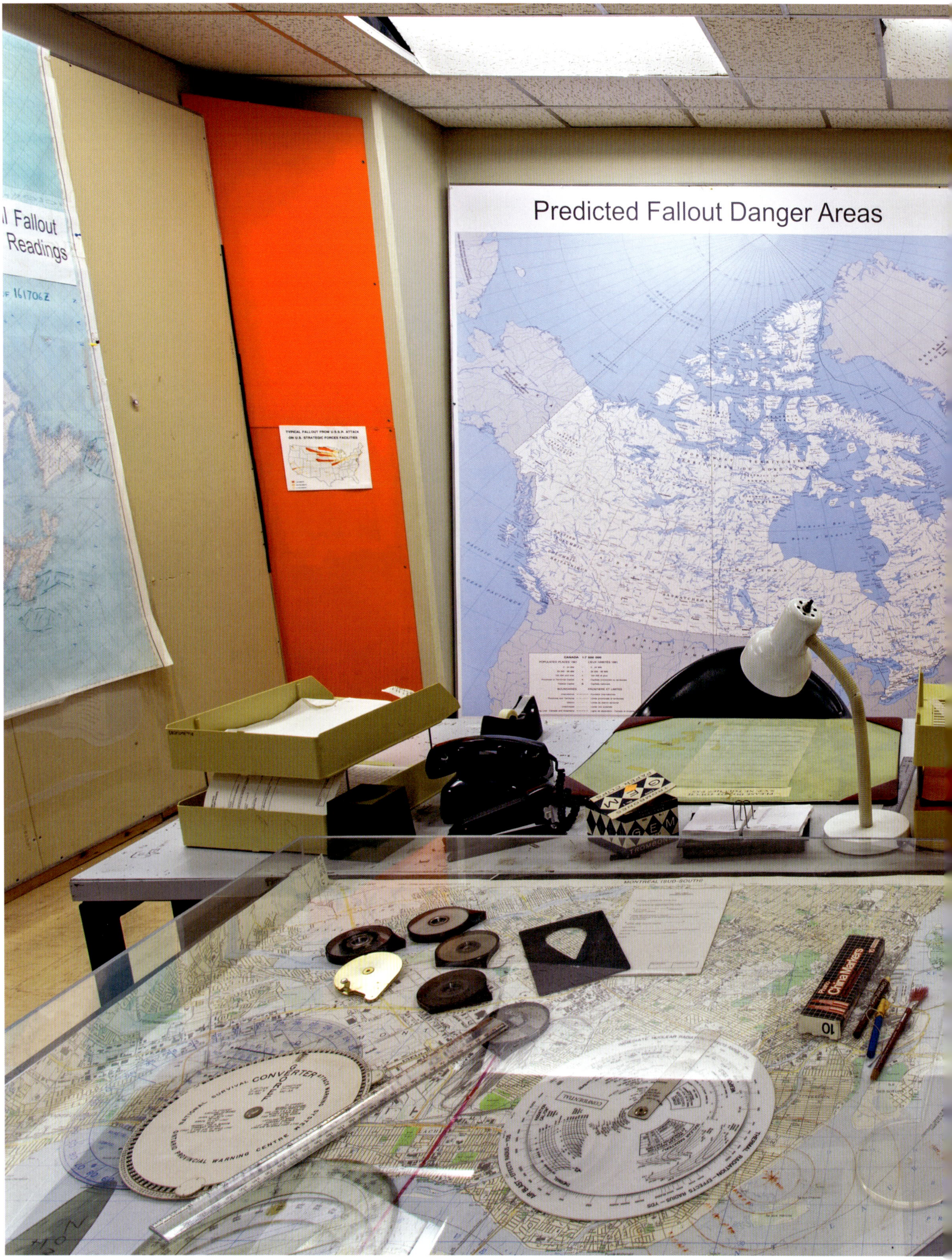

Predicted Fallout Danger Areas
Fallout
Readings

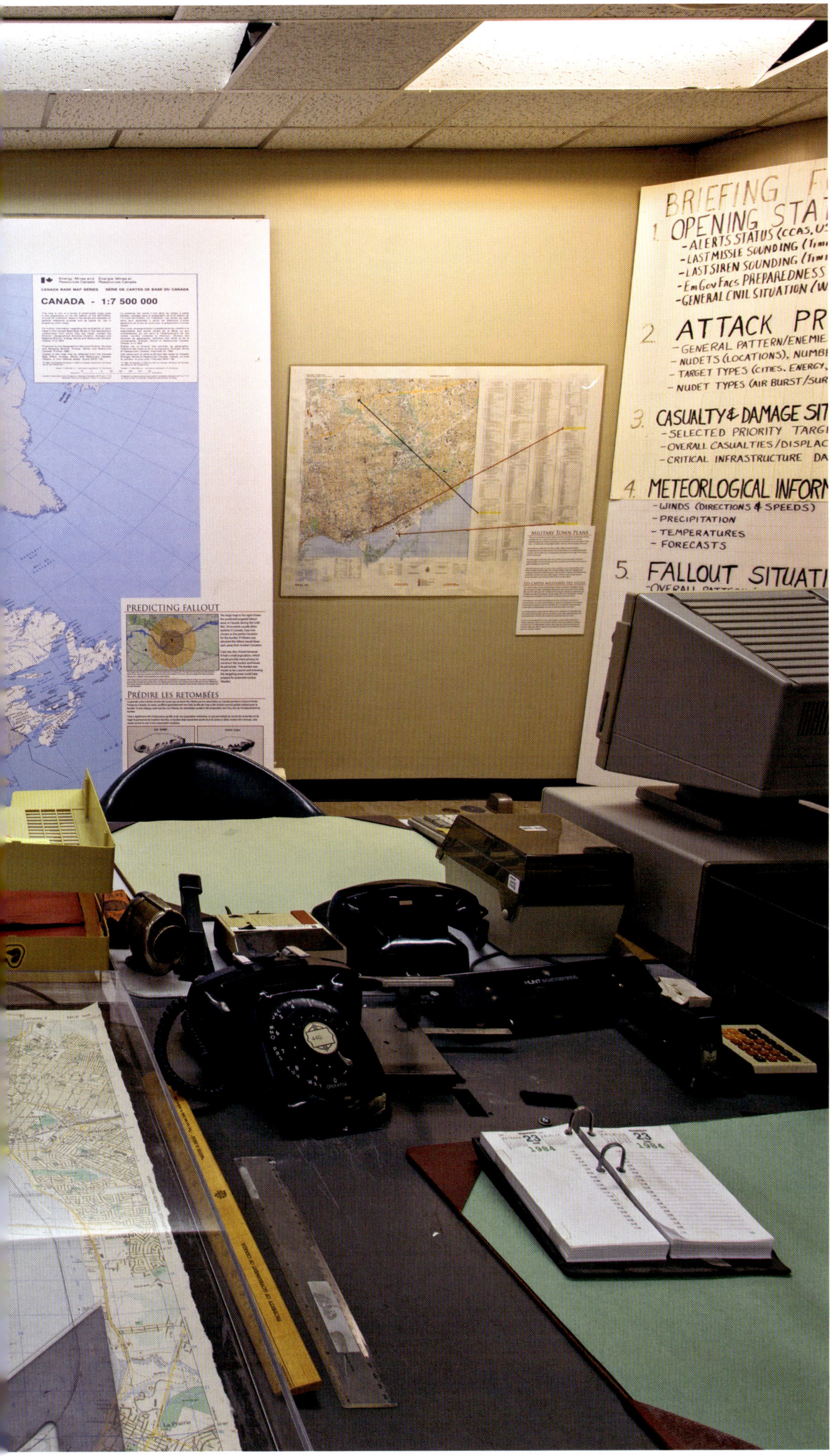

PREVIOUS PAGES:

Entrance Tunnel, Diefenbunker, CFS Carp, Ontario, Canada
Named after John Diefenbaker, the Prime Minister who ordered their construction, a series of over 50 emergency government headquarters were built in Canada during the 1950s and 1960s. The one at CFS Carp was the largest, with four storeys, but most were only one or two storeys and located outside major cities with sufficient supplies for a few weeks' stay underground.

LEFT:

War Room, Diefenbunker, CFS Carp, Ontario, Canada
Even if Canada was not directly targeted in the event of a nuclear war, her proximity to the US ensured that the risk of fallout spreading northwards, or eastwards from Alaska or westwards from Greenland, was significant. As Canada did not have nuclear weapons of its own, the focus of the war room was on minimizing casualties and maintaining government functions.

Diefenbunker, CFS Carp, Ontario, Canada
The shelter at CFS Carp had its own Canadian Broadcasting Corporation studio, as well as space for the prime minister and other key government ministers, medical and communications facilities, a conference room and a vault designed to hold Canada's gold reserves. The bunker cost $25 million to build, and approximately $450,000 a year to maintain. It was decommissioned in 1994.

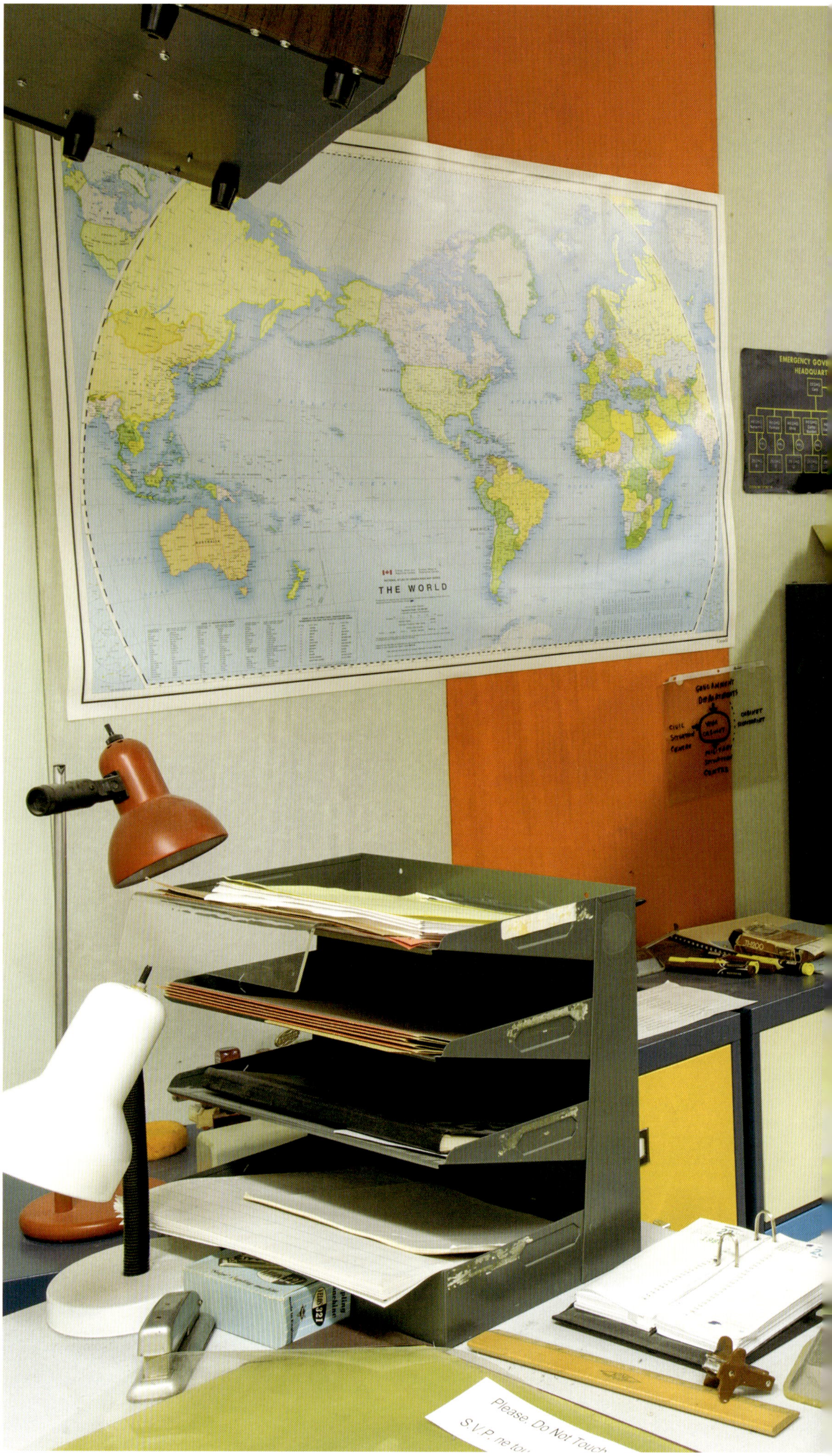

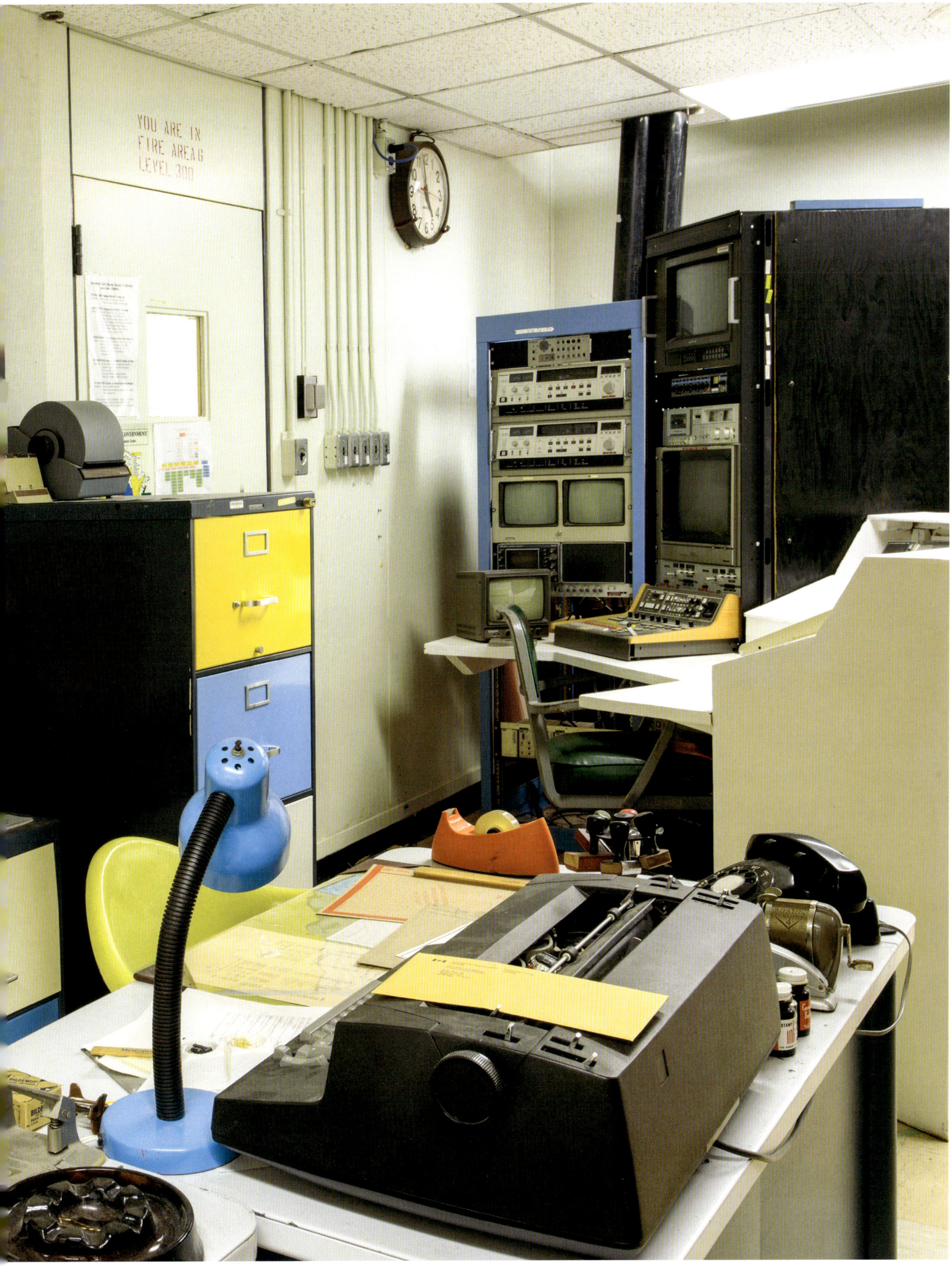
YOU ARE IN
FIRE AREA G
LEVEL 300

Former Lyndonville AFS, Vermont, USA
Established as a radar station in a remote part of Vermont in the 1950s, Lyndonville was designed to provide early warning of aircraft intruding on US airspace and guide interceptors to meet them. In 1962 it was upgraded with two concrete towers for new radars but only one was installed before the site was closed in 1963.

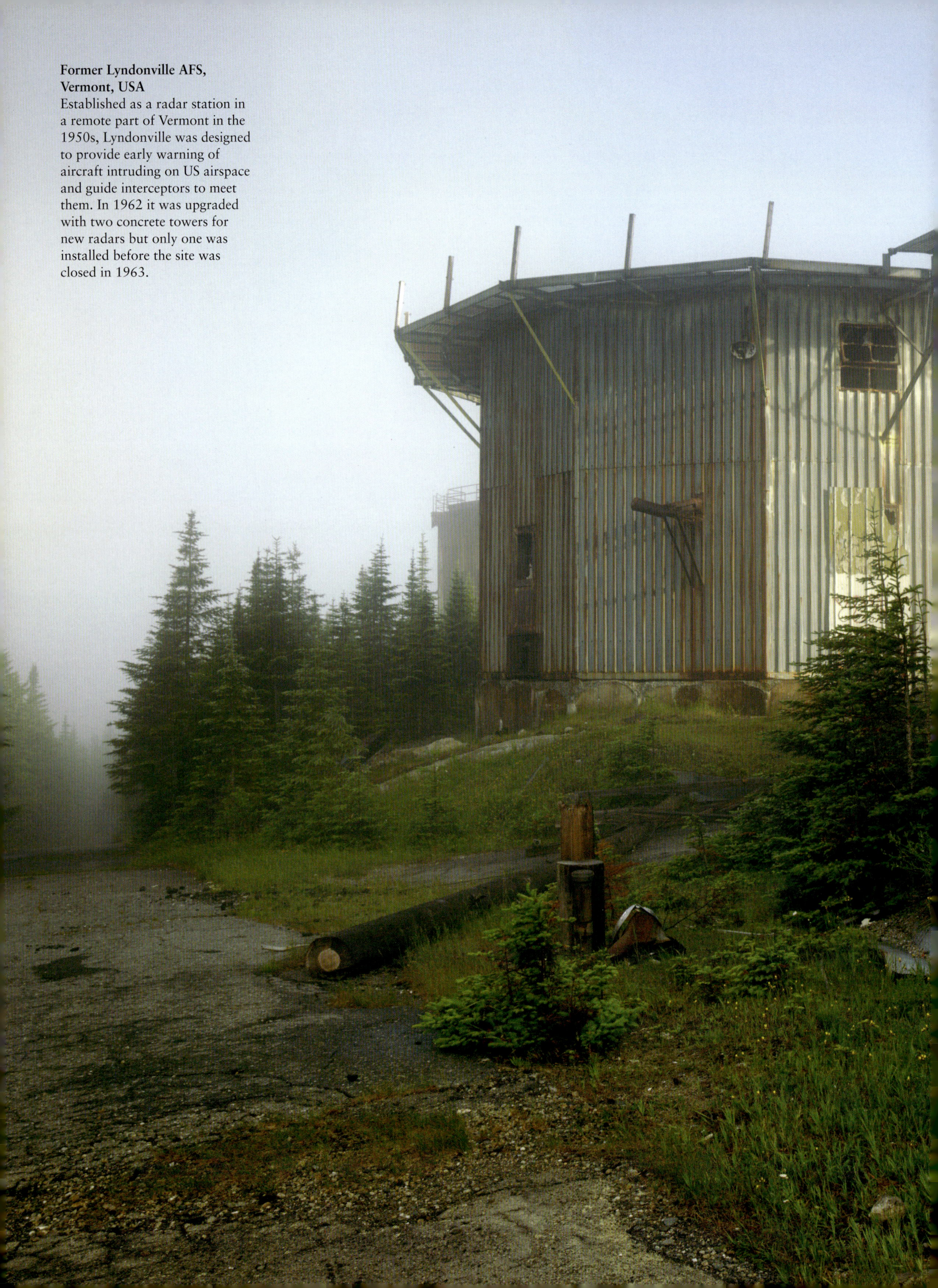

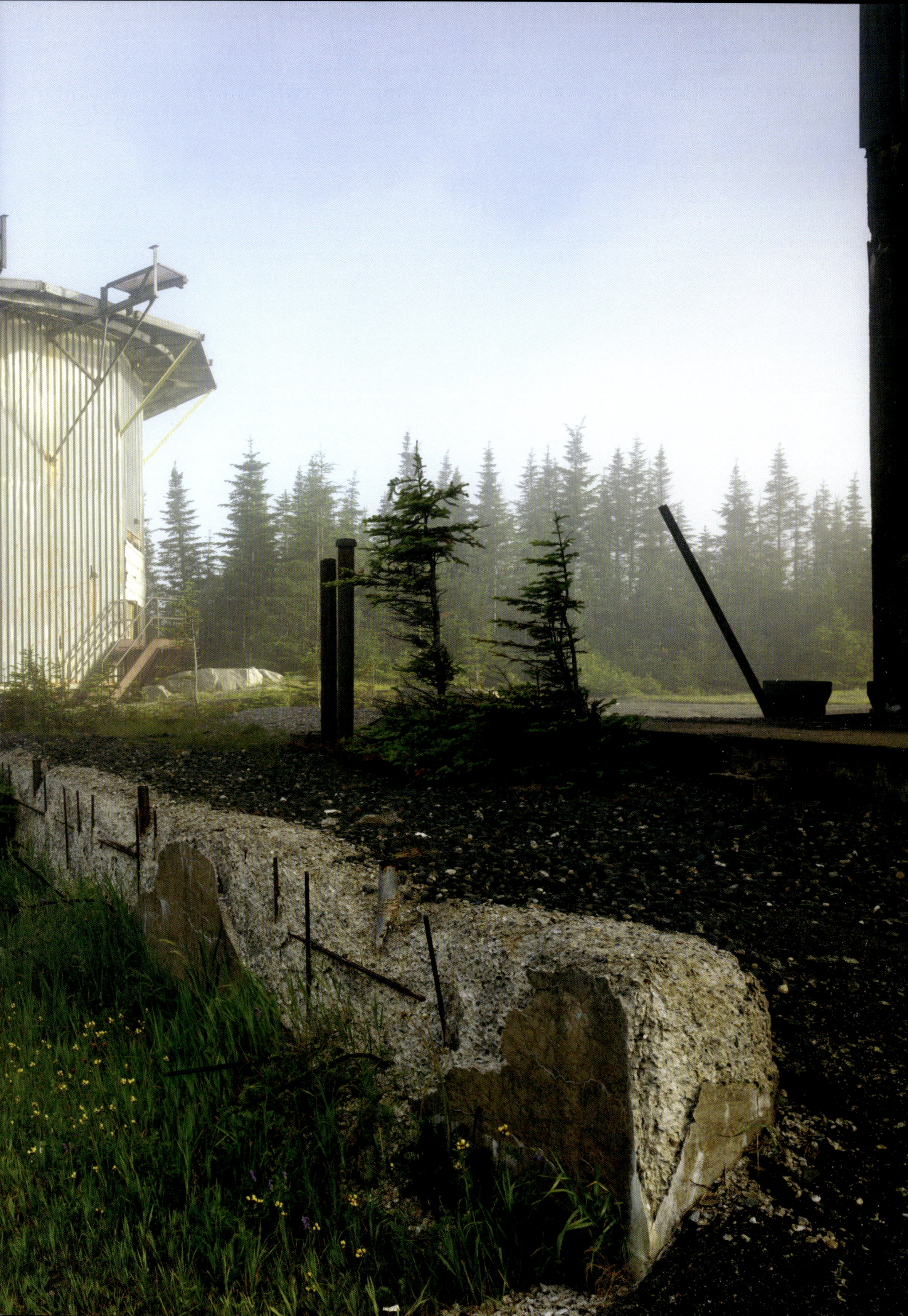

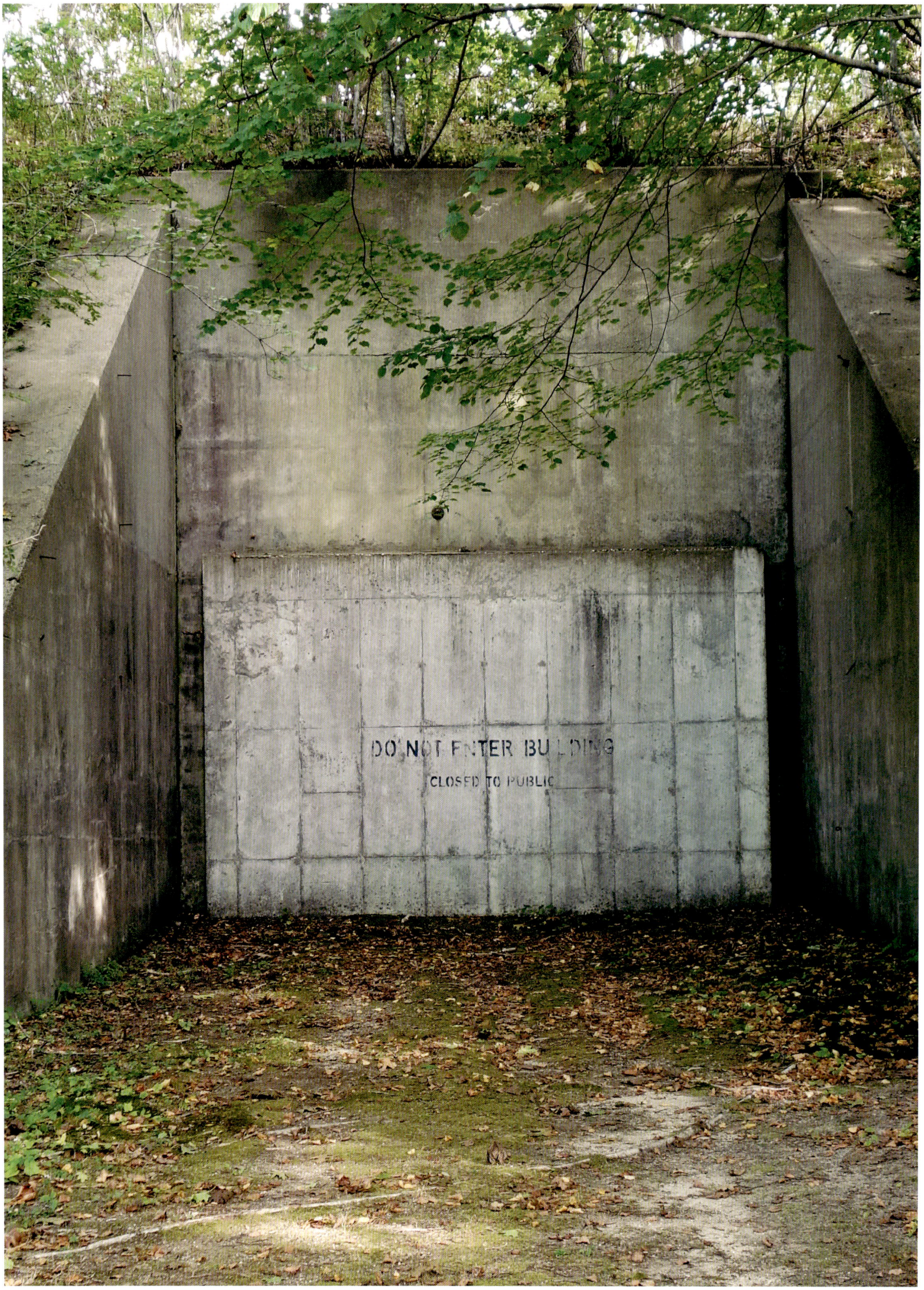
DO NOT ENTER BUILDING
CLOSED TO PUBLIC

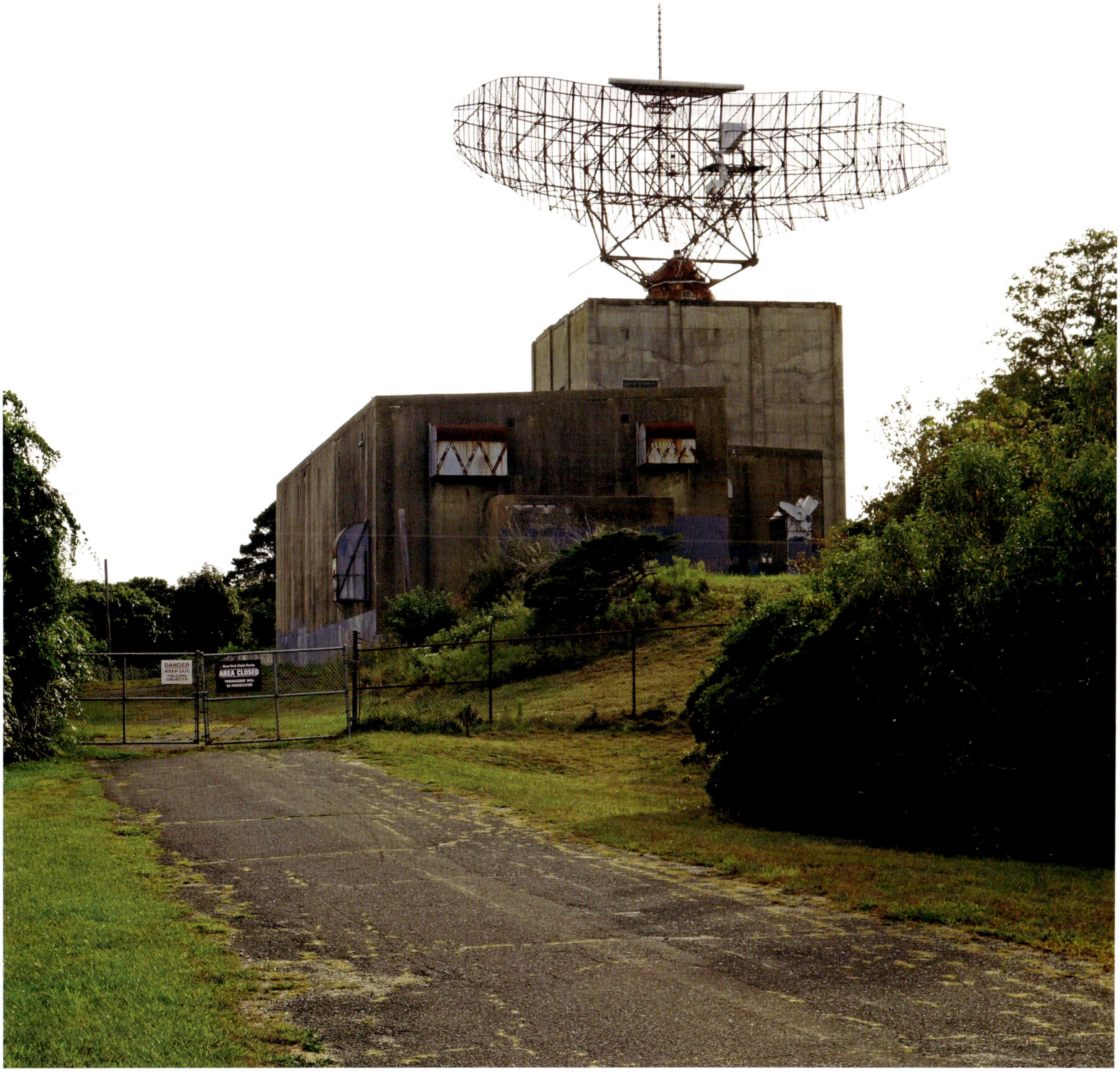

OPPOSITE:
Bunker Entrance, Montauk AFS, New York, USA
The threat of the Soviet Tupolev Tu-4 bomber, the first that could reach the US mainland, prompted the establishment of early warning radar systems across the country. One was built in 1948 at Fort Hero, a former World War II facility that included naval guns covering the Long Island Sound against potential German invasion.

ABOVE:
Radar Station, Montauk AFS, New York, USA
Montauk was upgraded several times during the Cold War, becoming a key part of the NORAD early warning system. This AN/FPS-35 radar was installed in 1960, and was powerful enough to cause interference to local radio and television equipment. It could detect aircraft over 320km (200 miles) away. The increased use of satellites resulted in Montauk being decommissioned in 1981, when it was decided to leave the antenna in place rather than attempt to remove it.

OVERLEAF:
Stanley R. Mickelsen Safeguard Complex, Nekoma, North Dakota, USA
Built to defend the nearby Minuteman ICBM silos, the complex was the first anti-ballistic site in the US. The Missile Site Radar (the pyramid structure) would guide Spartan (long range) and Sprint (short range) missiles to incoming warheads. Viewed as a failure, the complex was closed in 1976 after less than a year.

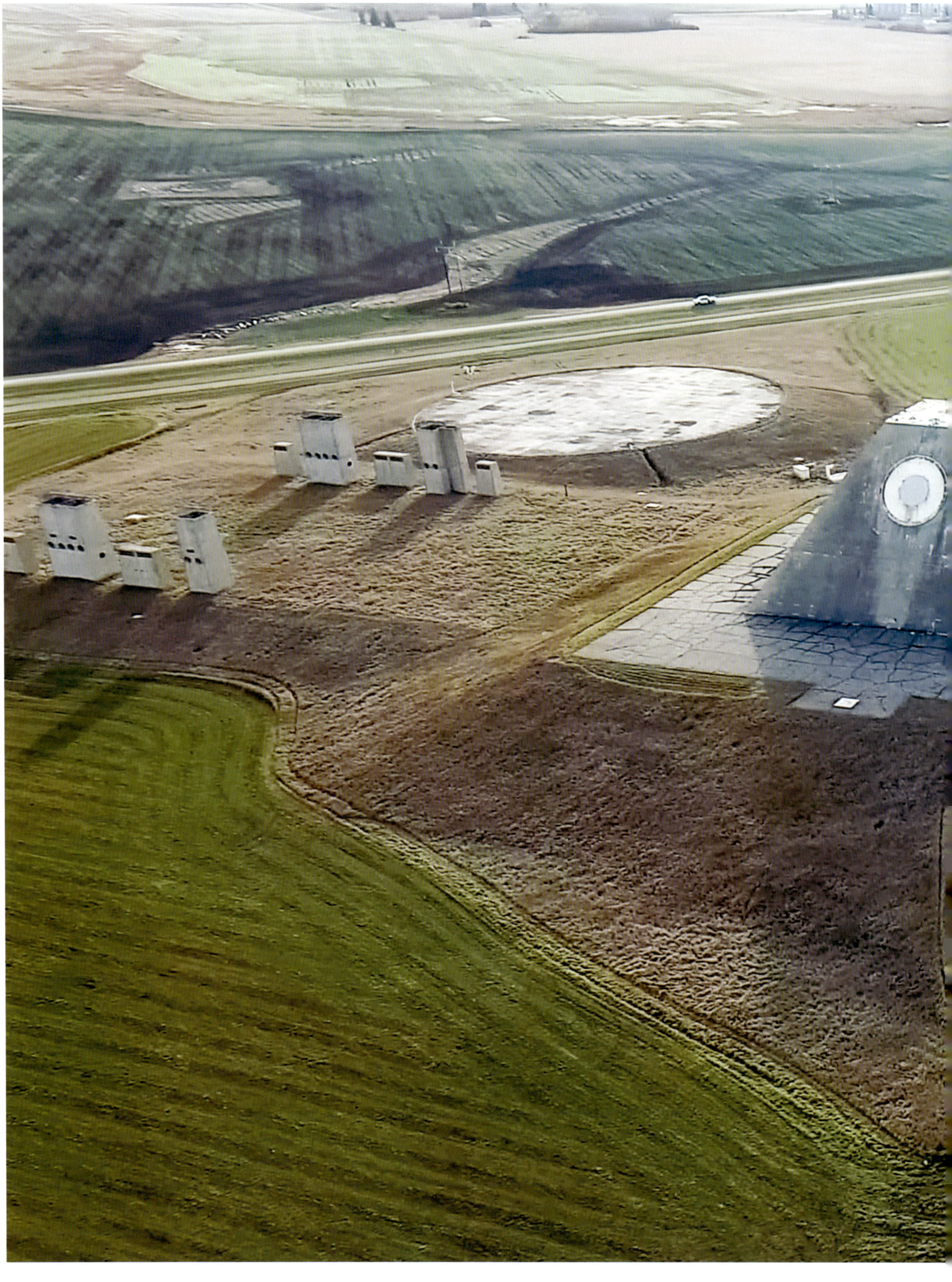

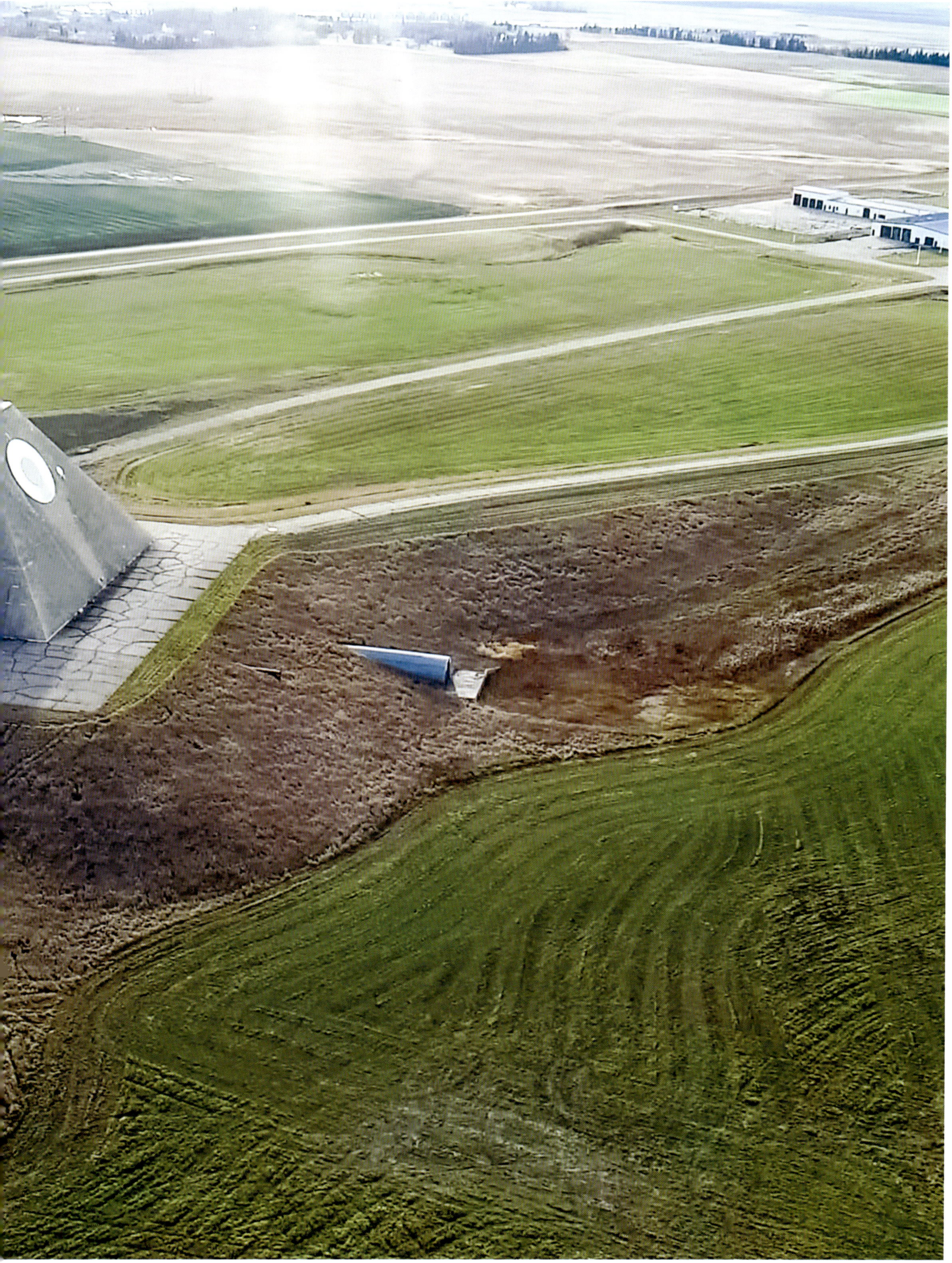

Former US Army Black Hills Ordnance Depot, South Dakota, USA
Opened in 1942, the depot was used for the storage and maintenance of various types of ordnance. Chemical weapons such as mustard gas and sarin were tested on site. The depot was closed in 1967. In 2016, it was bought by a developer who has converted some of the 575 bunkers into shelters that allow up to 24 people to survive for one year.

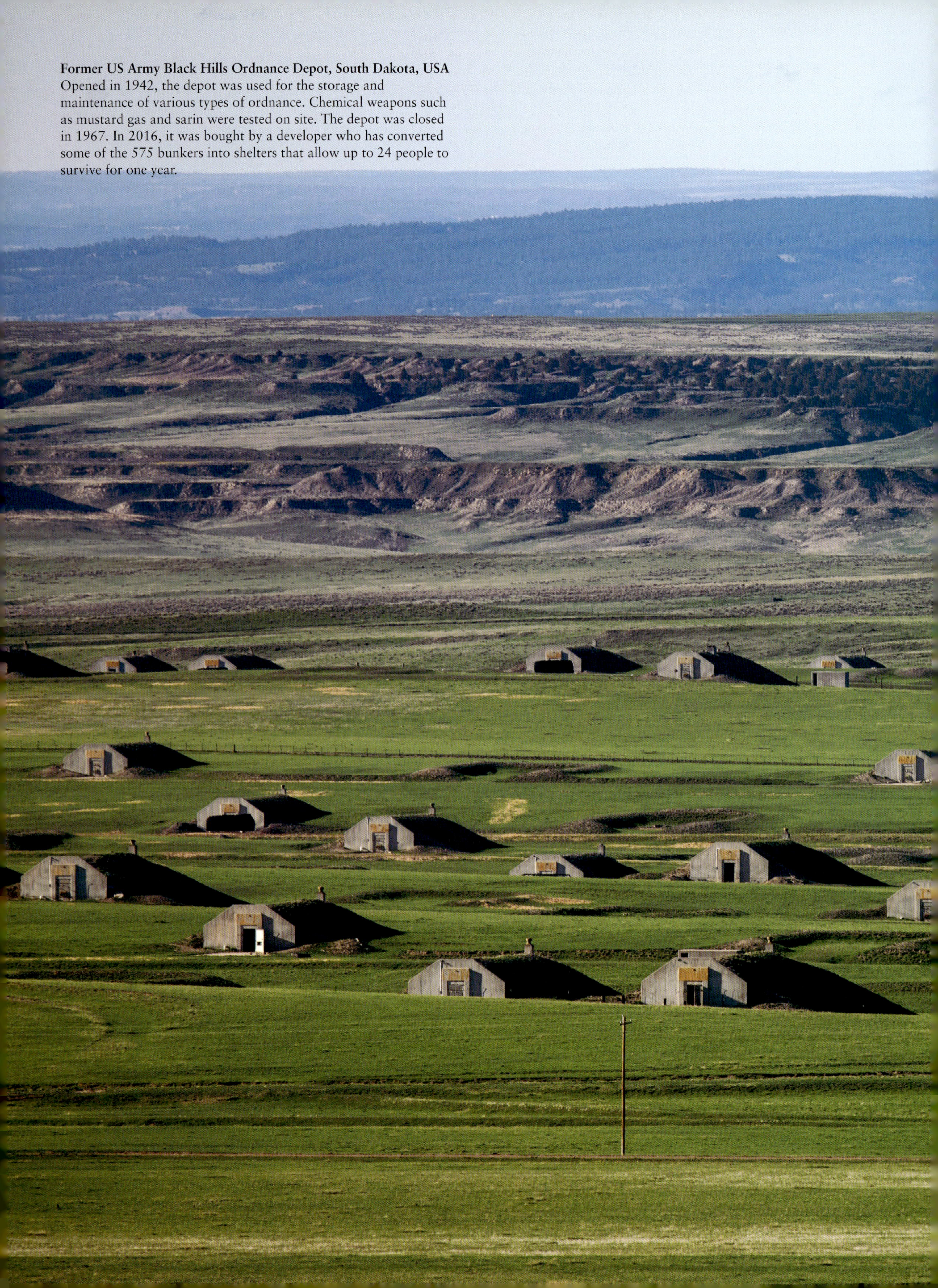

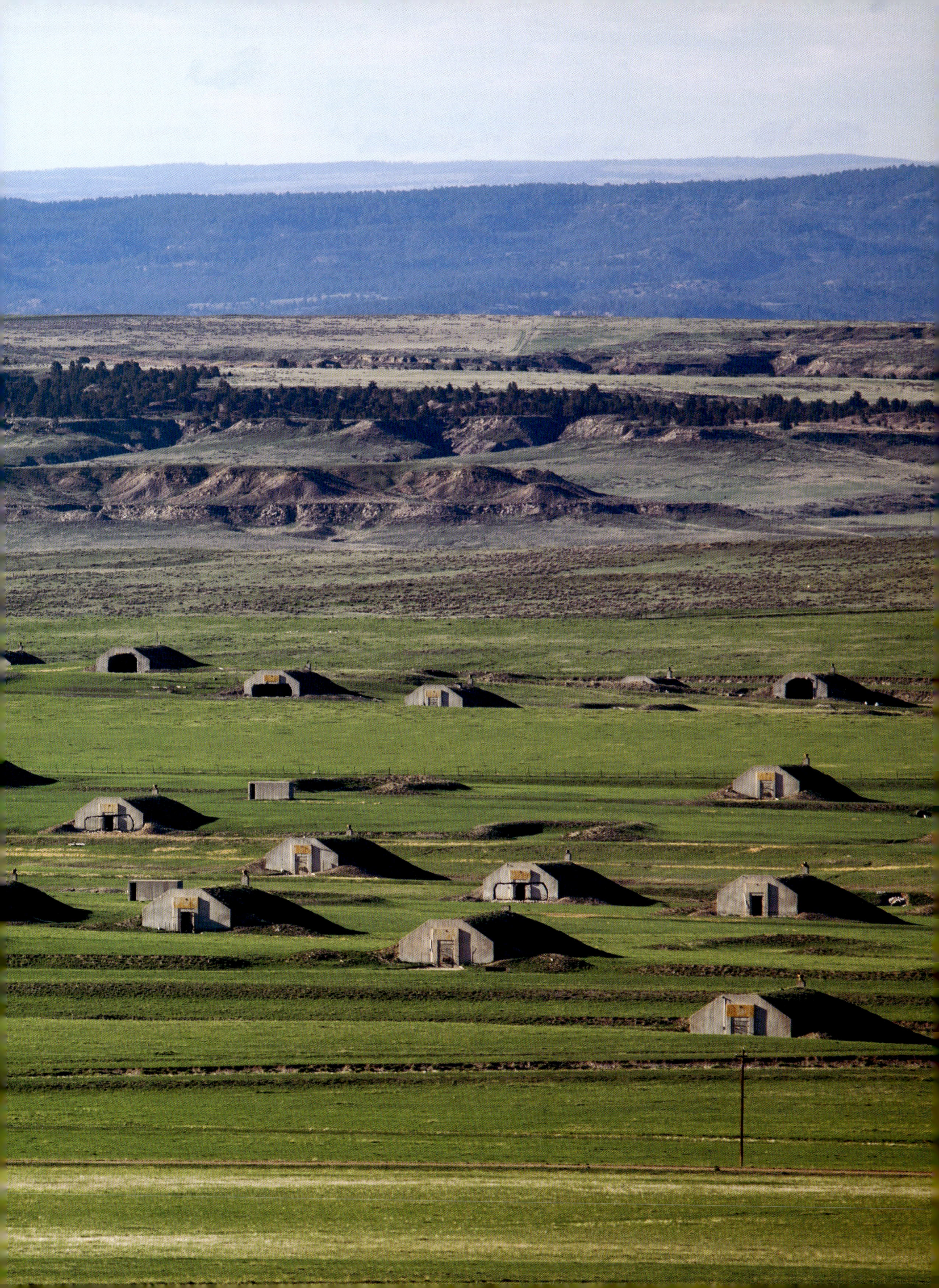

Bathroom, Former Peacekeeper Silo, Wyoming, USA
Conditions for the men who operated the Peacekeeper missiles at Francis E. Warren Air Force Base (AFB) from 1986 to 2005 were far from luxurious. Under development since the early 1970s, the Peacekeeper carried up to ten warheads with a range of around 9660km (6000 miles). The operators' bunker was approximately 30m (100ft) underground.

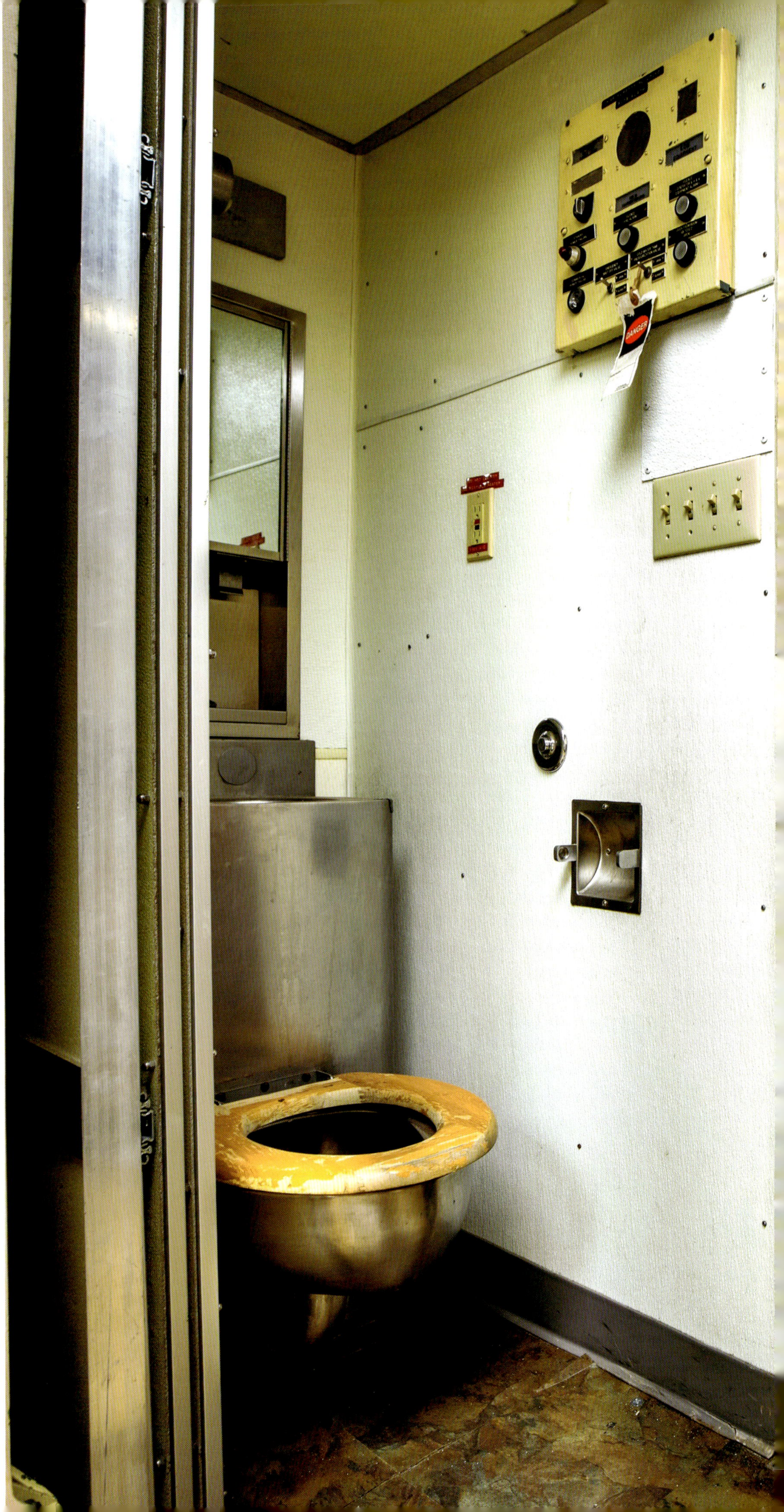

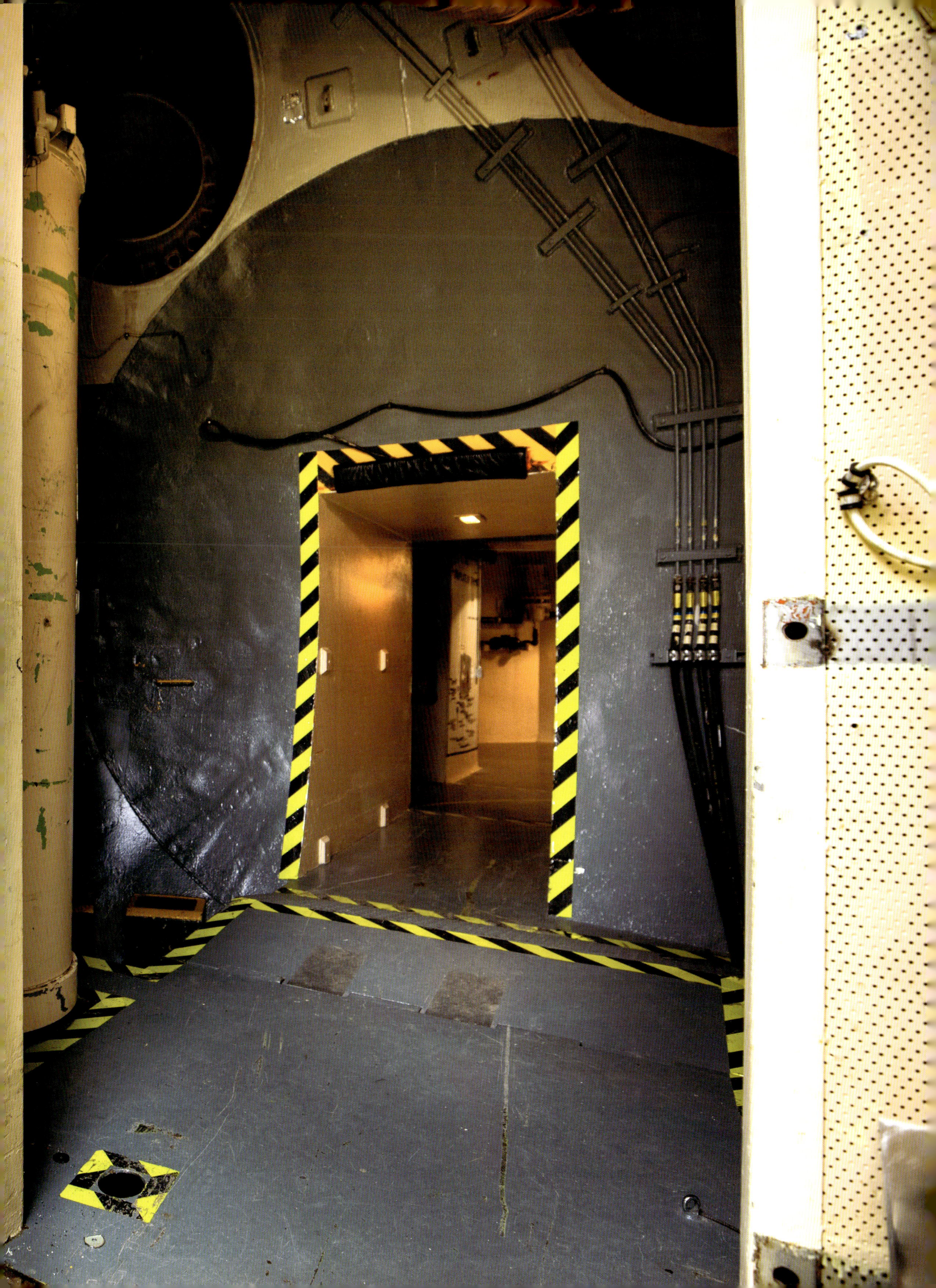

RIGHT:

Corridor, Greenbrier Bunker, West Virginia, USA

Codenamed 'Casper', the House of Representatives had a nuclear shelter built for them under a new extension to the Greenbrier Hotel in White Sulphur Springs that was completed in 1962. With its own generators and air purification, the shelter could accommodate over 1000 people, and the food supplies were regularly replenished. As the bunker was owned by the hotel, the Federal Government paid an annual rent of $25,000.

OPPOSITE:

Blast Door, Greenbrier Bunker, West Virginia, USA

The blast doors at Greenbrier were so heavy that two of them were shipped from Ohio by train, but they were balanced so that they could be opened easily by one person. The existence of the bunker was kept secret so that only a few Congressmen and the 12 staff required to maintain it knew, until a *Washington Post* journalist discovered it in 1992, and the bunker – never deployed in anger – was then decommissioned.

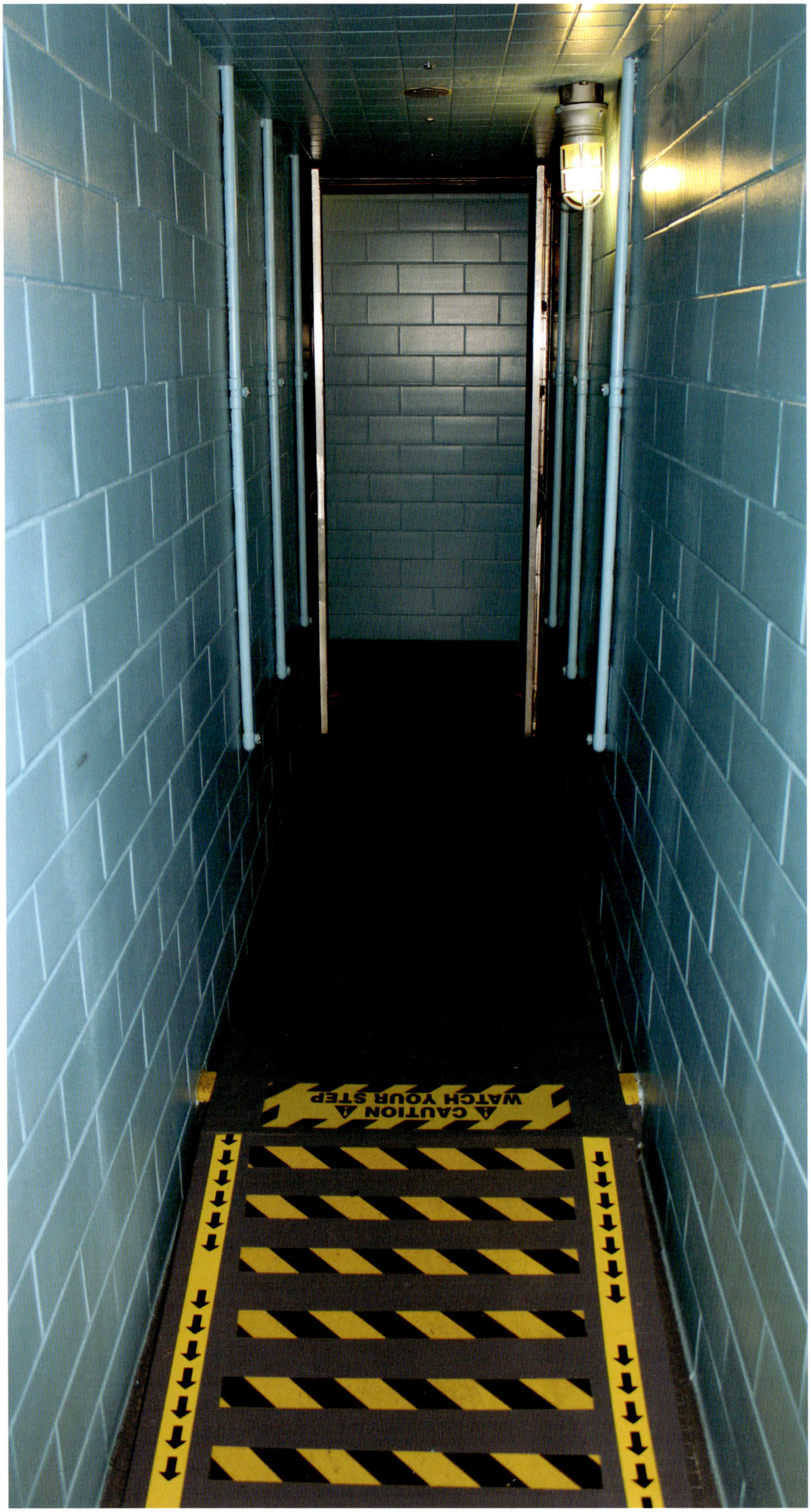

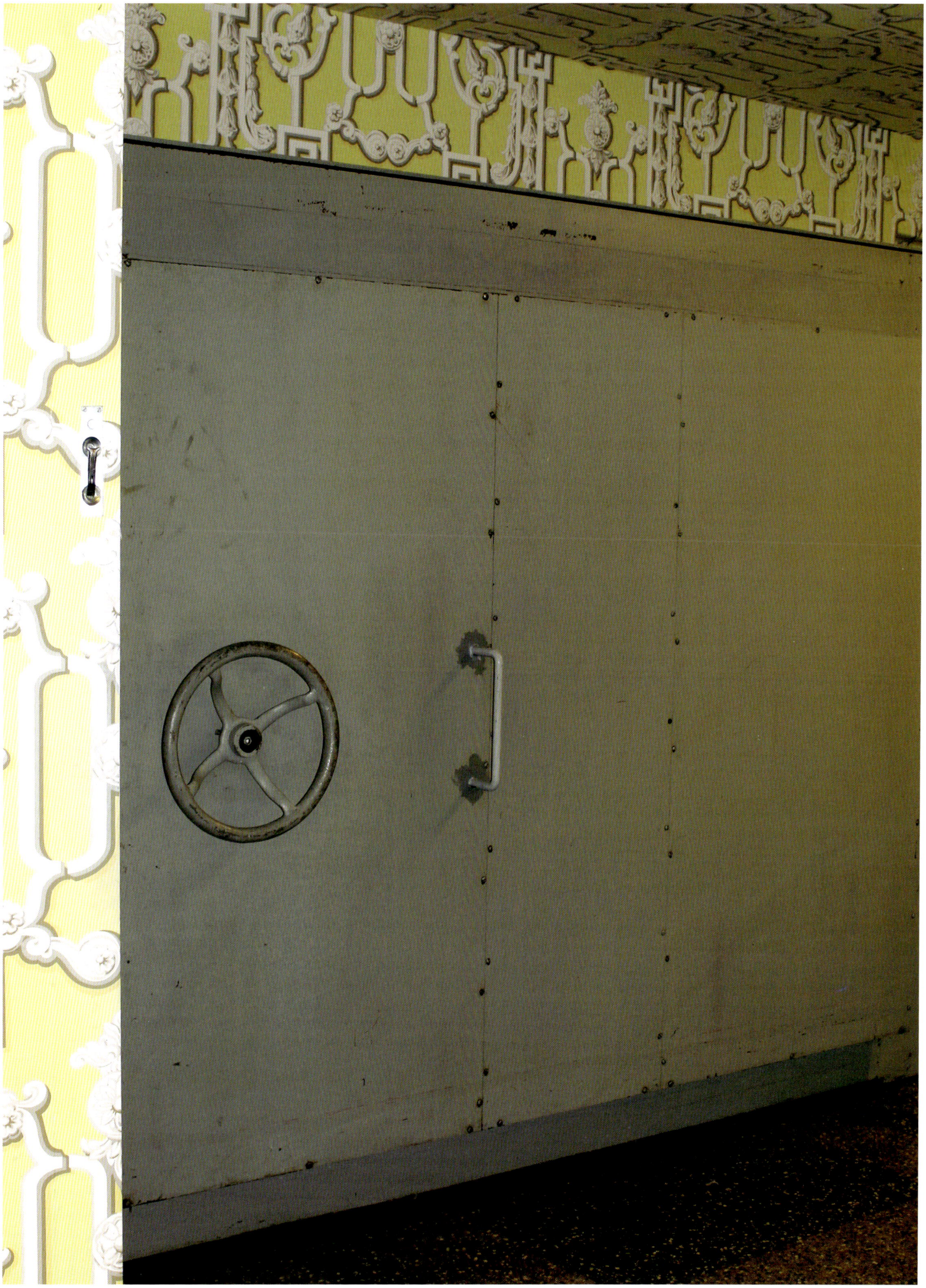

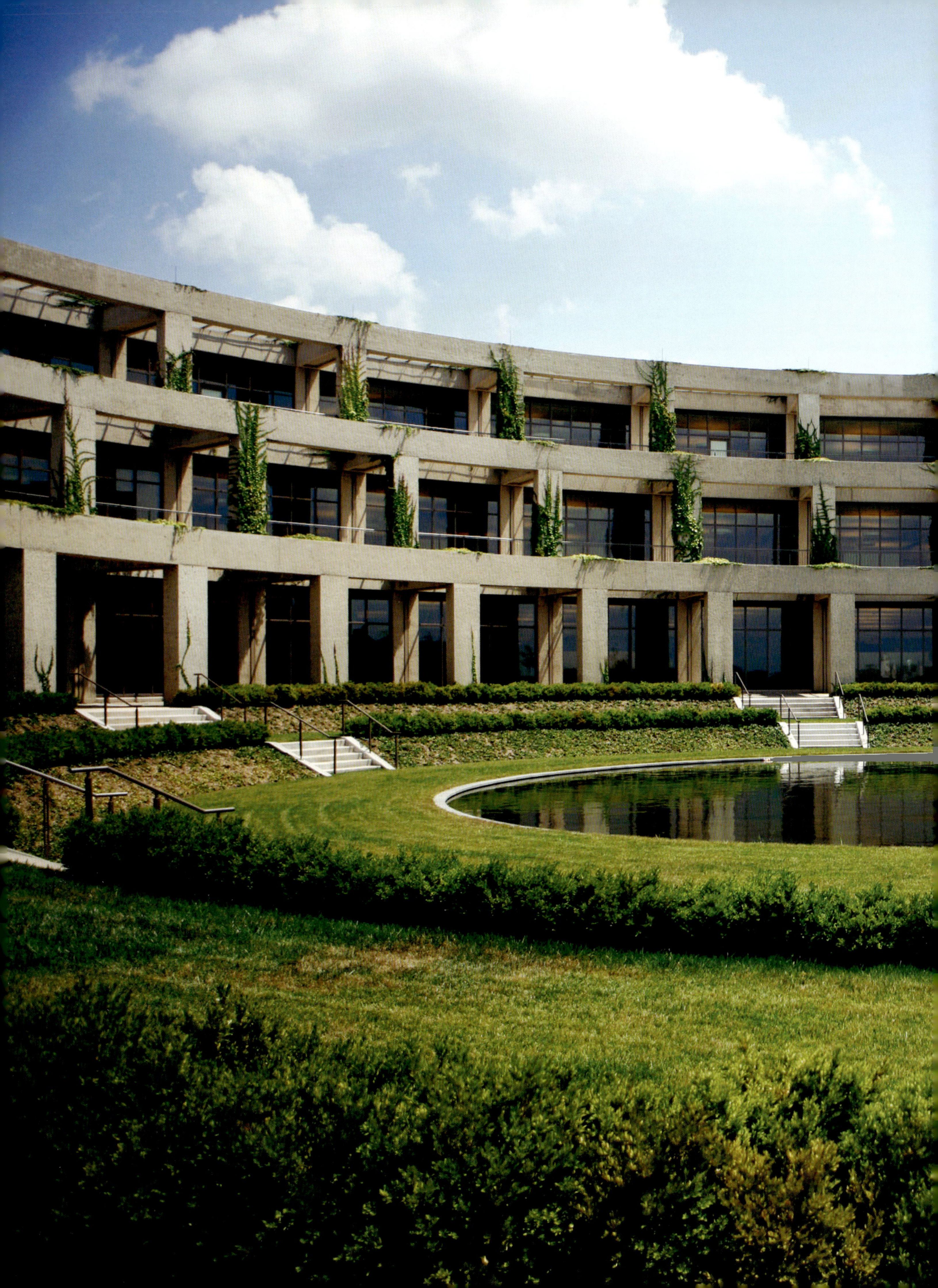

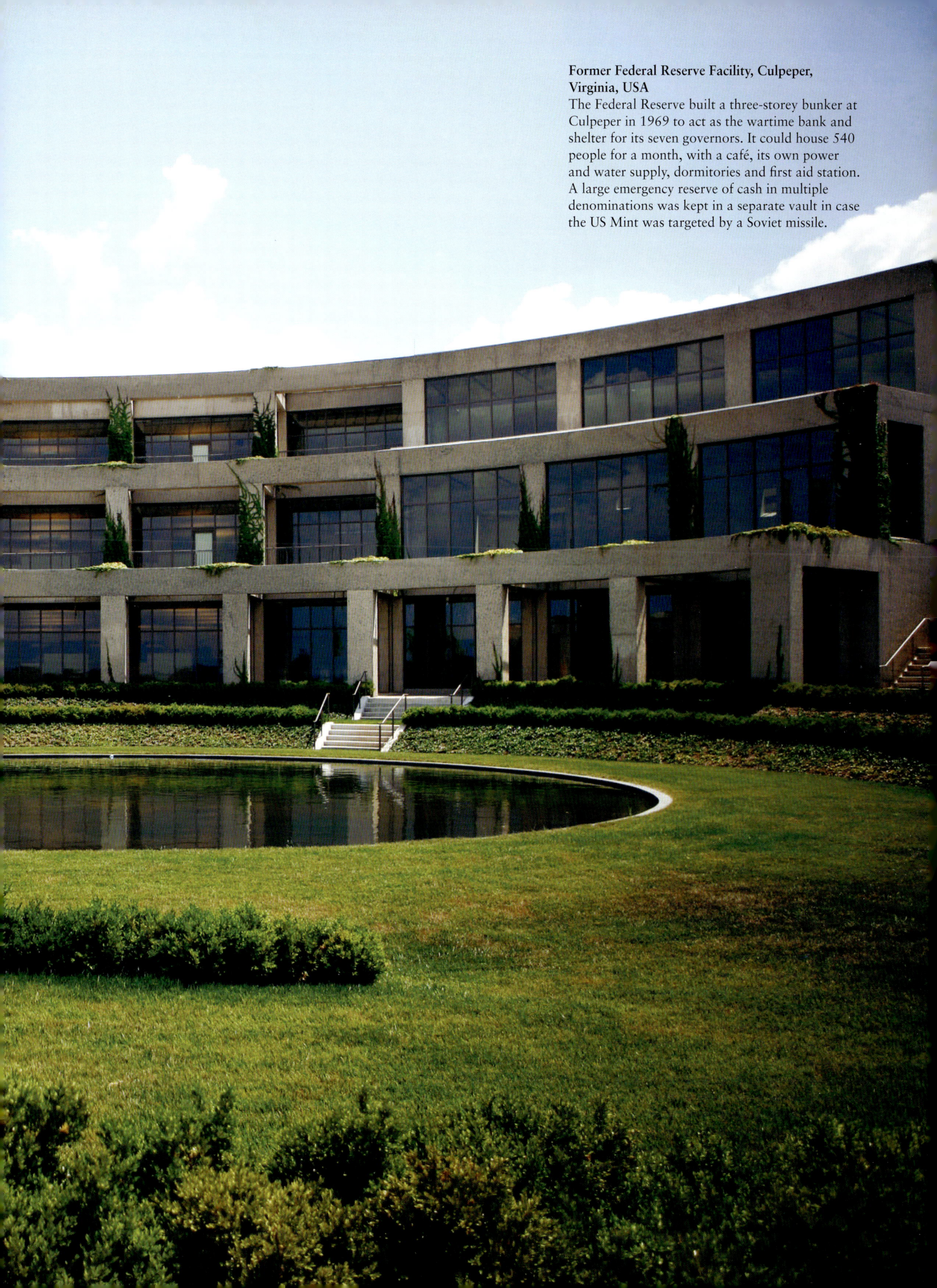

Former Federal Reserve Facility, Culpeper, Virginia, USA
The Federal Reserve built a three-storey bunker at Culpeper in 1969 to act as the wartime bank and shelter for its seven governors. It could house 540 people for a month, with a café, its own power and water supply, dormitories and first aid station. A large emergency reserve of cash in multiple denominations was kept in a separate vault in case the US Mint was targeted by a Soviet missile.

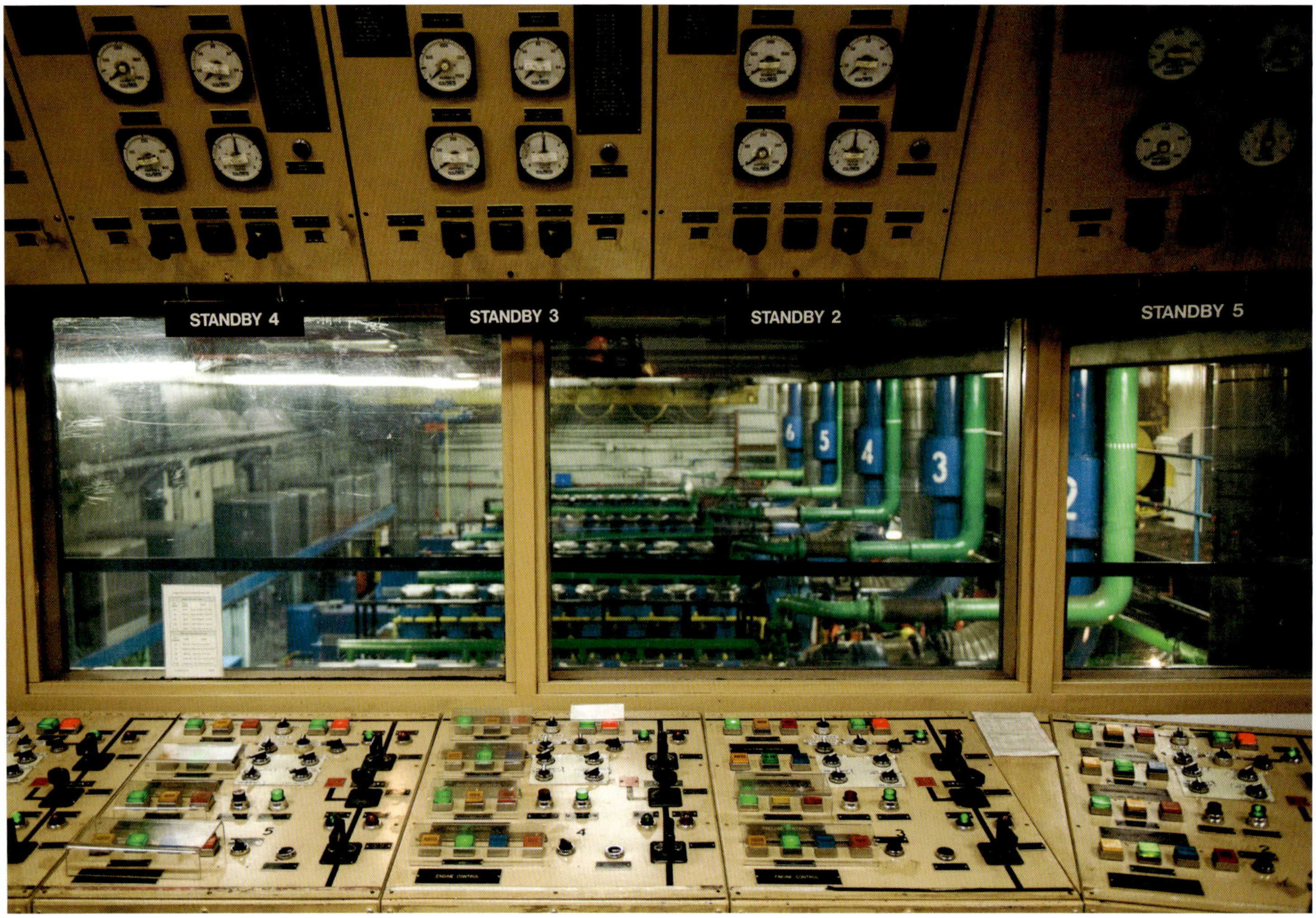
STANDBY 4
STANDBY 3
STANDBY 2
STANDBY 5

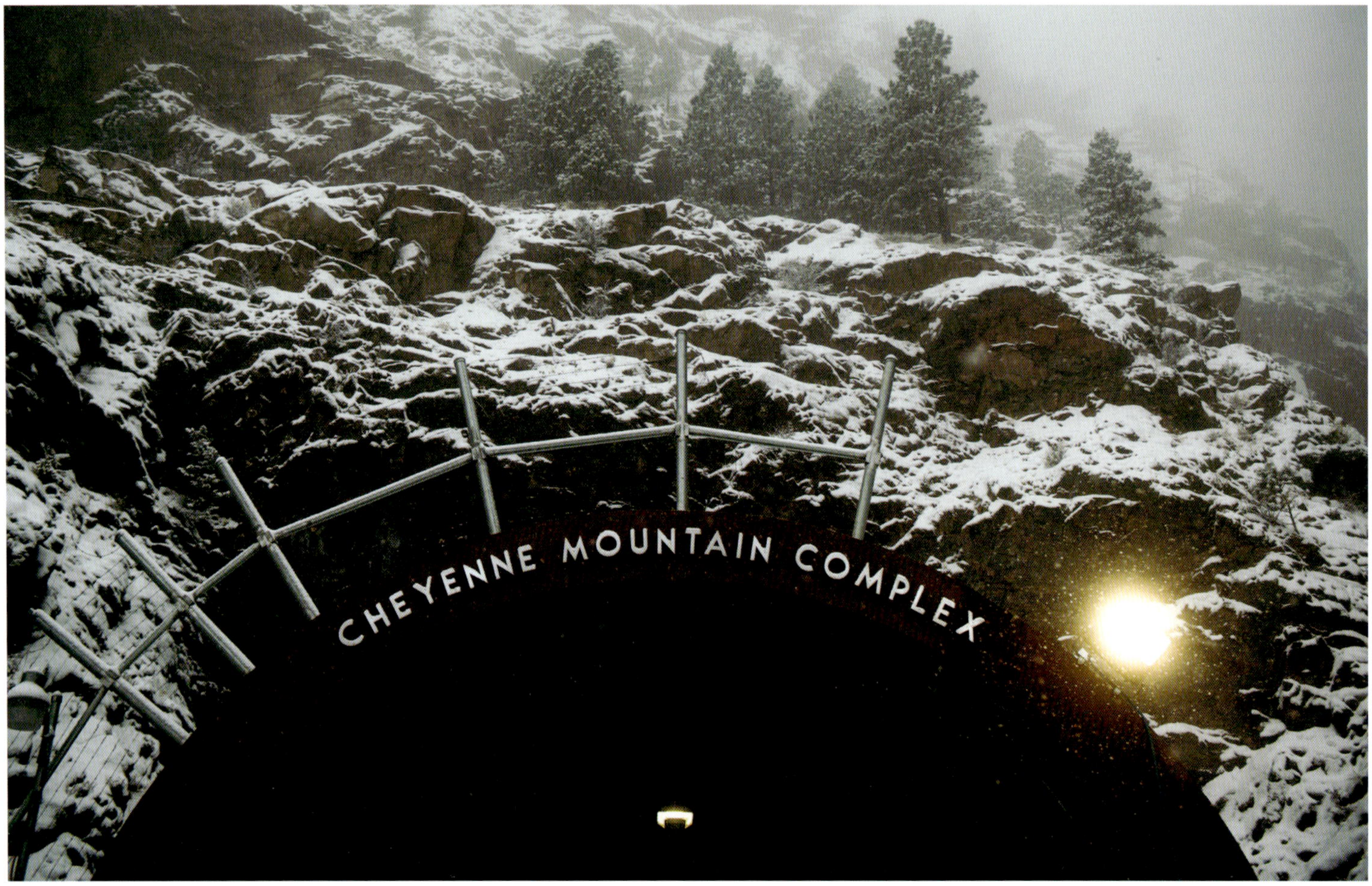
CHEYENNE MOUNTAIN COMPLEX

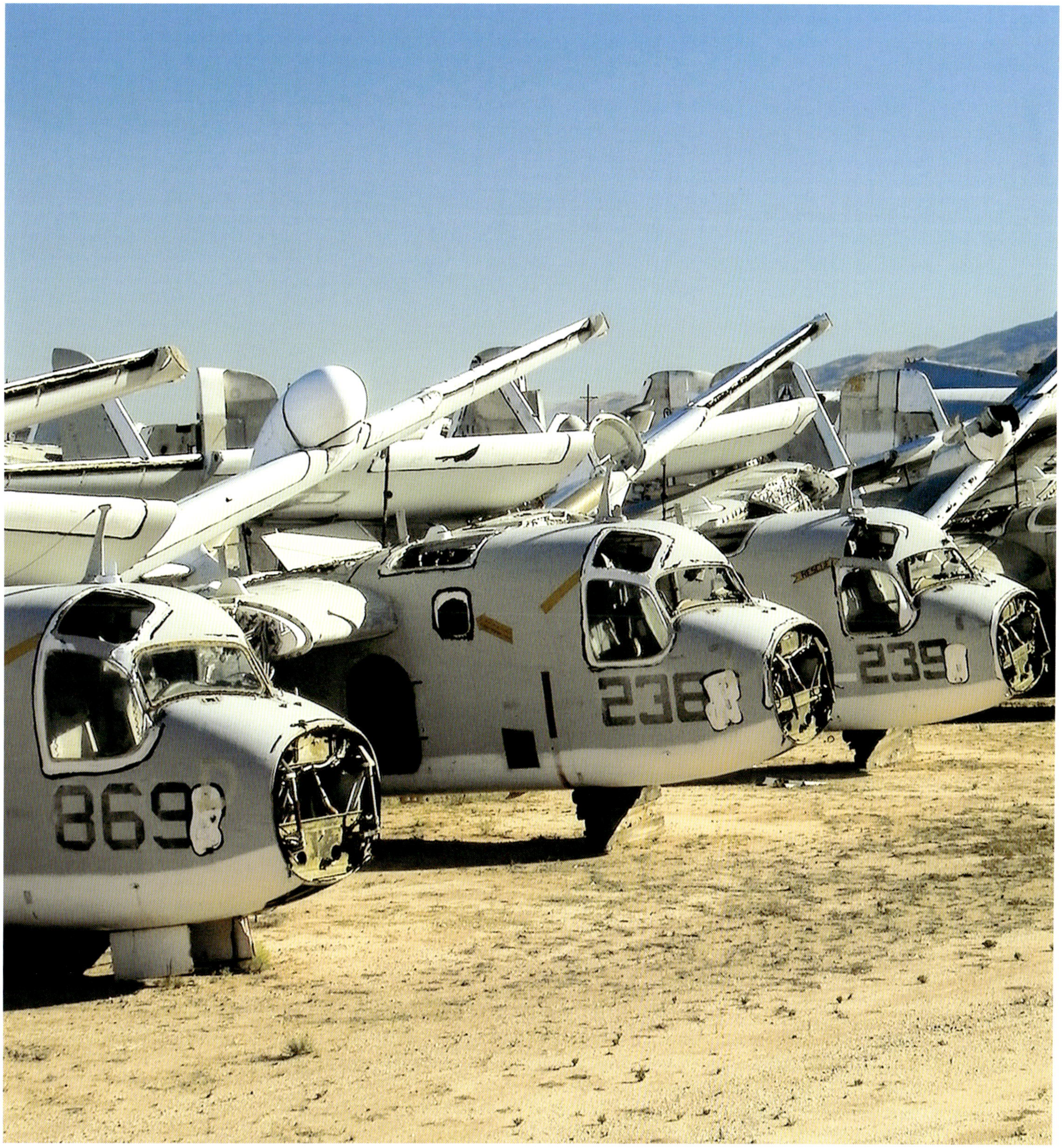

OPPOSITE TOP AND BOTTOM:
Cheyenne Mountain Complex, Wyoming, USA
Construction on the North American Air Defense Command (NORAD) site began in 1961, with the complex becoming operational five years later. Over 1000 US and Canadian personnel worked there by the 1970s. The size of a small town, the site had six emergency diesel generators ready to generate up to 500 kilowatts should the regular power supply fail.

It was expected that the NORAD complex would be targeted by Soviet missiles should war break out. The entrance had two 25 ton doors 15m (50ft) apart to protect the buildings inside from any blast effect, with concrete-covered steel ribs on the outside to divert any dislodged rocks from blocking access to the interior.

ABOVE:
S-2 Tracker Storage Area, Davis-Monthan AFB, Tucson, Arizona, USA
The dry climate and alkali soil around Tucson provide perfect conditions for the storage of surplus or retired aircraft. The 309th Aerospace Maintenance and Regeneration Group are responsible for the 'boneyard' for ballistic missiles and aircraft like the S-2 Tracker anti-submarine aircraft shown.

OVERLEAF:
B-52 Storage Area, Davis-Monthan AFB, Tucson, Arizona, USA
Approximately 4000 aircraft are stored at Davis-Monthan. Most are eventually scrapped or raided for spares, but two B-52s have been restored to active service to replace those lost in accidents. The latest took four months to become airworthy after being in storage for over ten years.

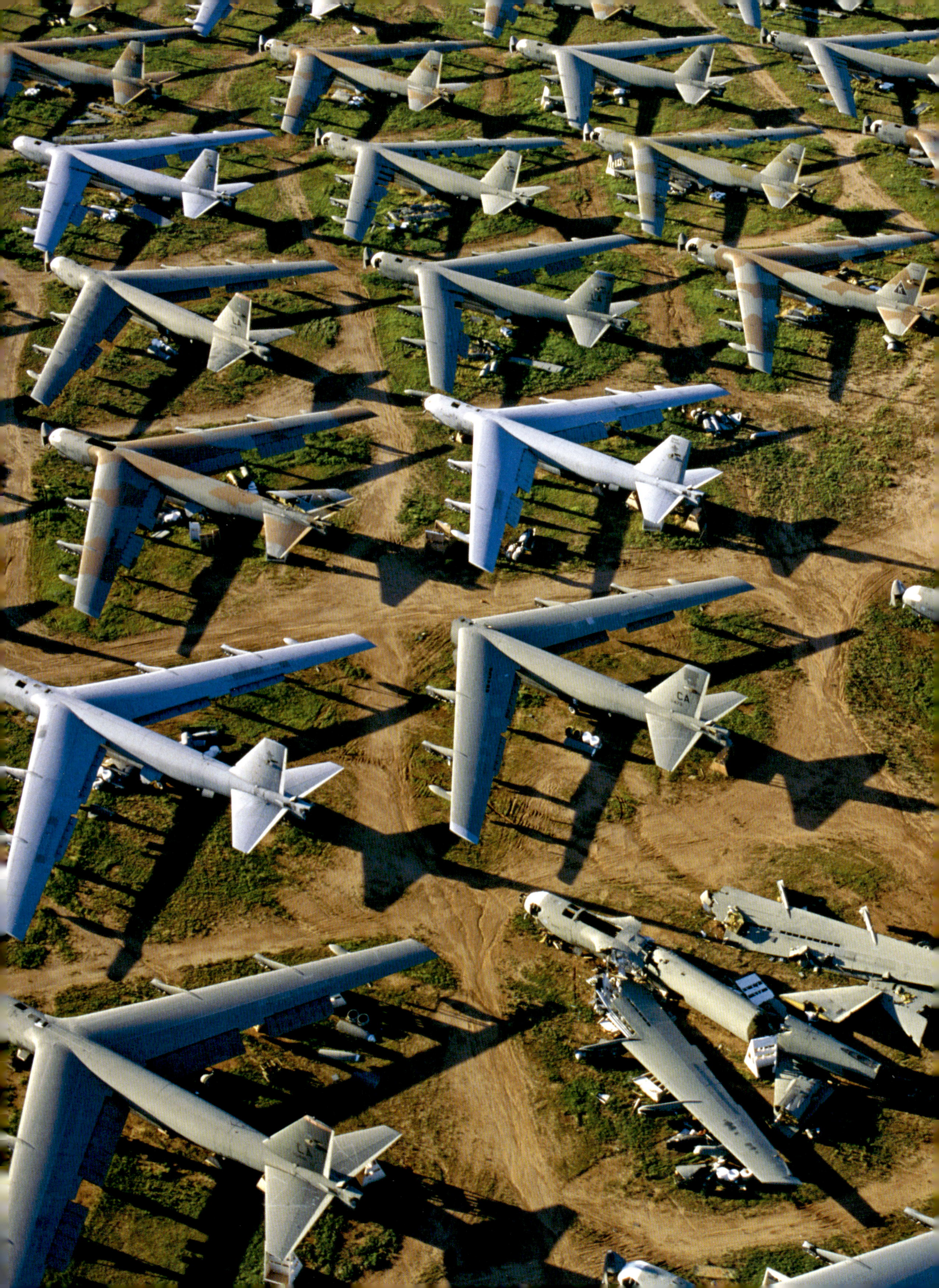

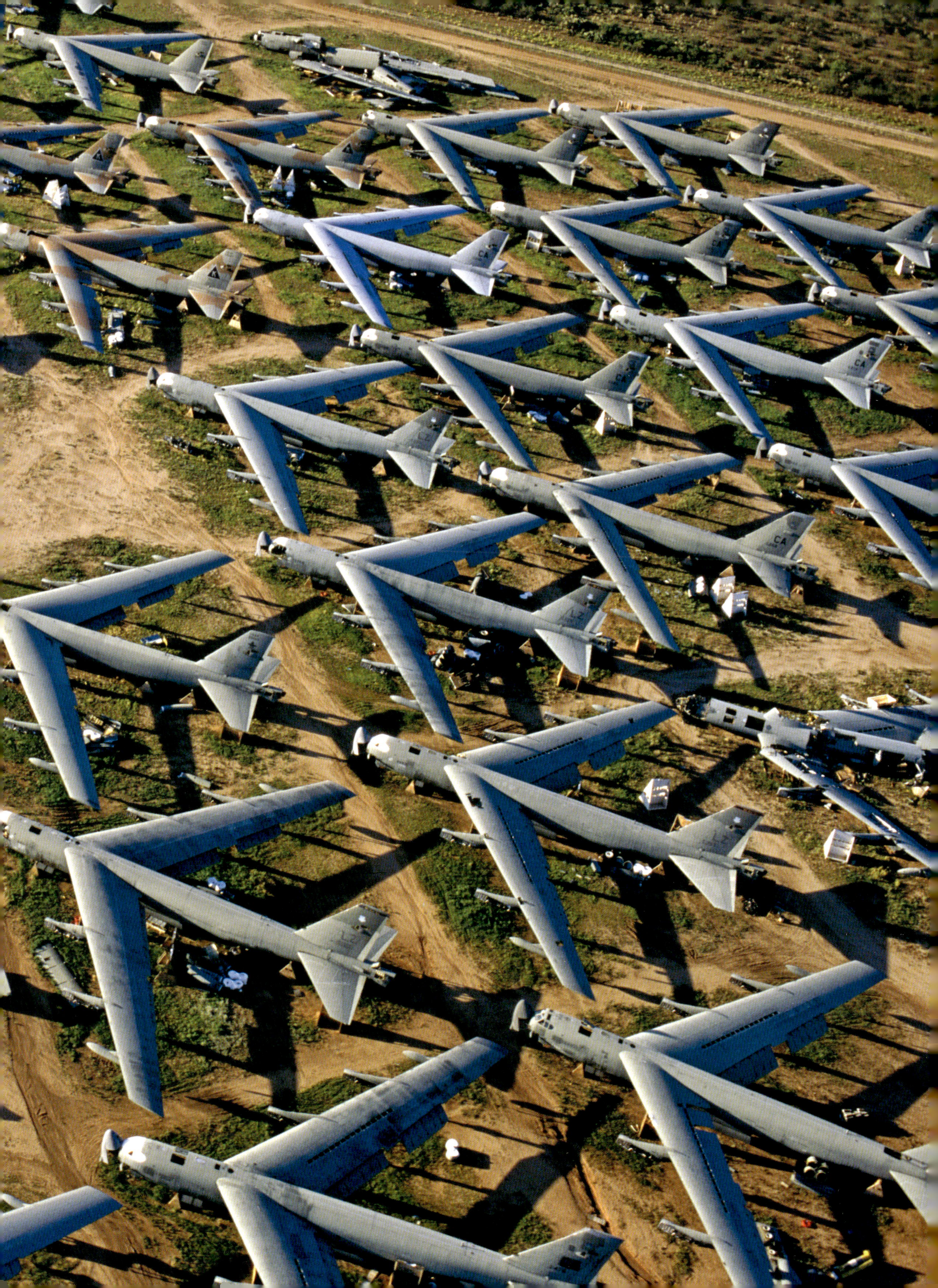

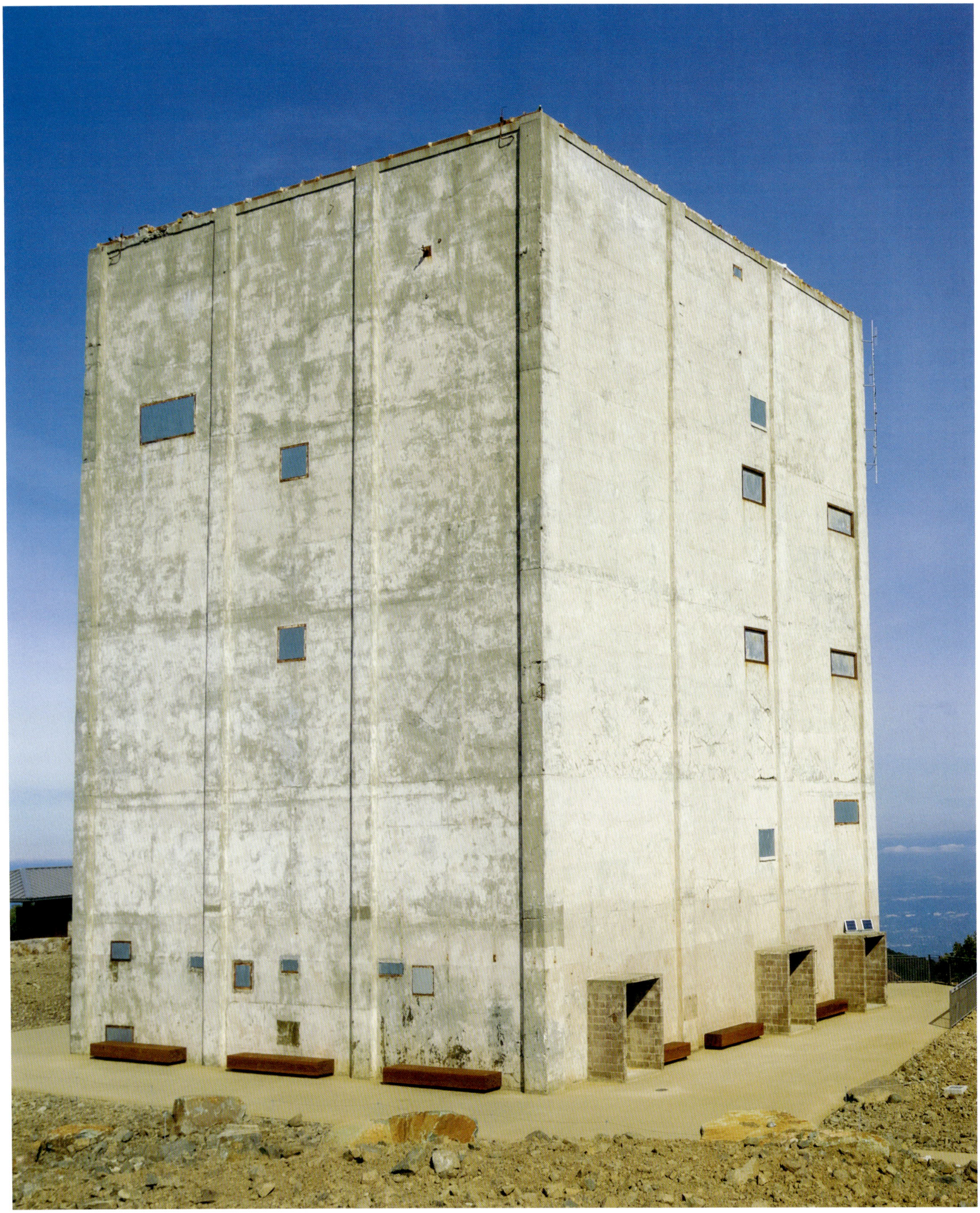

ABOVE:

Radar, Sierra Azul OSP, Santa Clara, California, USA

Almaden Air Force Station was a radar surveillance post on the summit of Mount Umunhum. The concrete box contained a radar that was so powerful it affected radios and televisions for miles around. Operating from 1958 to 1980, the site has now been given historic status.

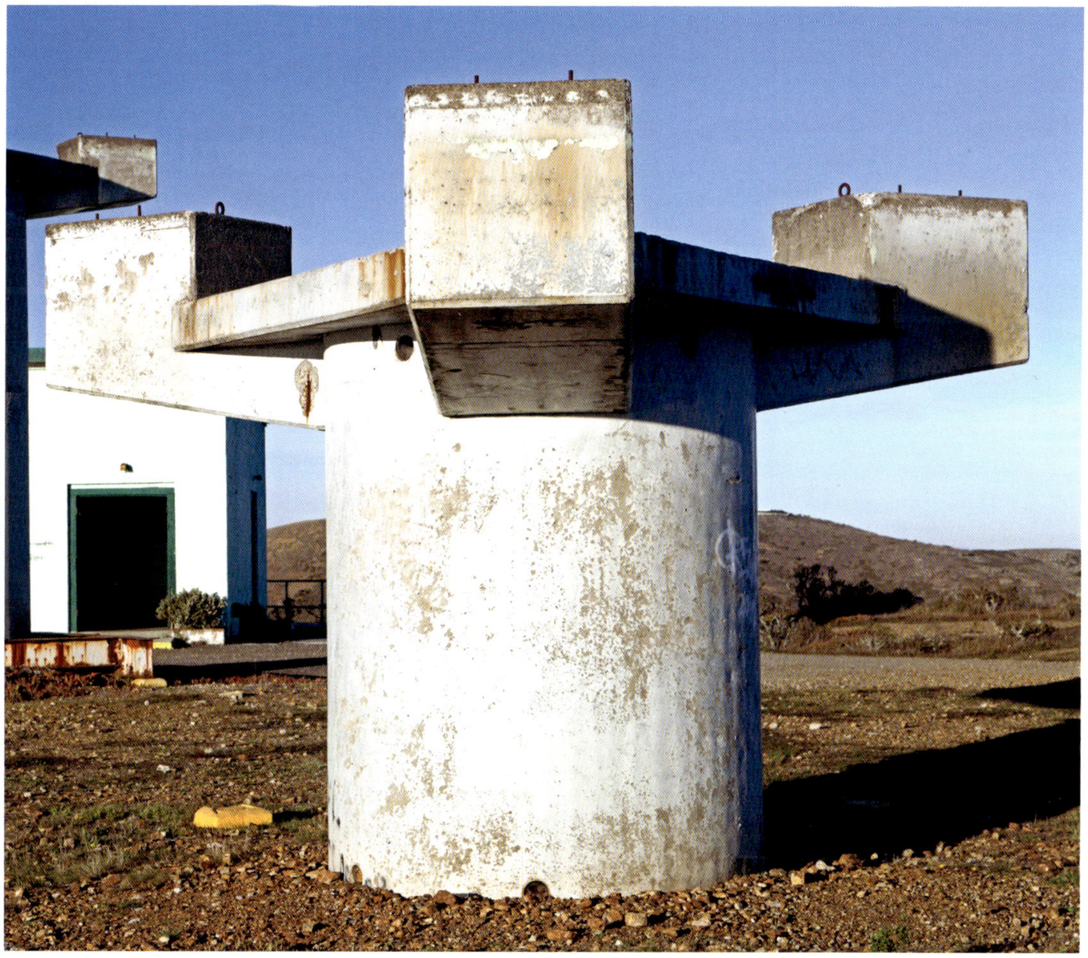

LEFT:
Former Radar Platform, Fort Barry, California, USA
A Nike anti-aircraft missile site was built here in 1954 as one of twelve sites to protect the San Francisco Bay Area from Soviet bomber attack. The radar formed part of the missile battery's integrated fire control system. The installation was closed in 1974 but preserved under the terms of the SALT treaty as a historic site.

BELOW:
Nike Missile Magazine, Angel Island, San Francisco, California USA
The metal doors shown cover what would have been the magazine where the Nike Ajax anti-aircraft missiles operated from this site were stored. The battery was only in operation between 1955 and 1961.

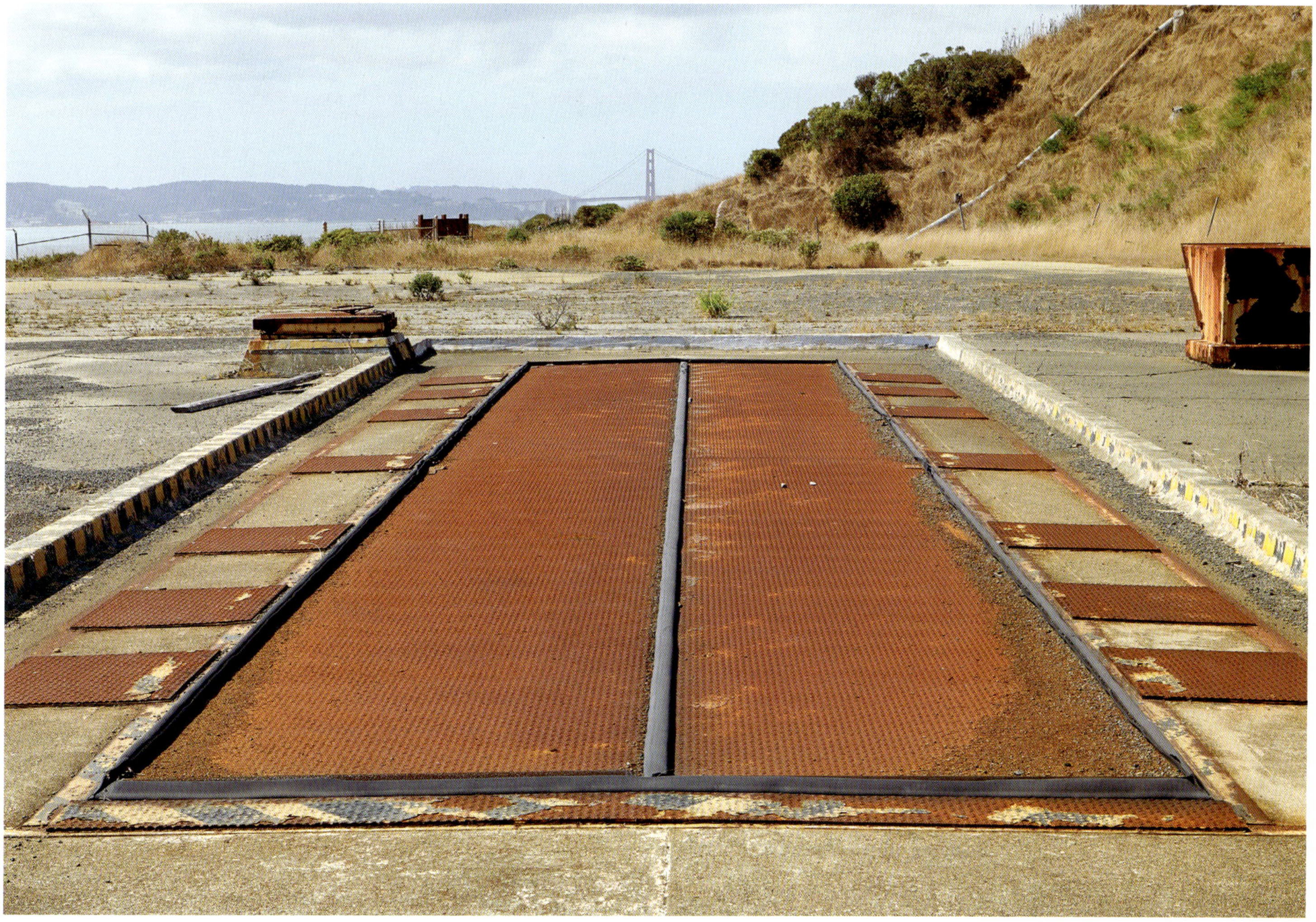

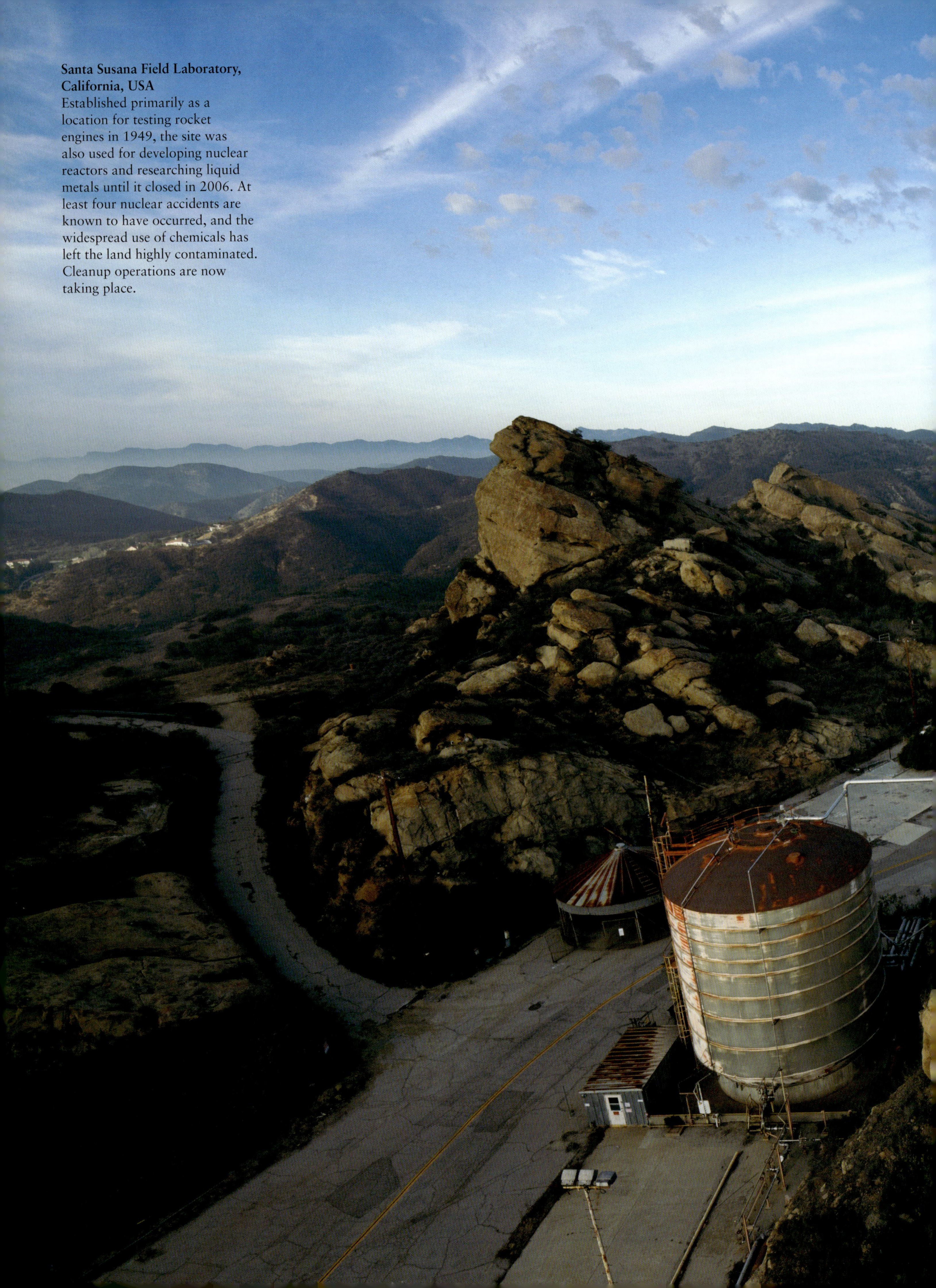

Santa Susana Field Laboratory, California, USA
Established primarily as a location for testing rocket engines in 1949, the site was also used for developing nuclear reactors and researching liquid metals until it closed in 2006. At least four nuclear accidents are known to have occurred, and the widespread use of chemicals has left the land highly contaminated. Cleanup operations are now taking place.

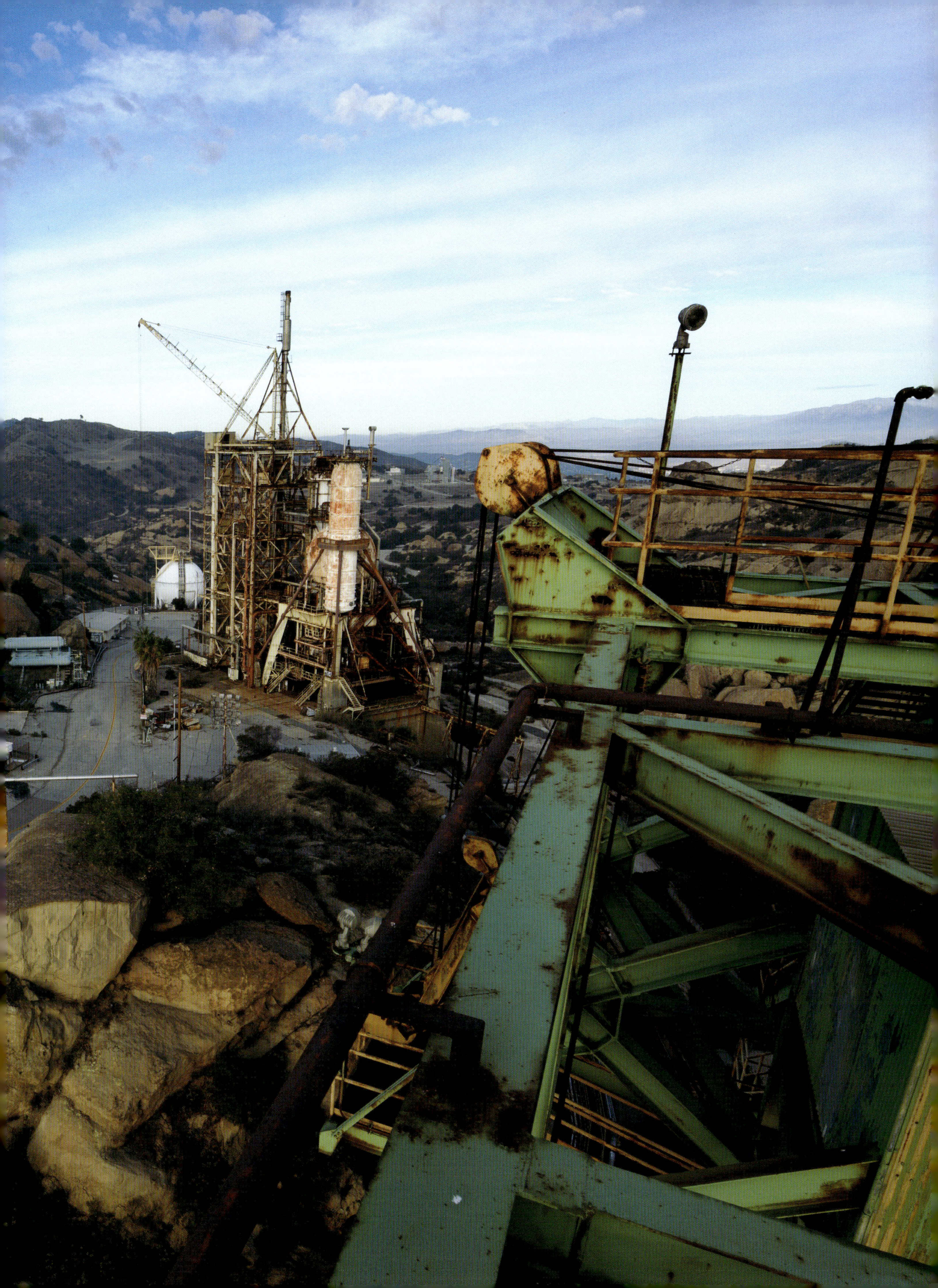

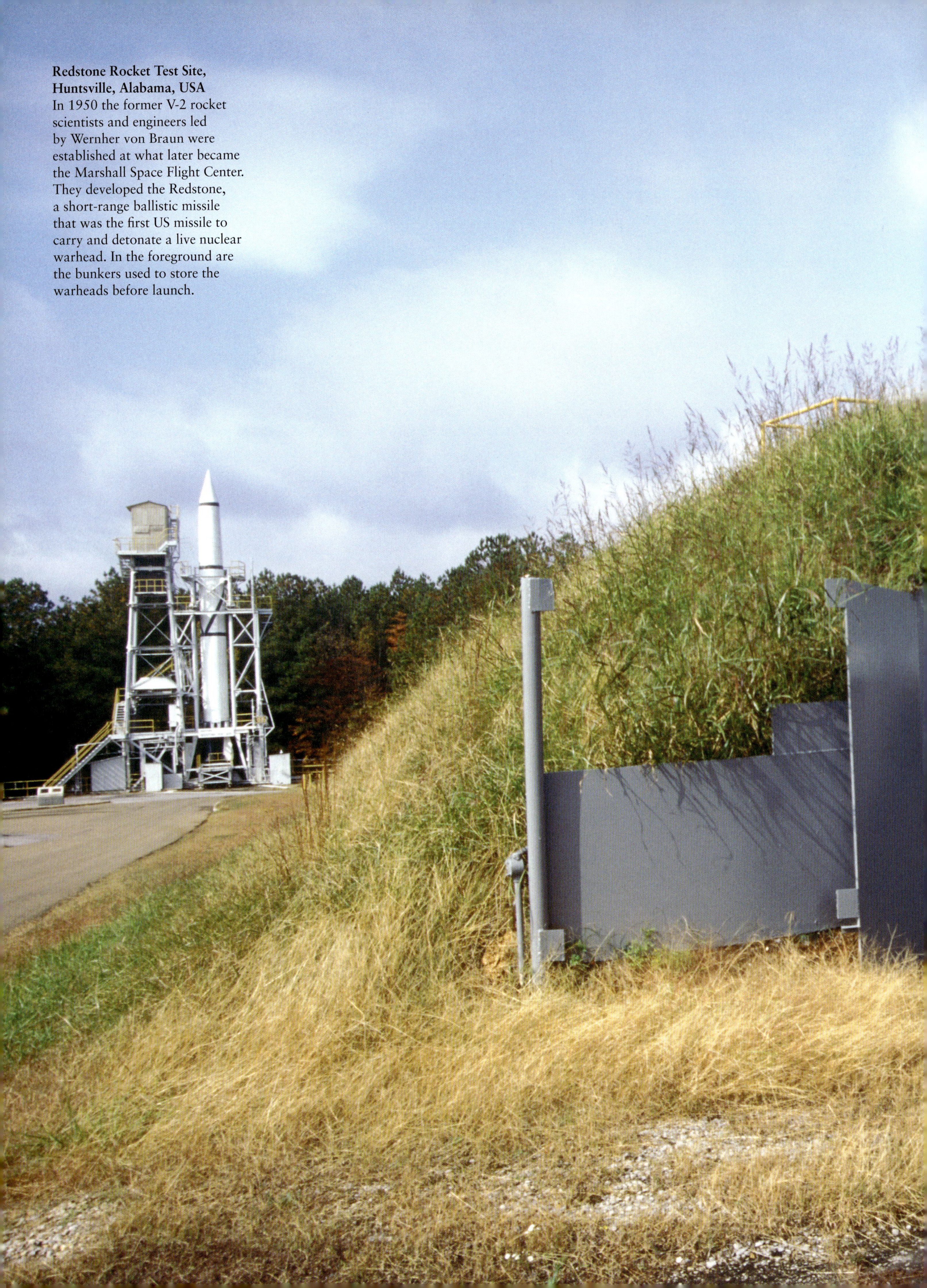

Redstone Rocket Test Site, Huntsville, Alabama, USA
In 1950 the former V-2 rocket scientists and engineers led by Wernher von Braun were established at what later became the Marshall Space Flight Center. They developed the Redstone, a short-range ballistic missile that was the first US missile to carry and detonate a live nuclear warhead. In the foreground are the bunkers used to store the warheads before launch.

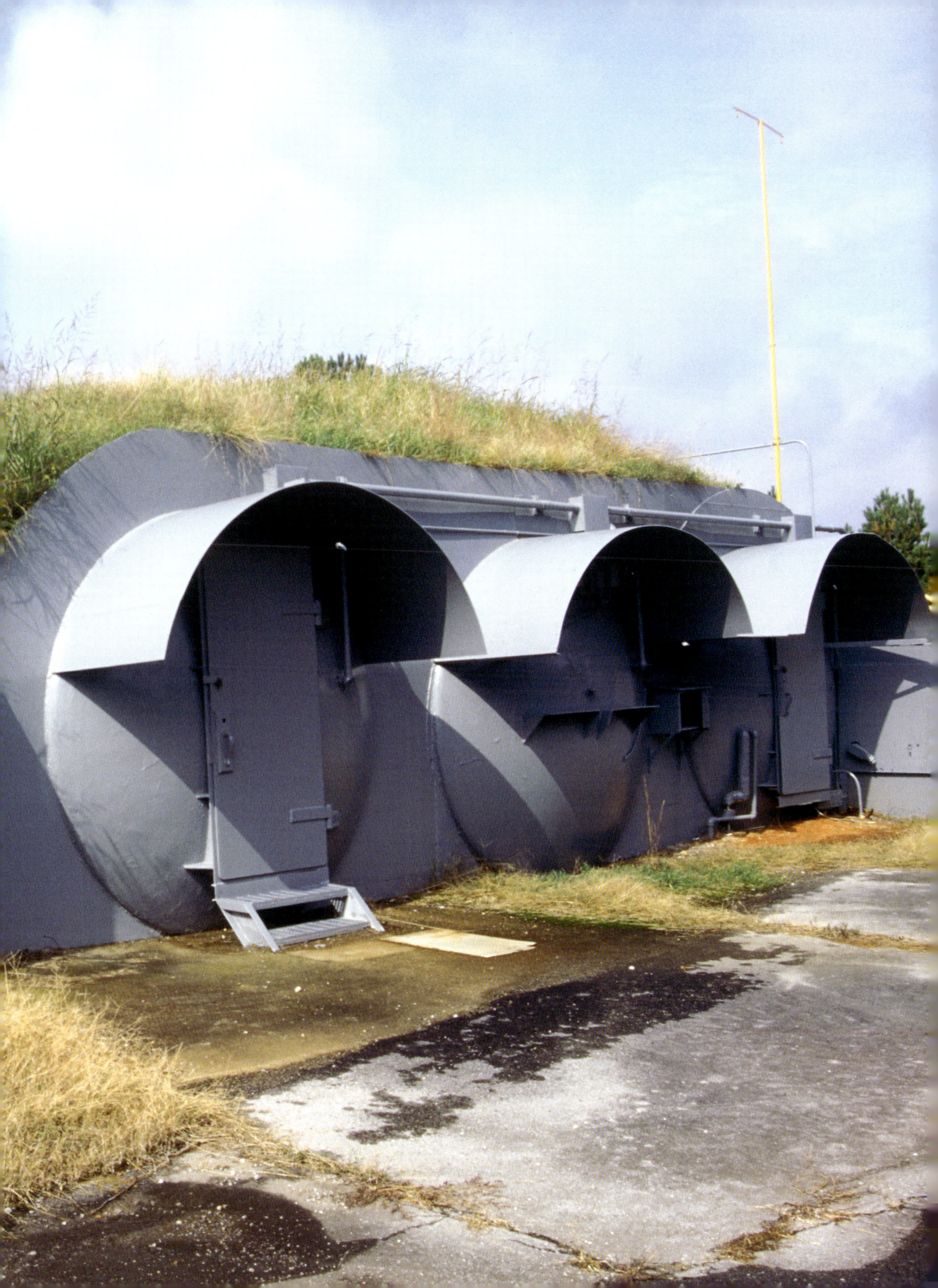

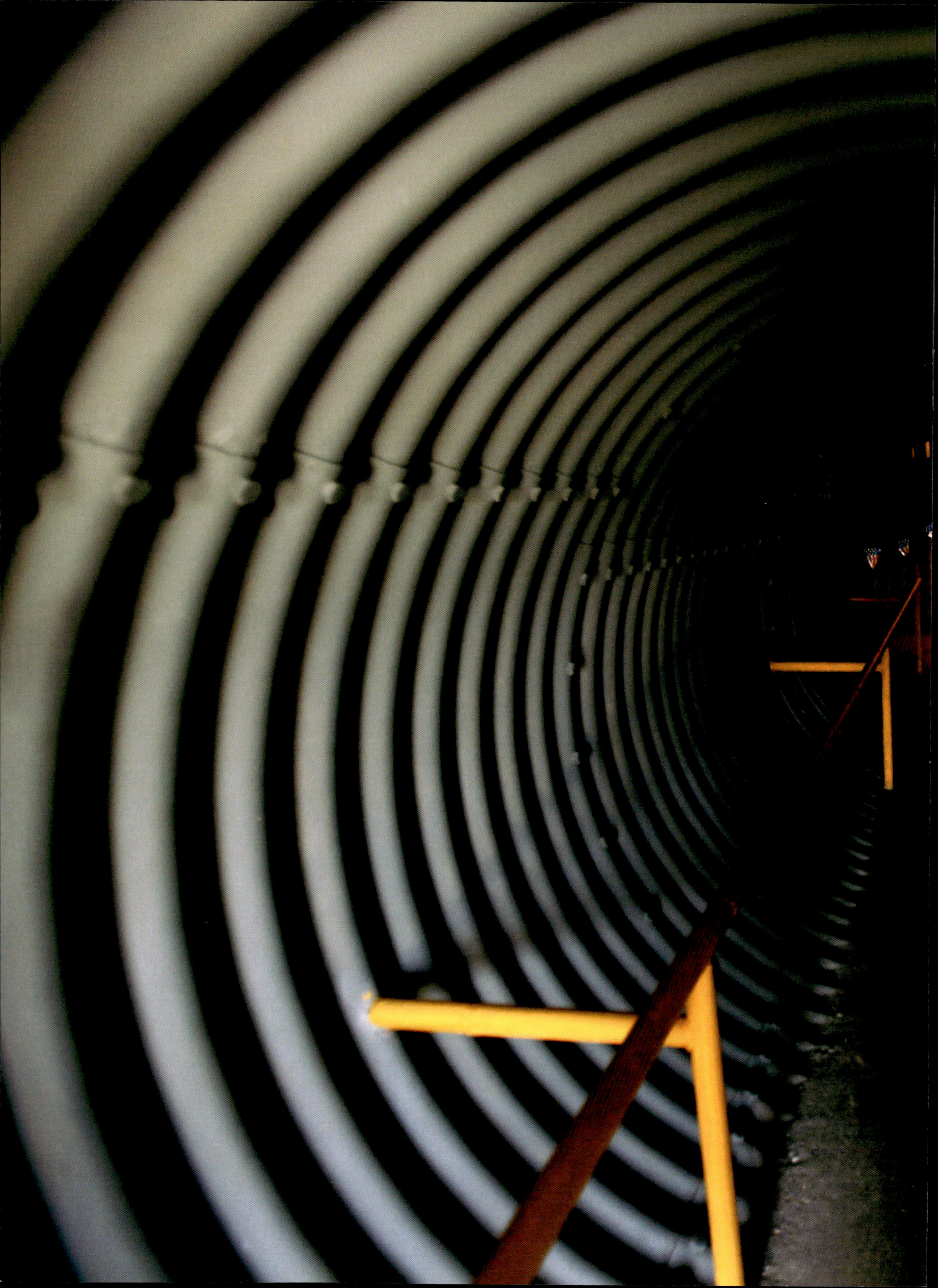

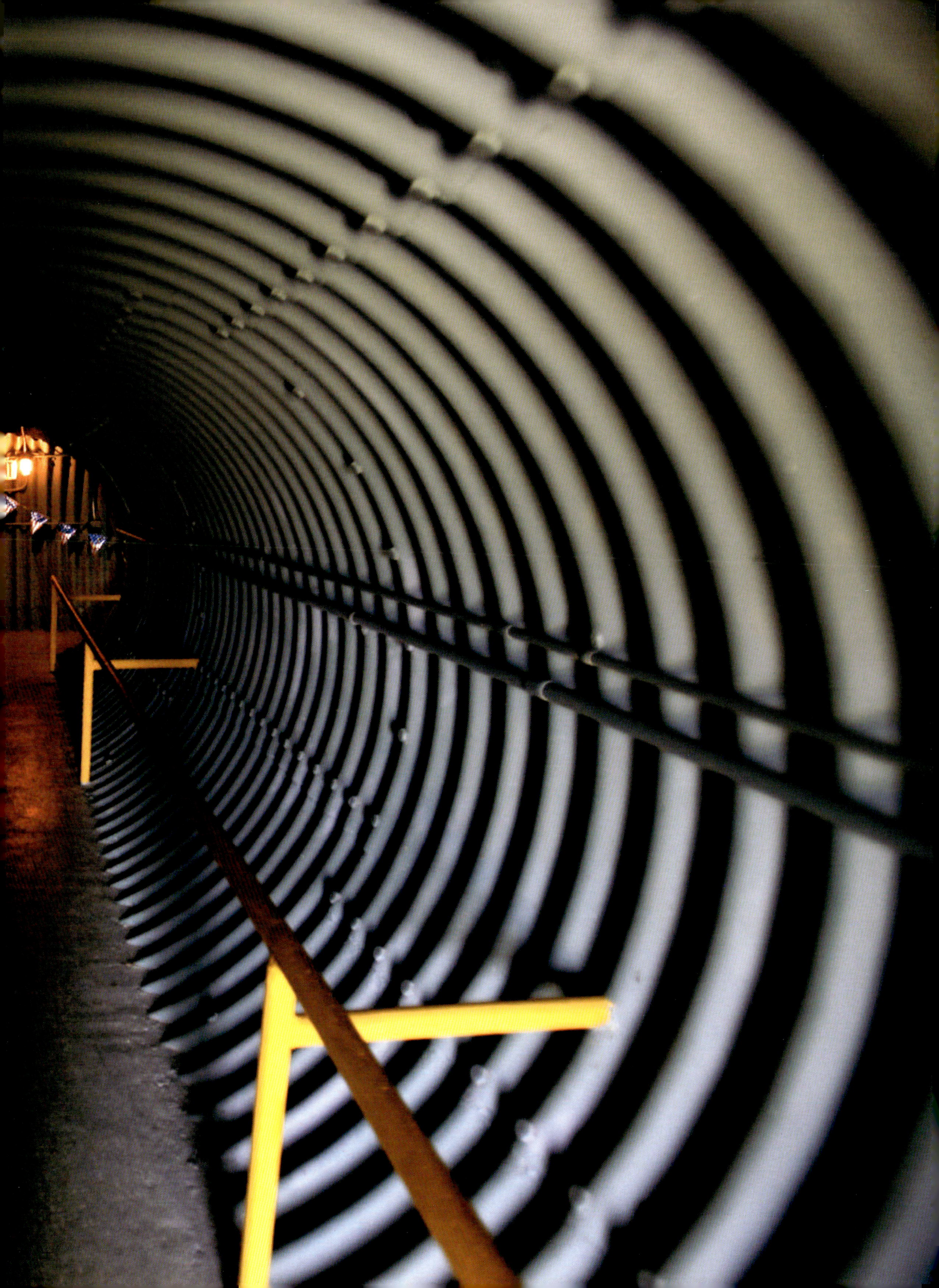

PREVIOUS PAGES:

Entrance, Detachment Hotel Bunker, Peanut Island, Florida, USA

Shortly after his election in 1960, a bunker was built within two weeks for President Kennedy near his family home in Palm Beach. In an emergency a helicopter would transfer the President to the bunker in five minutes. It could house 30 people for up to a month.

RIGHT:

Detachment Hotel Bunker, Peanut Island, Florida, USA

The Presidential Seal is clearly visible on the floor of the bunker, which had its own radio room and decontamination area. The roof was protected with lead sheets under several metres of soil. After Kennedy's assassination in 1963, the bunker was closed and subsequently declassified in 1974.

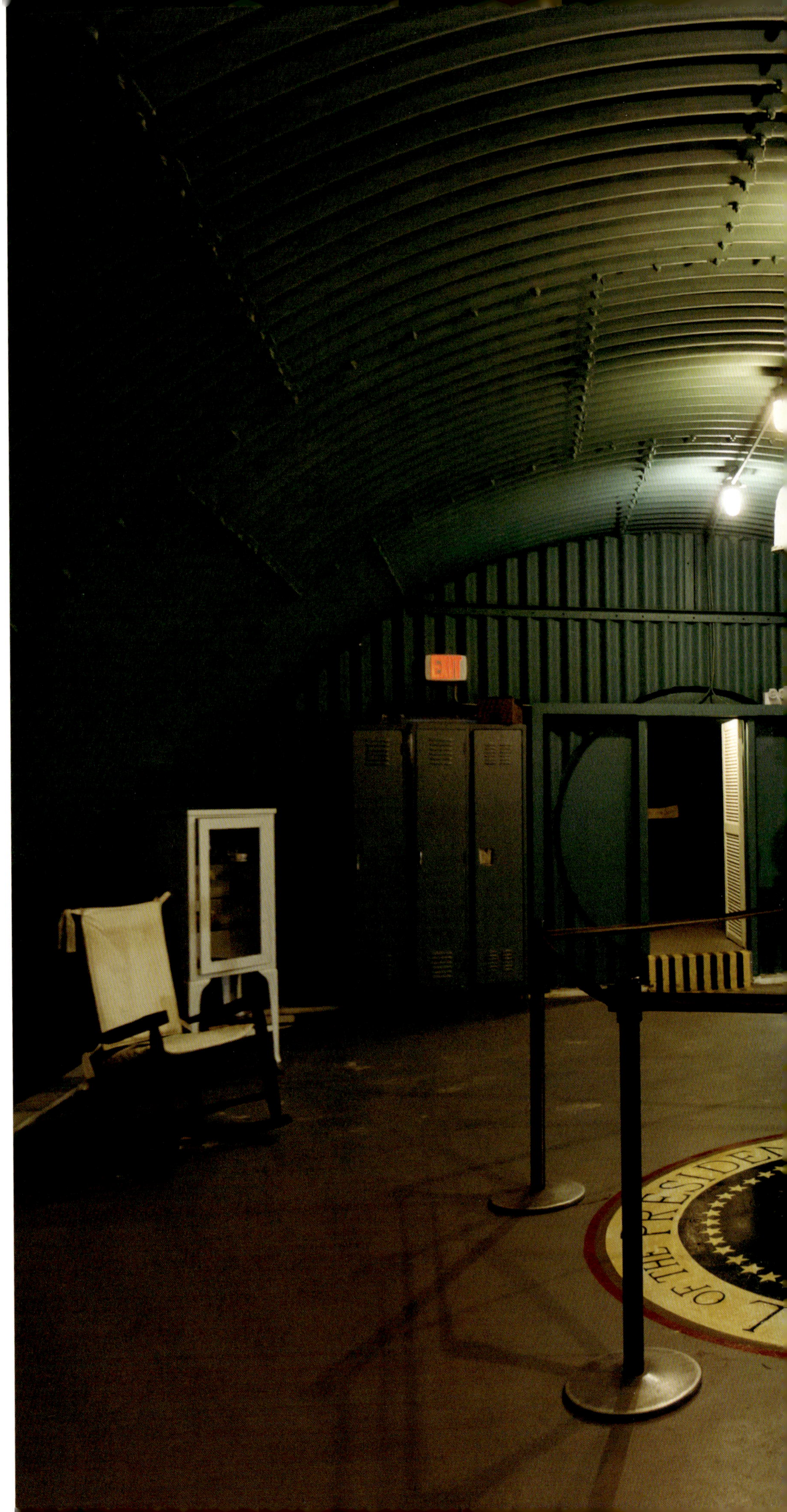

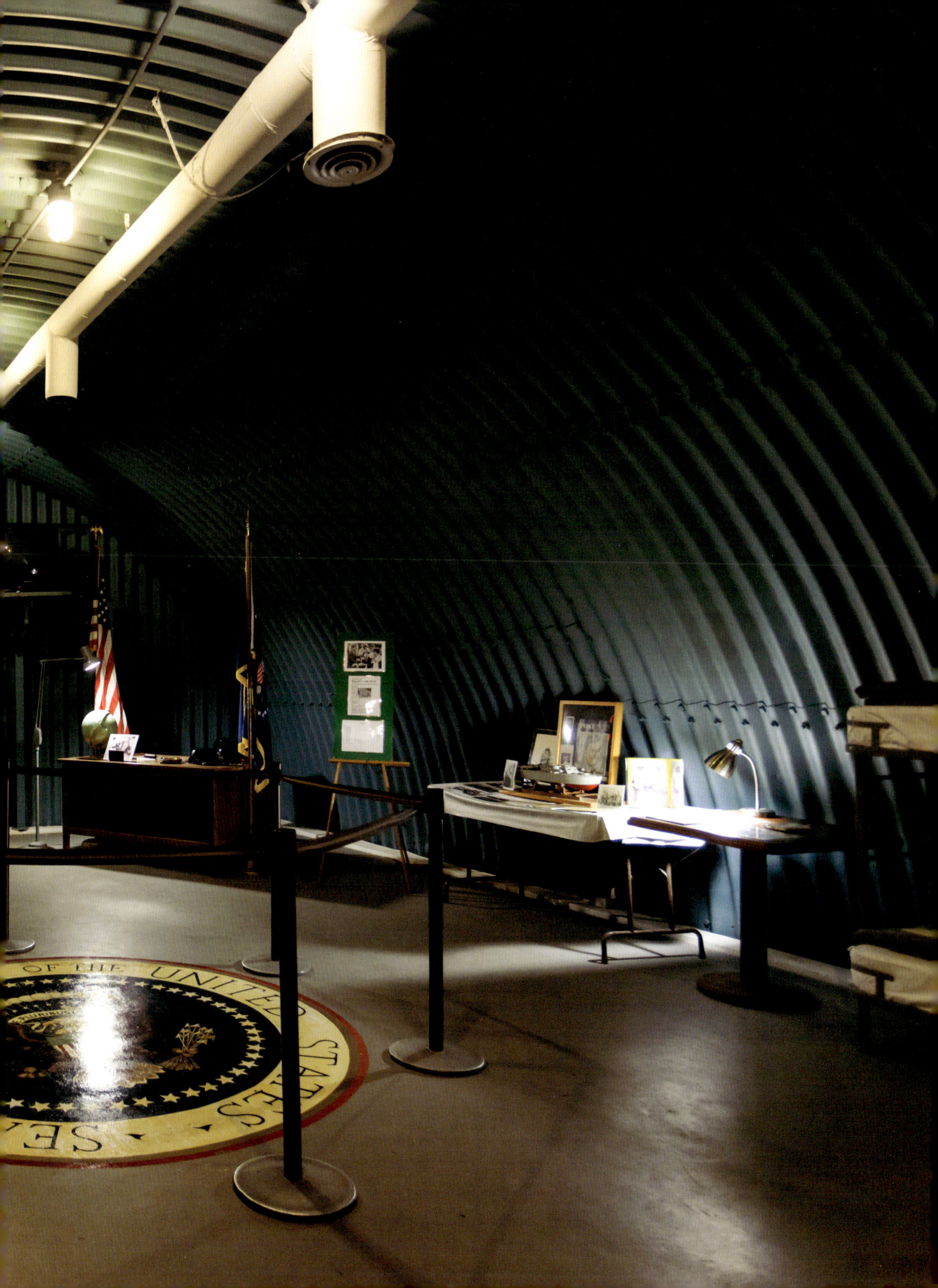
OF THE UNITED STATES

RIGHT:
Nike Missile Site, Everglades, Florida, USA
Part of the Homestead-Miami Defense Area created to defend Miami, this Nike Hercules anti-aircraft missile site became operational in 1964. The battery had three launch units with four missiles each. The site closed in 1979, and this telephone has likely not rung since.

BELOW:
Abandoned Radar Station, USA
The 1962 Cuban Missile Crisis (when the Soviet attempt to base nuclear missiles on Cuba brought the world to the brink of war) prompted the rapid improvement of anti-aircraft defences across the US, with radar sites and missile batteries built to protect major cities.

OPPOSITE TOP:
Cubana Airlines An-6 Wreck, Pearls Airport, Grenada
Pearls Airport was taken over by the US Marines during the 1983 invasion after a coup by a faction in the Marxist government. This Soviet-built An-6 was operated by Cubana Airlines. A significant number of Cubans were on the island, some of whom resisted the US forces.

OPPOSITE BOTTOM:
Lourdes Listening Station, Havana, Cuba
Shortly before the Cuban Missile Crisis in 1962, work began on this Soviet-funded facility built to listen to sensitive US communications. Over 1500 intelligence operatives from the KGB and GRU as well as other Warsaw Pact agencies worked there.

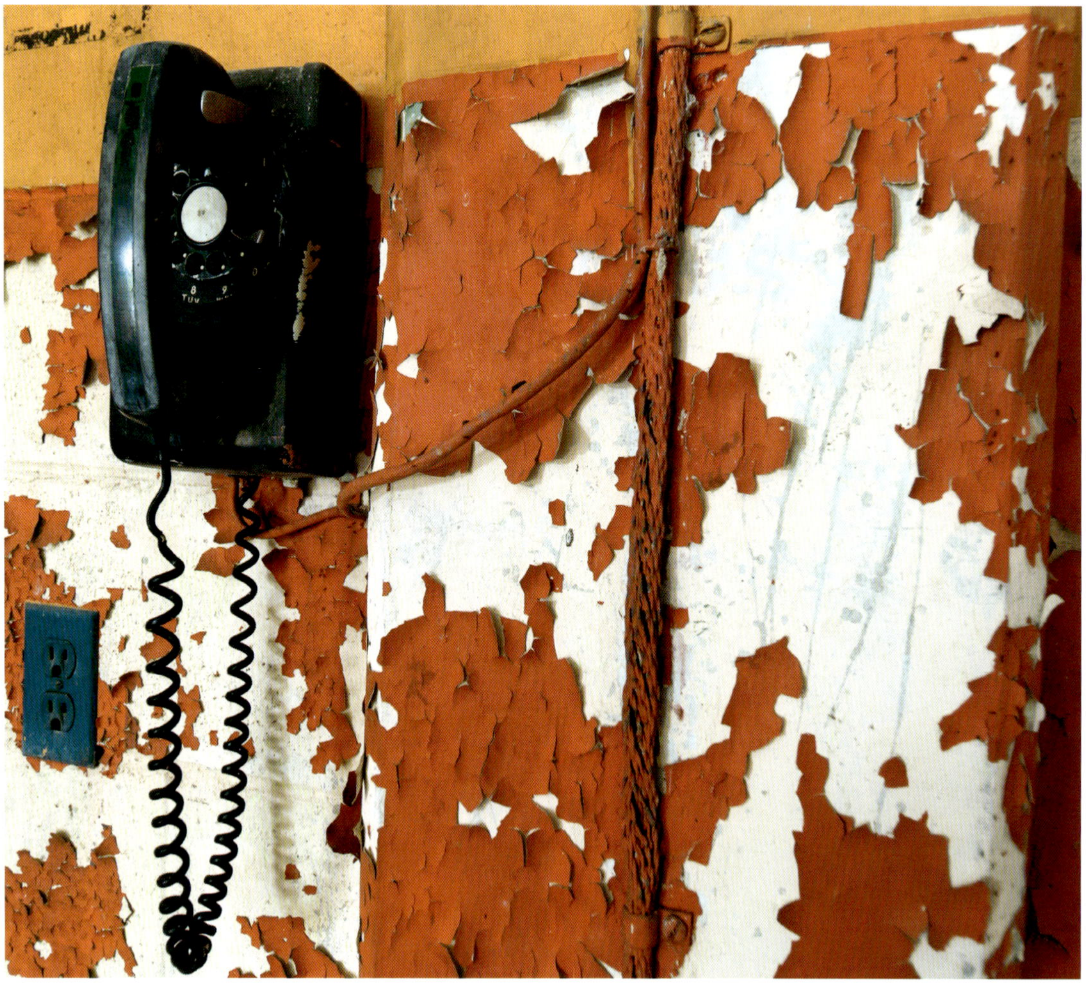

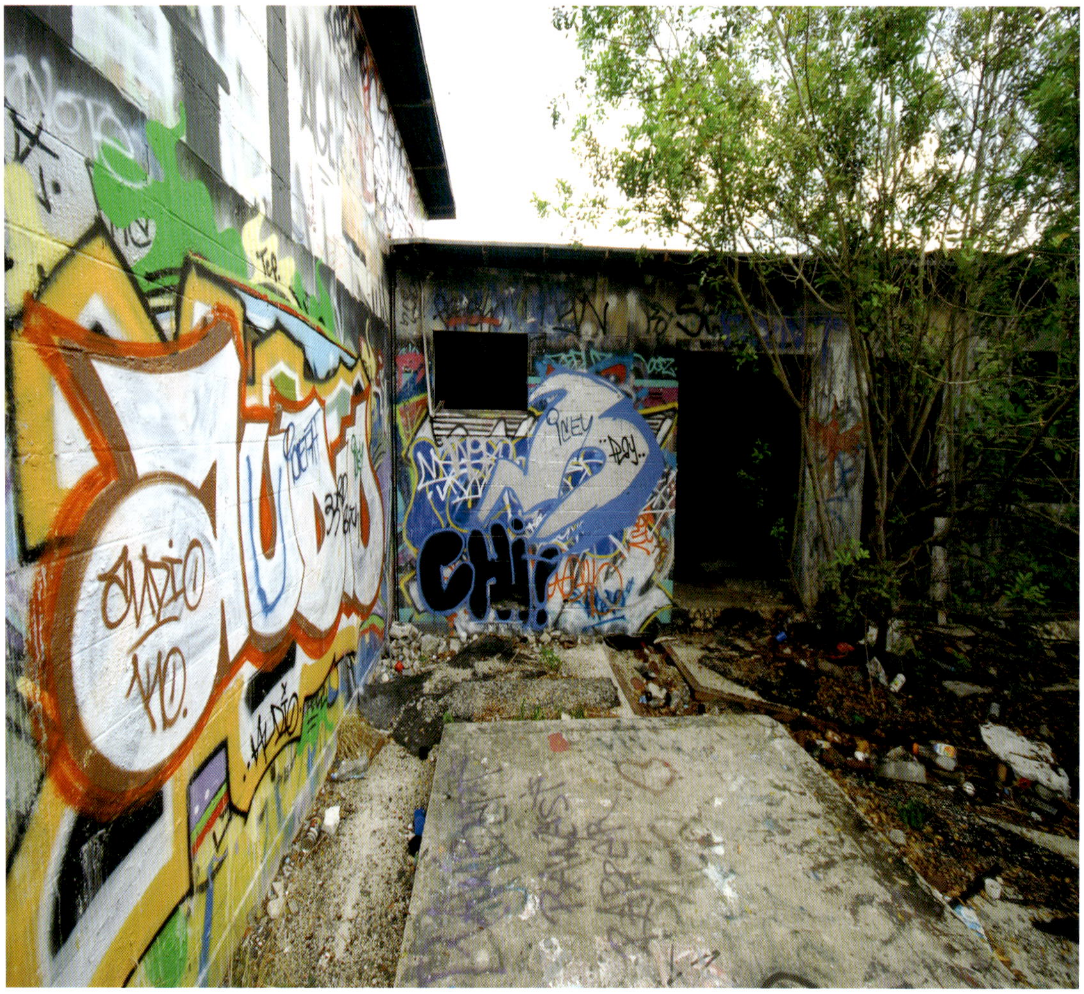

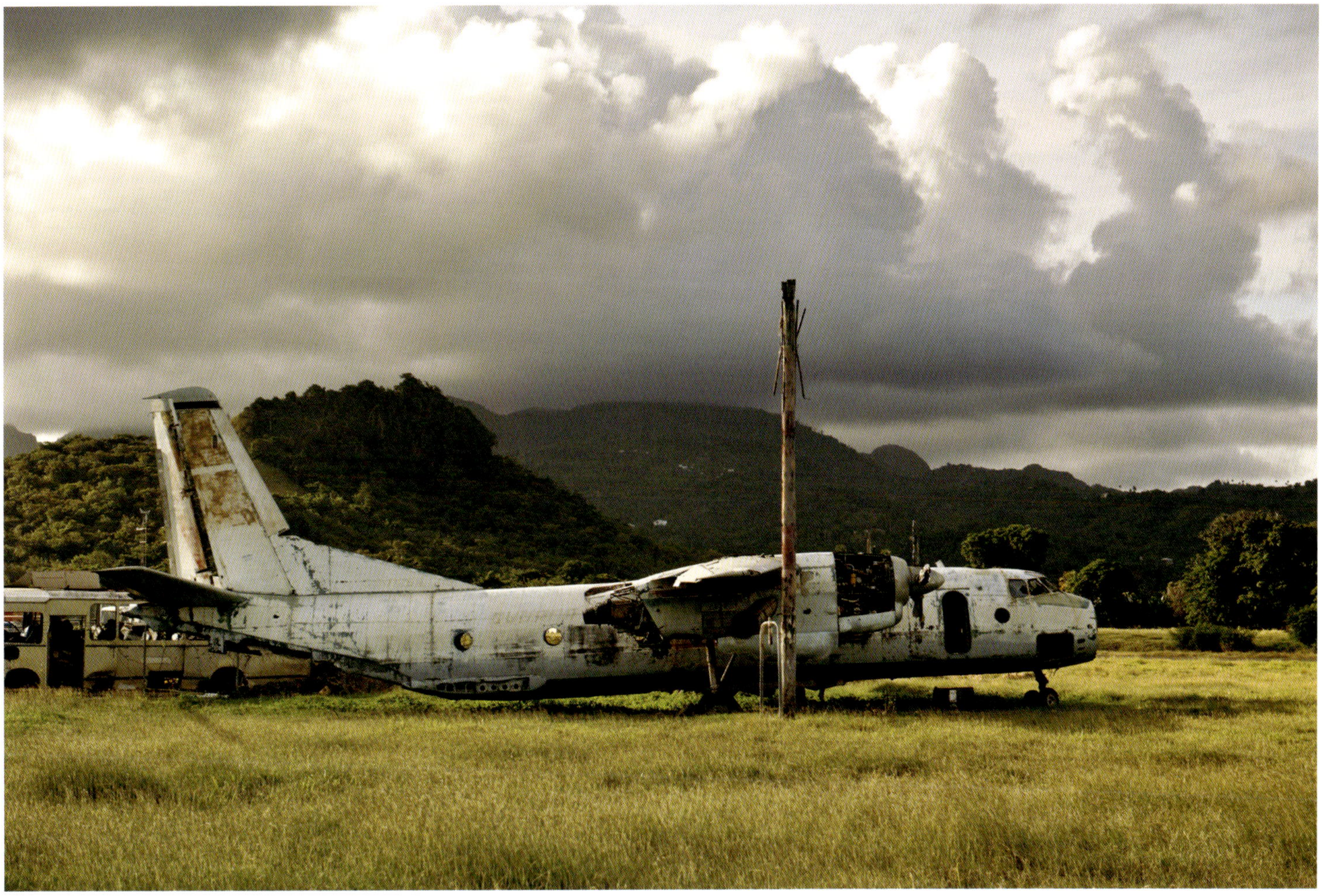

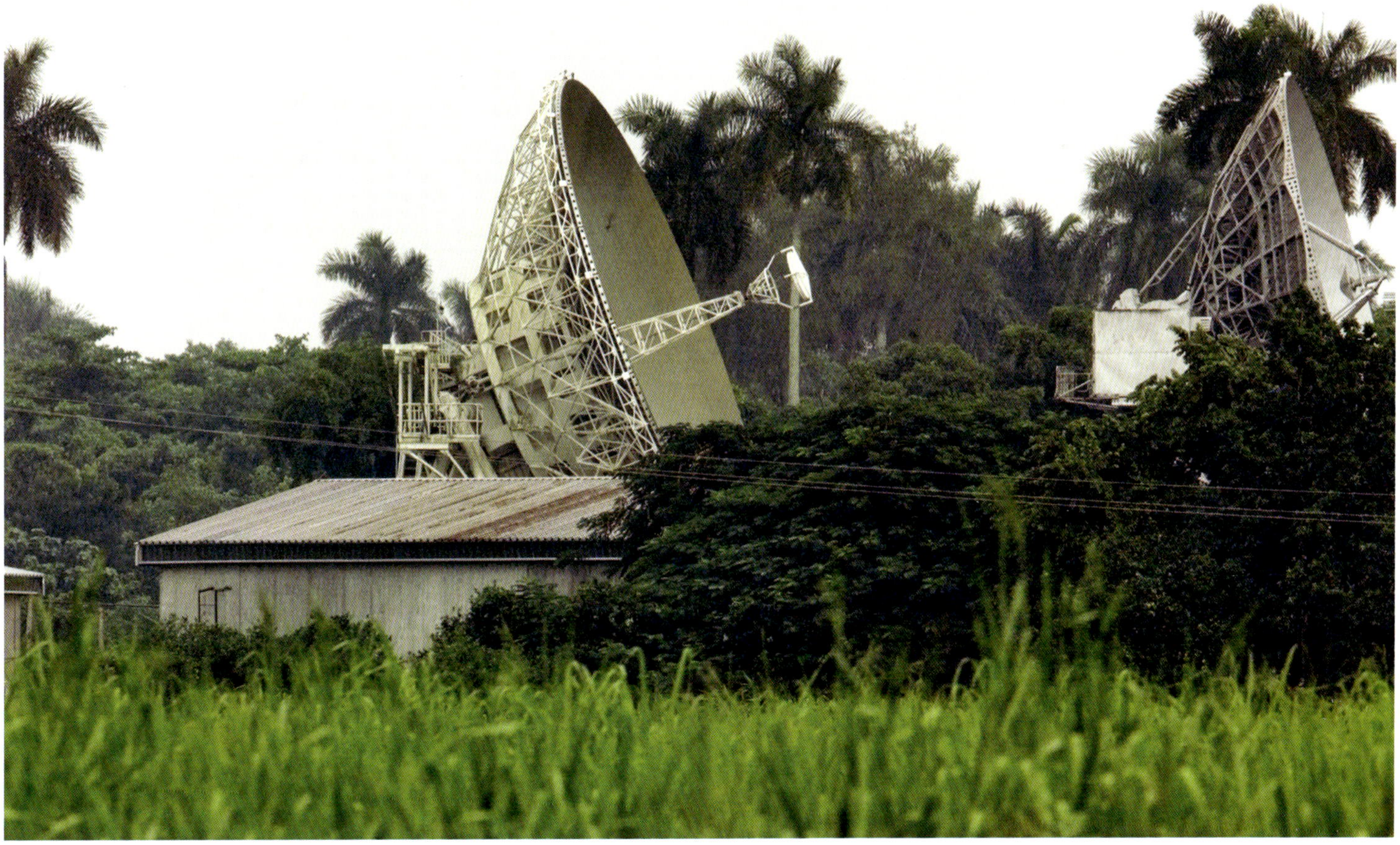

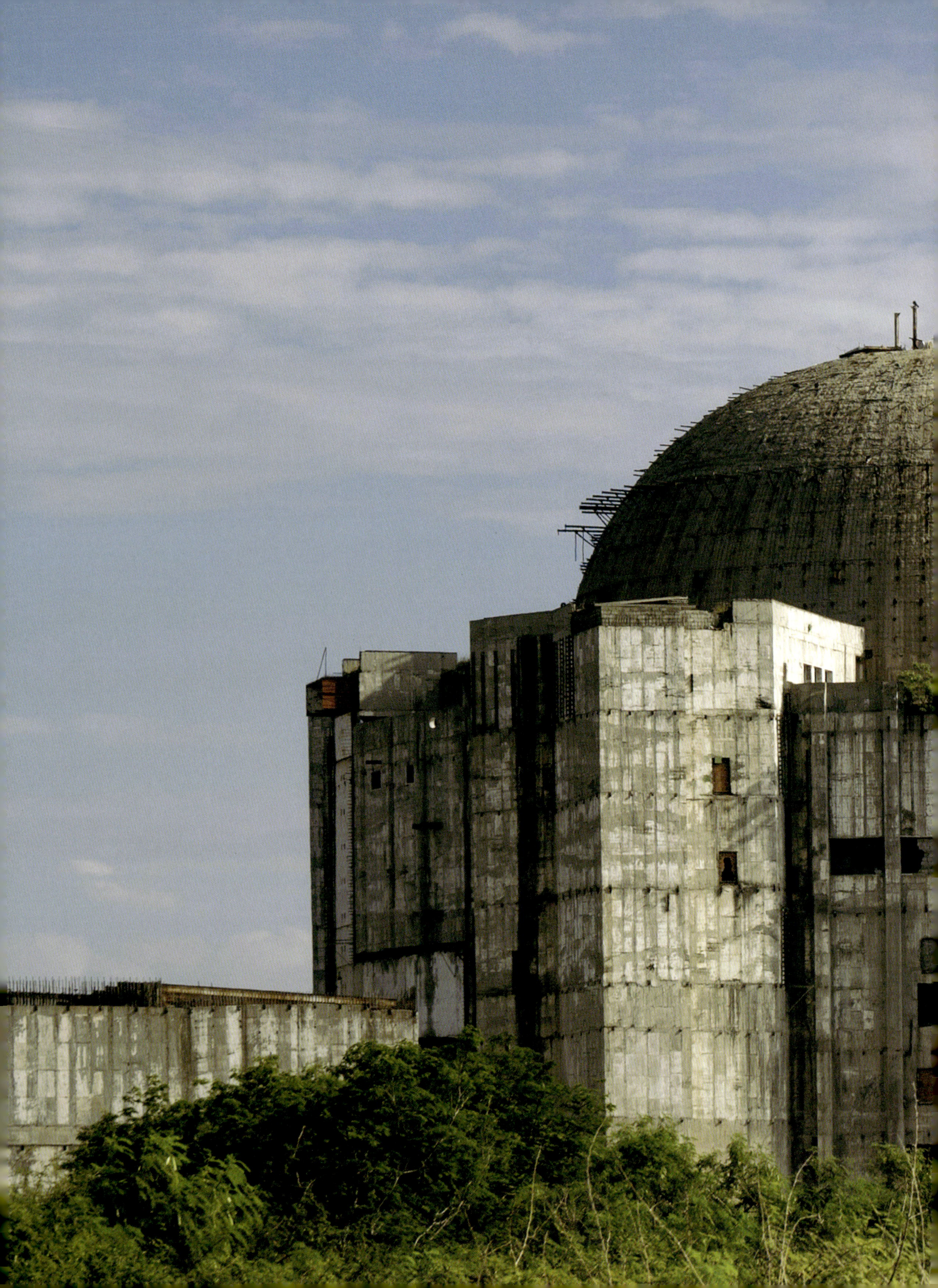

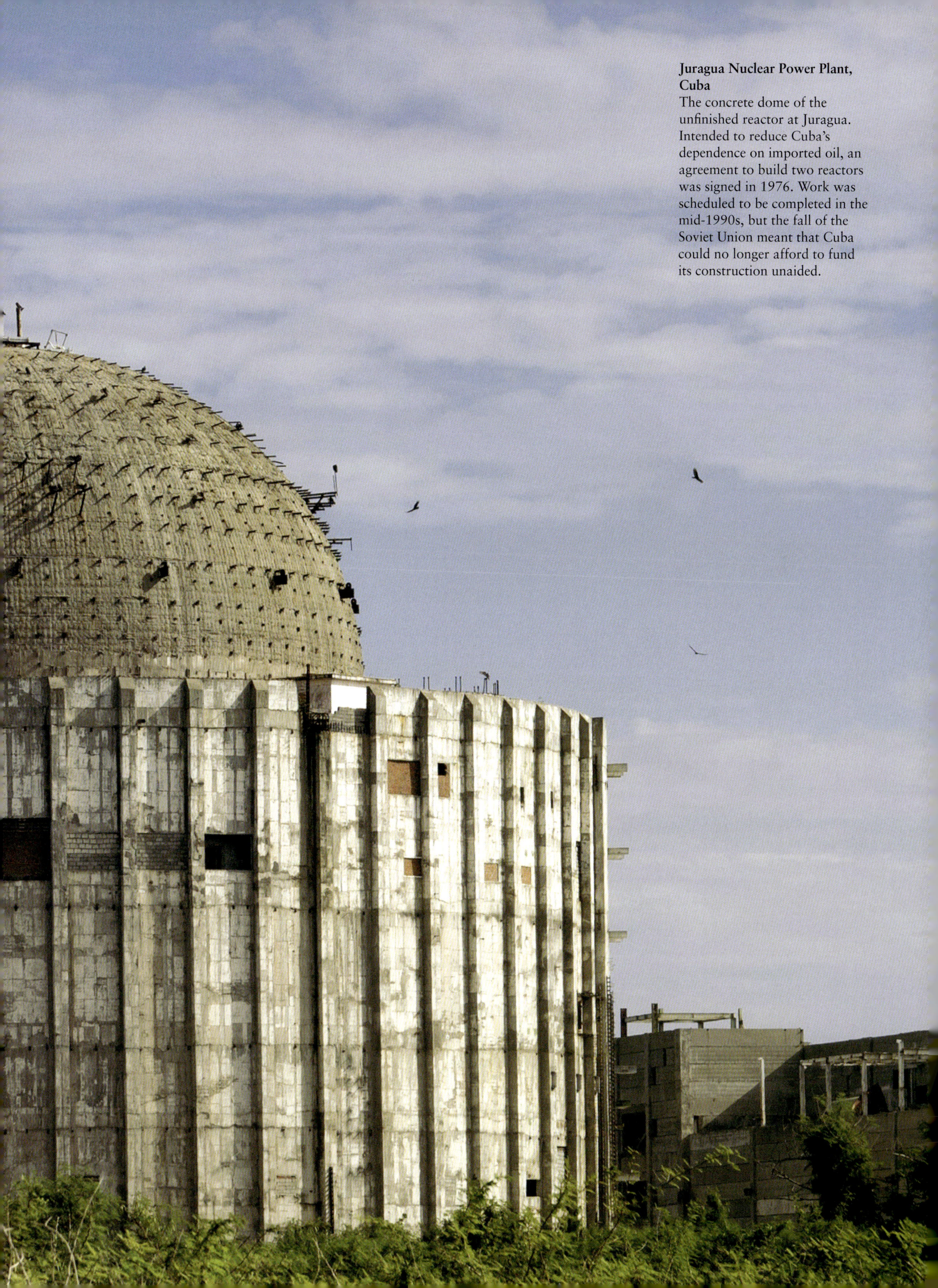

Juragua Nuclear Power Plant, Cuba
The concrete dome of the unfinished reactor at Juragua. Intended to reduce Cuba's dependence on imported oil, an agreement to build two reactors was signed in 1976. Work was scheduled to be completed in the mid-1990s, but the fall of the Soviet Union meant that Cuba could no longer afford to fund its construction unaided.

RIGHT:

Argentine Aircraft Wreck, Pebble Island, Falkland Islands

After almost four decades since its destruction, the wreckage of this aircraft is unrecognisable. Shortly after their invasion, Argentine forces established an airbase on Pebble Island with Pucara and Mentor aircraft that were destroyed by the British SAS in a night raid. Alternatively it could belong to a Dagger fighter bomber shot down by a British Sea Harrier.

BELOW:

Argentine Helicopter Wreck, East Falkland, Falkland Islands

Ten Argentine helicopters were destroyed during the war, but only four Puma helicopters were shot down in the air, two by gunfire, one by a handheld Stinger missile and one by a ship-launched Sea Dart missile. The others were destroyed on the ground. Thirteen more were captured by the British at the war's end.

OPPOSITE:

Argentine Chinook Helicopter Wreck, East Falkland, Falkland Islands

The Chinook served on both sides during the war, but after the loss of the *Atlantic Conveyor* the British only had one aircraft, with the call sign 'Bravo November', which survived the war and went on to serve in both Afghanistan and Iraq. The crew of this aircraft landed and fled after being seen by a Sea Harrier, which then destroyed the Chinook on the ground.

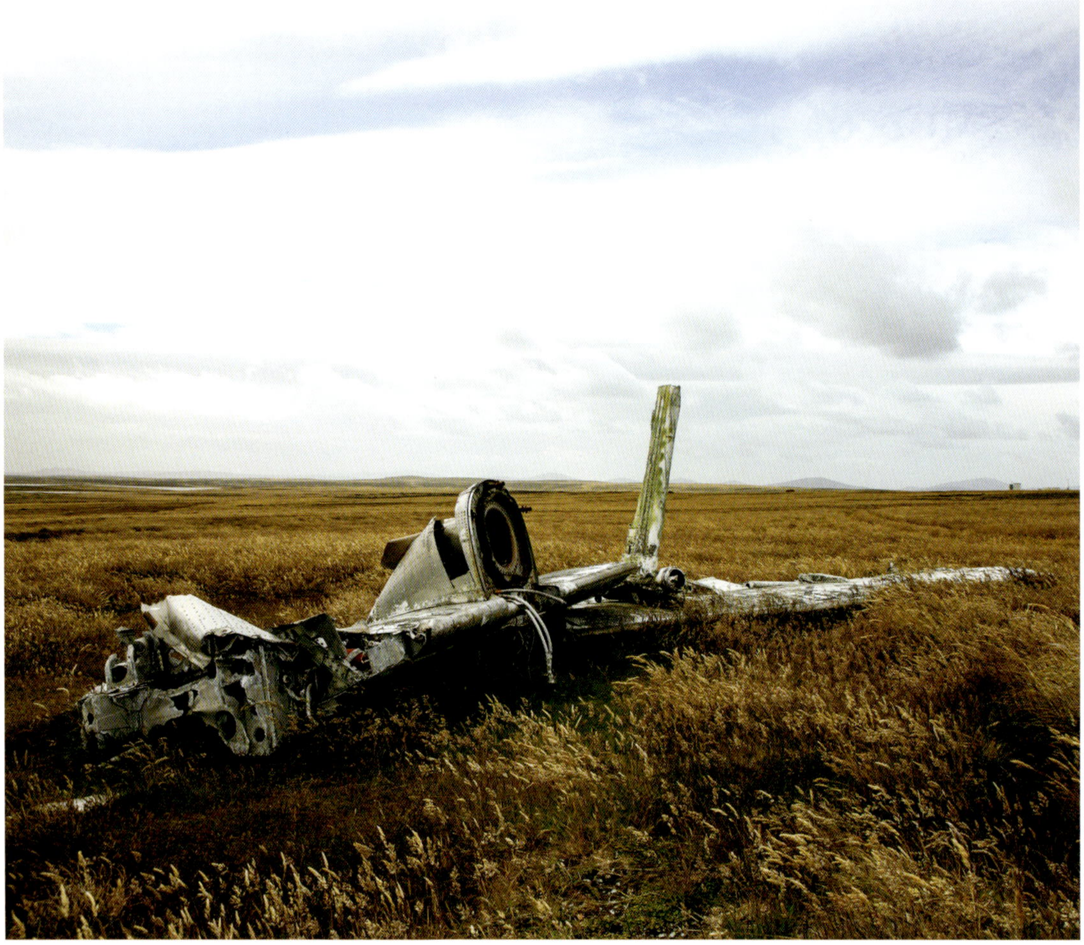

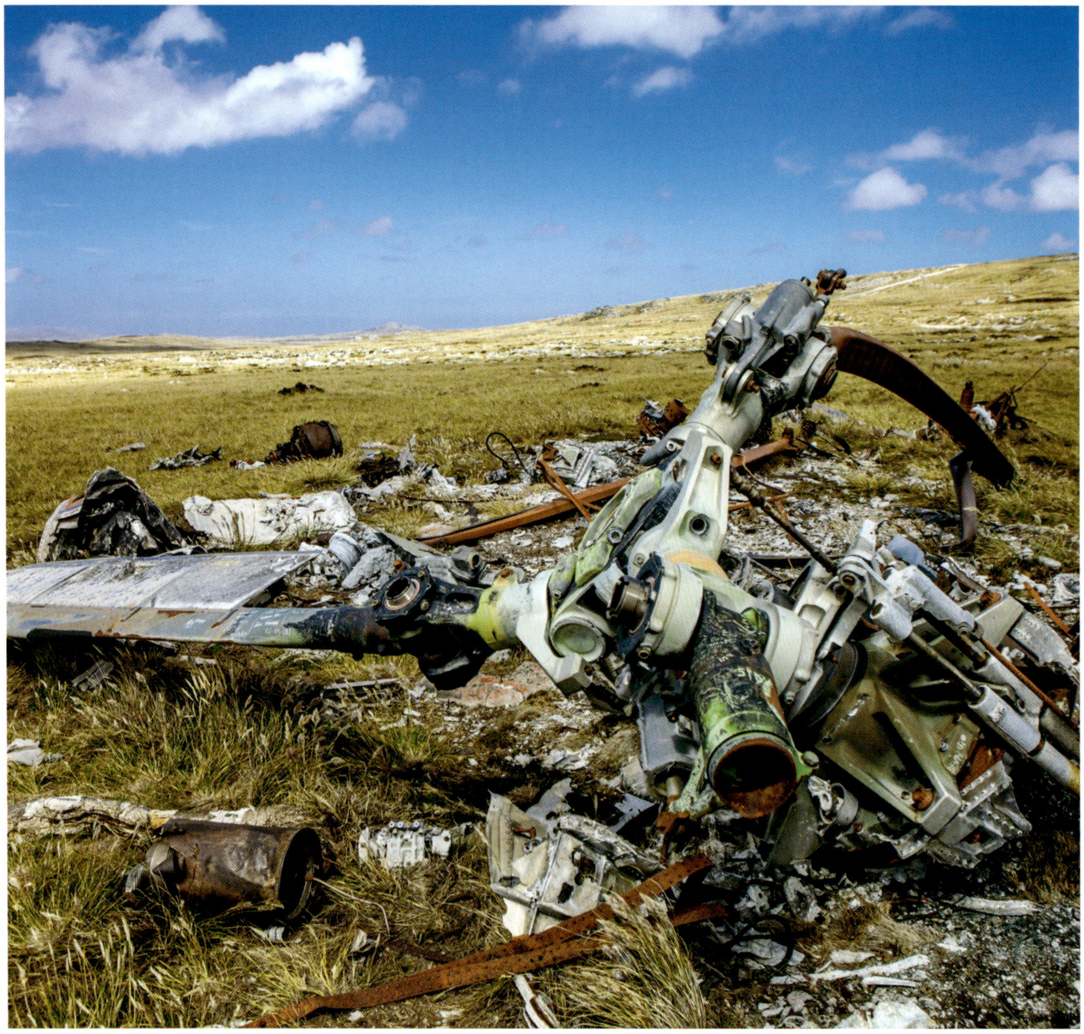

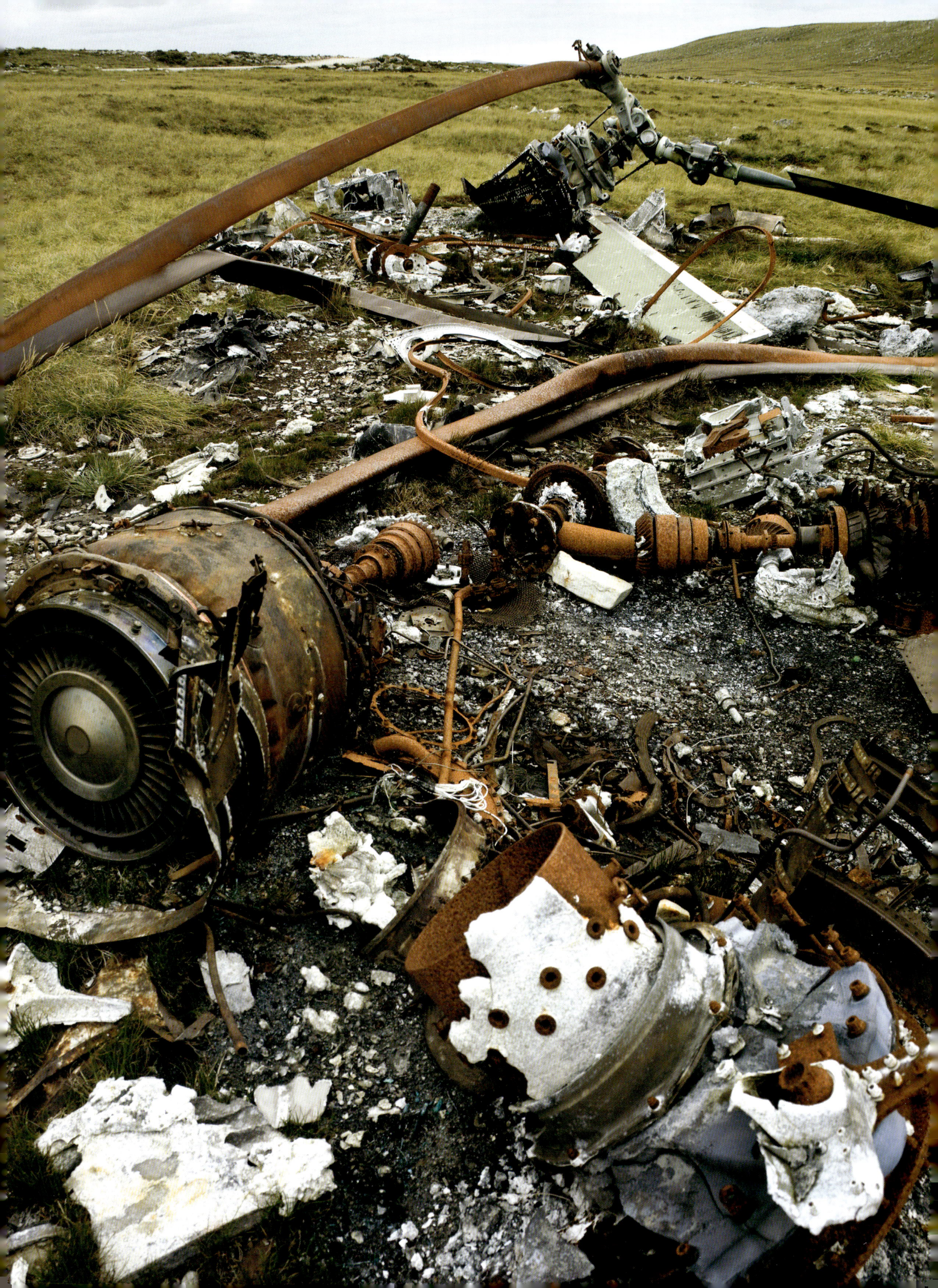

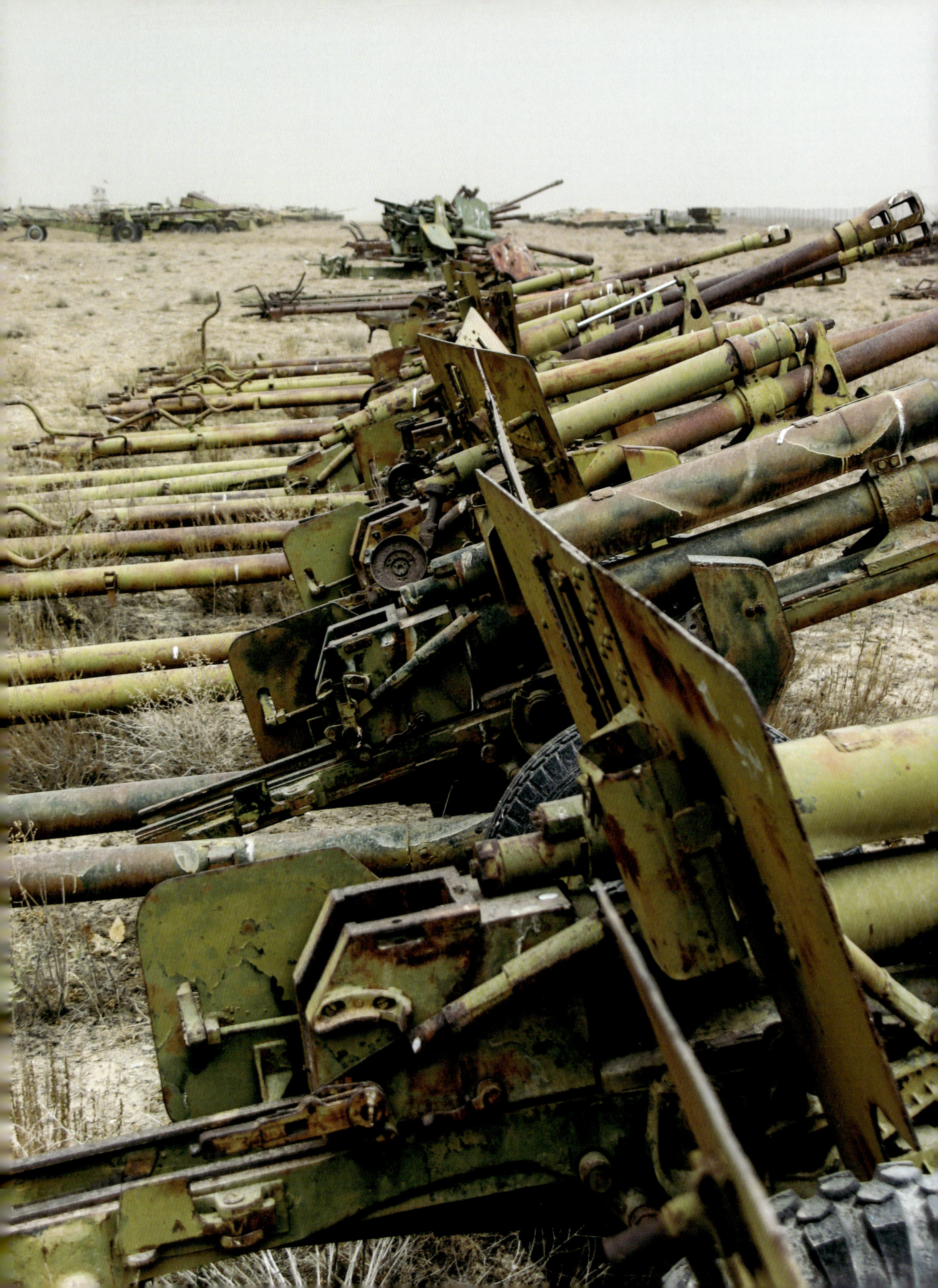

Asia, Middle East and Africa

The Cold War may have begun as a struggle for control of Europe, but it rapidly spread to other areas of the globe as the two sides vied to increase their spheres of influence. The Korean War saw Soviet pilots flying Chinese MiG-15 jets in defence of Communist North Korea against the United Nations pilots, while the North Vietnamese also received significant assistance from the Soviet Union during their conflict with the South Vietnamese and their US allies. For its part, the US provided financial aid and military equipment to the Mujahideen guerrillas fighting the Soviet occupation of Afghanistan in 1979, a war often referred to as the 'Soviet Vietnam'. In the Middle East, the newly-established state of Israel fought numerous wars with US, French and British-designed tanks and aircraft against its Arab neighbours using the latest in Soviet weaponry, while the US-supplied Pakistan and Soviet-supported India likewise sparred regularly over their disputed border in Kashmir. In Africa, cheap Soviet equipment and investment was welcomed by new post-colonial governments like Mozambique keen to distance themselves from the West. Even in the remote Pacific, traces of the Cold War remain at Bikini Atoll, where the radiation levels left by 1950s US bomb tests remain too high for humans to return.

OPPOSITE:
Soviet-era Artillery, Afghanistan
The Soviet Union was supplying arms to Afghanistan long before its invasion of 1979, so it is impossible to say when and by whom these weapons were taken into the country. The field guns in the foreground appear to be World War II-era ZiS-3 76mm (3in) pieces, but armoured cars, rocket launchers and other artillery types are also visible.

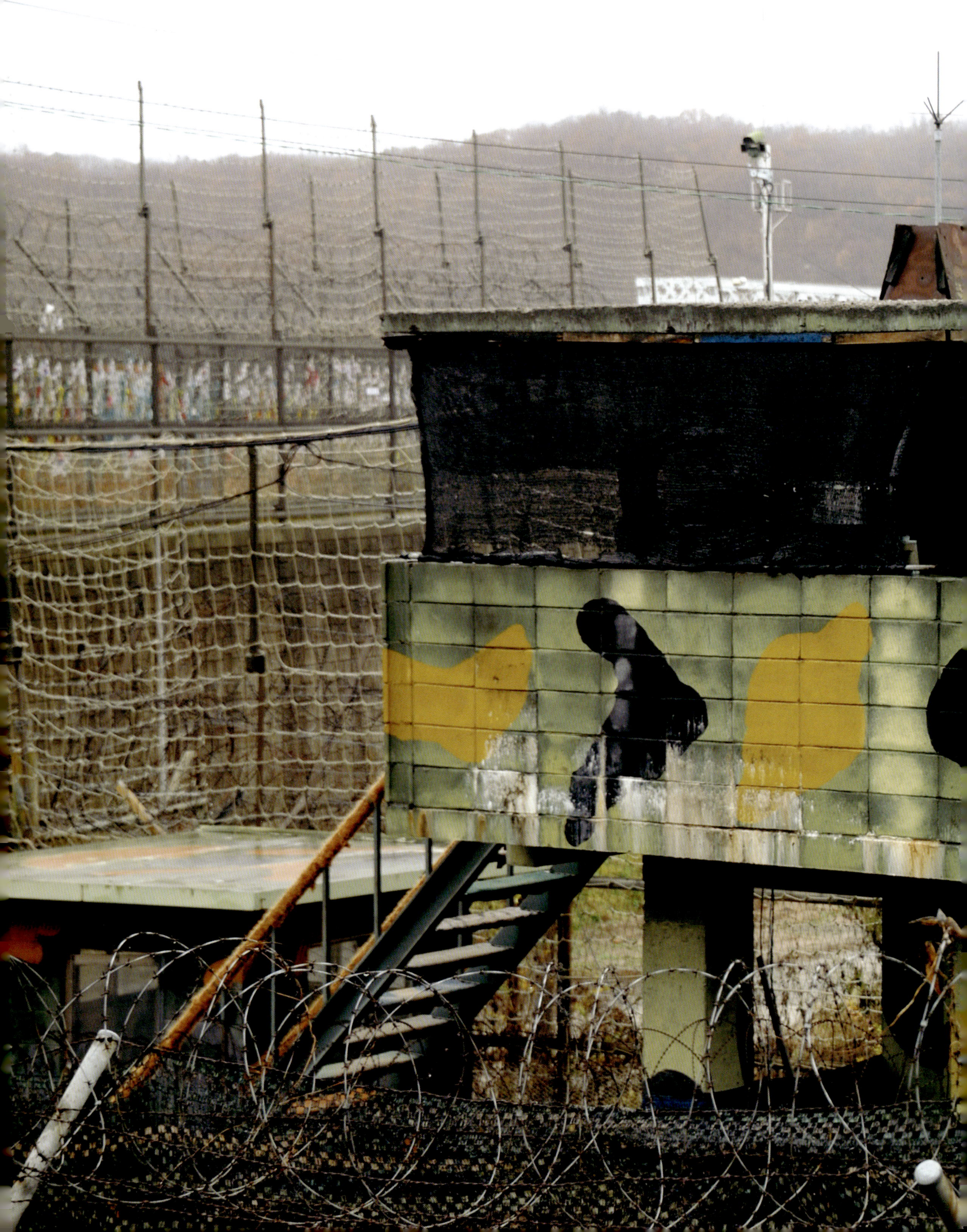

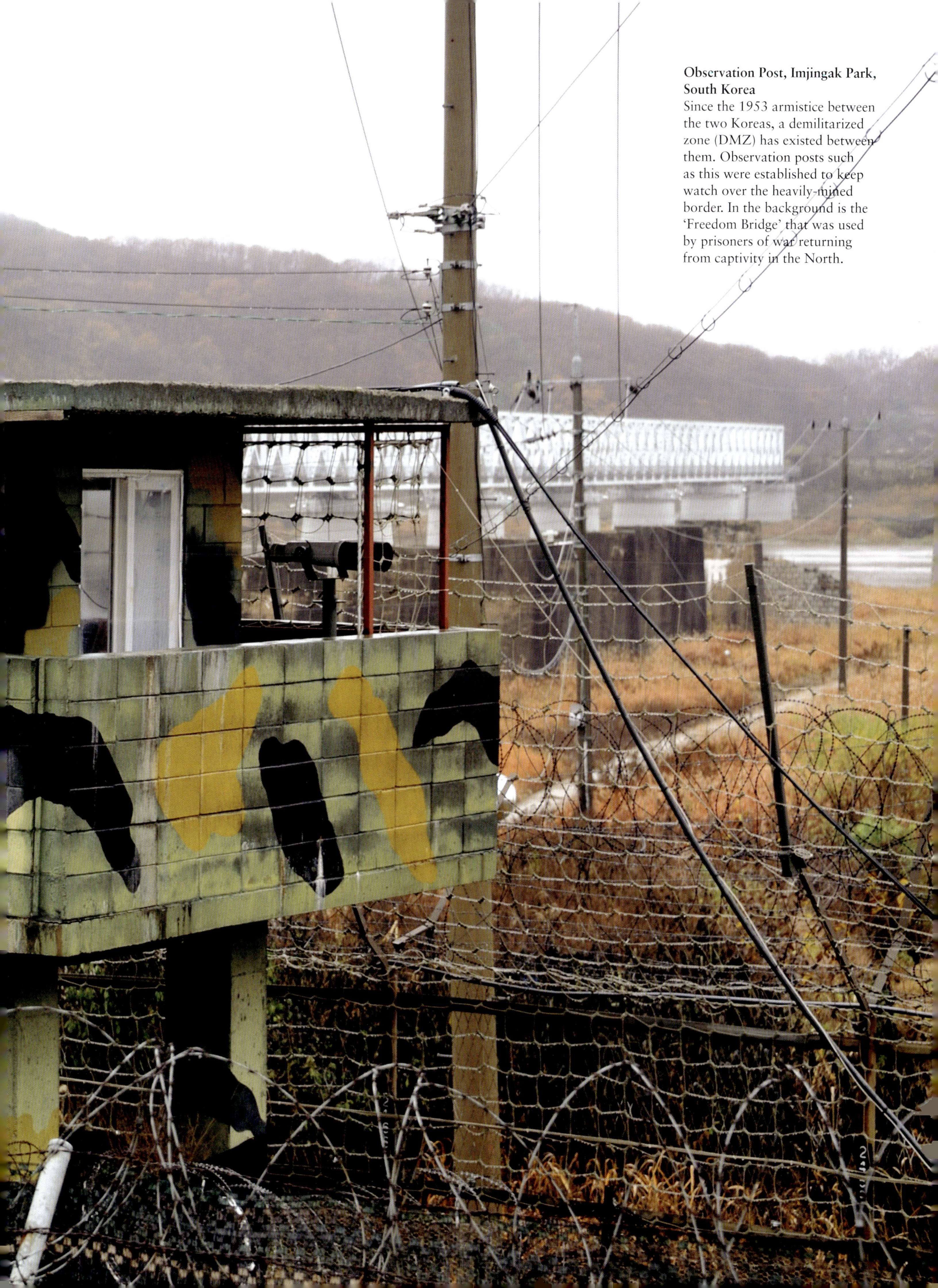

Observation Post, Imjingak Park, South Korea
Since the 1953 armistice between the two Koreas, a demilitarized zone (DMZ) has existed between them. Observation posts such as this were established to keep watch over the heavily-mined border. In the background is the 'Freedom Bridge' that was used by prisoners of war returning from captivity in the North.

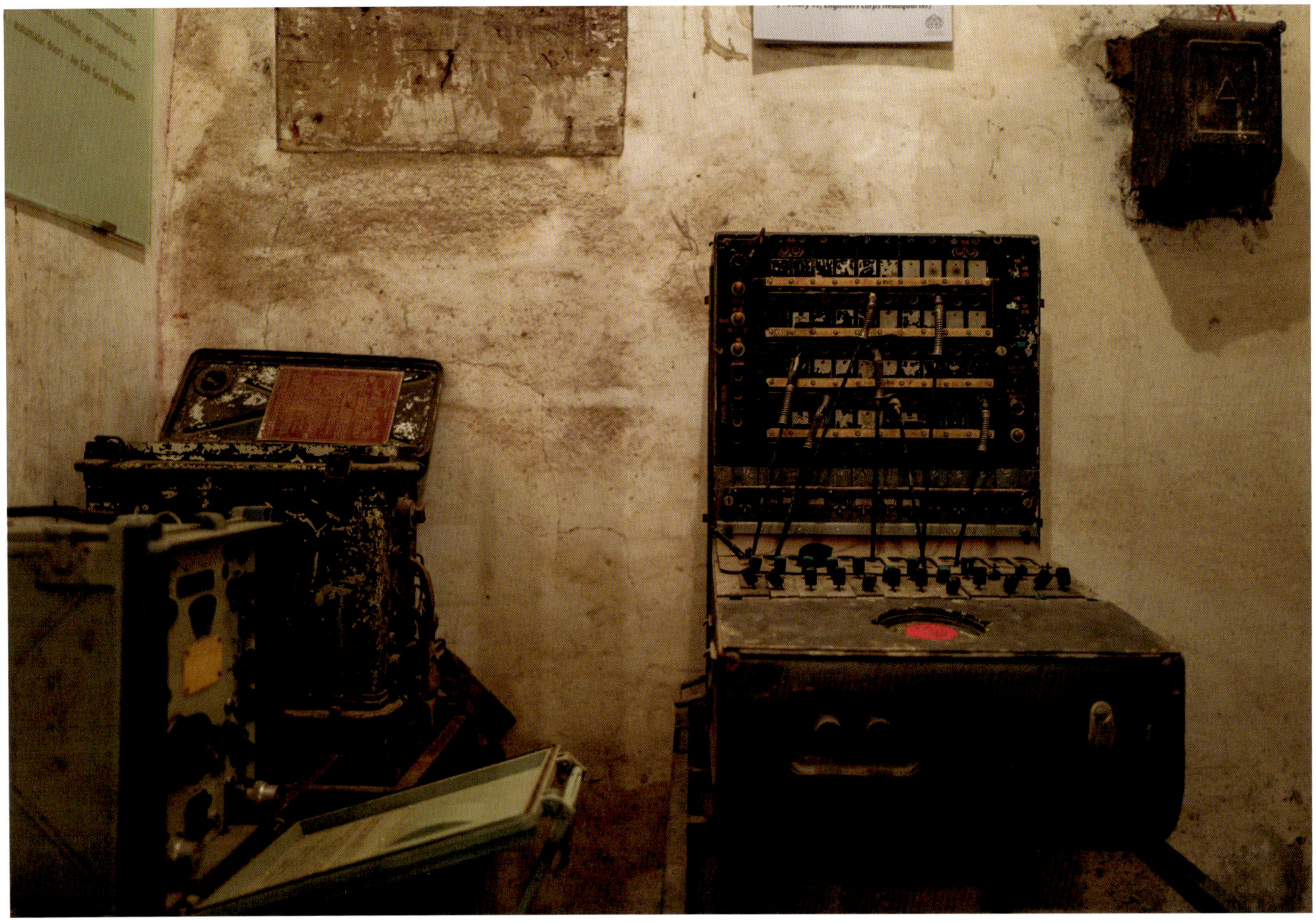

ABOVE:
Communications Equipment, Bunker, Thang Long Citadel, Hanoi, Vietnam
The site of the former royal citadel, seat of the ruling dynasty until 1810, was used as the headquarters of the People's Army of Vietnam from 1954 to 1975 with the codename D67.

RIGHT:
Blast Door, Bunker, Thang Long Citadel, Hanoi, Vietnam
D67 was built to protect the North Vietnamese general staff against the threat of American bombing of Hanoi during the Vietnam War.

OPPOSITE:
UH-1 Huey Helicopter, Ho Chi Minh City, Vietnam
The Huey was synonymous with the US involvement in Vietnam, and made famous by its regular appearance in television news coverage. It was used in many roles, including as a gunship, transport, casualty evacuation and search and rescue.

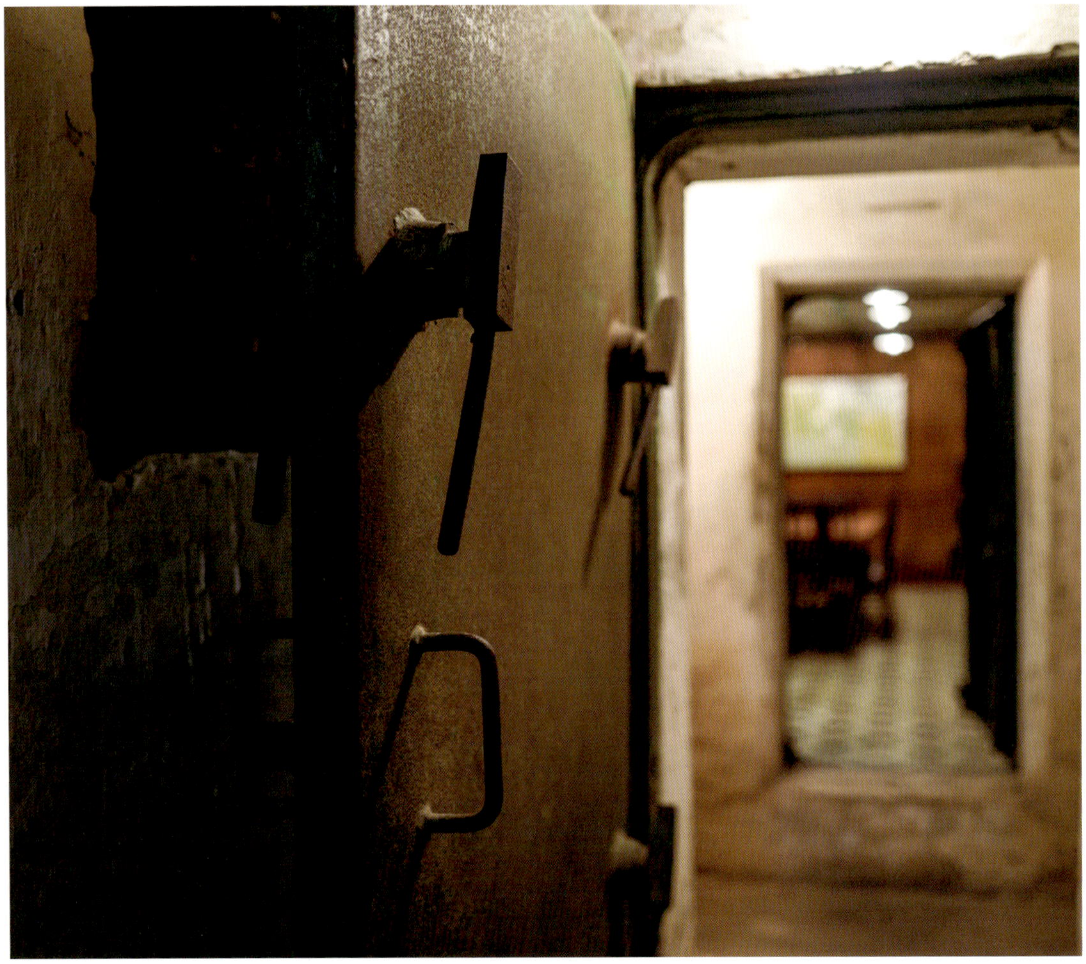

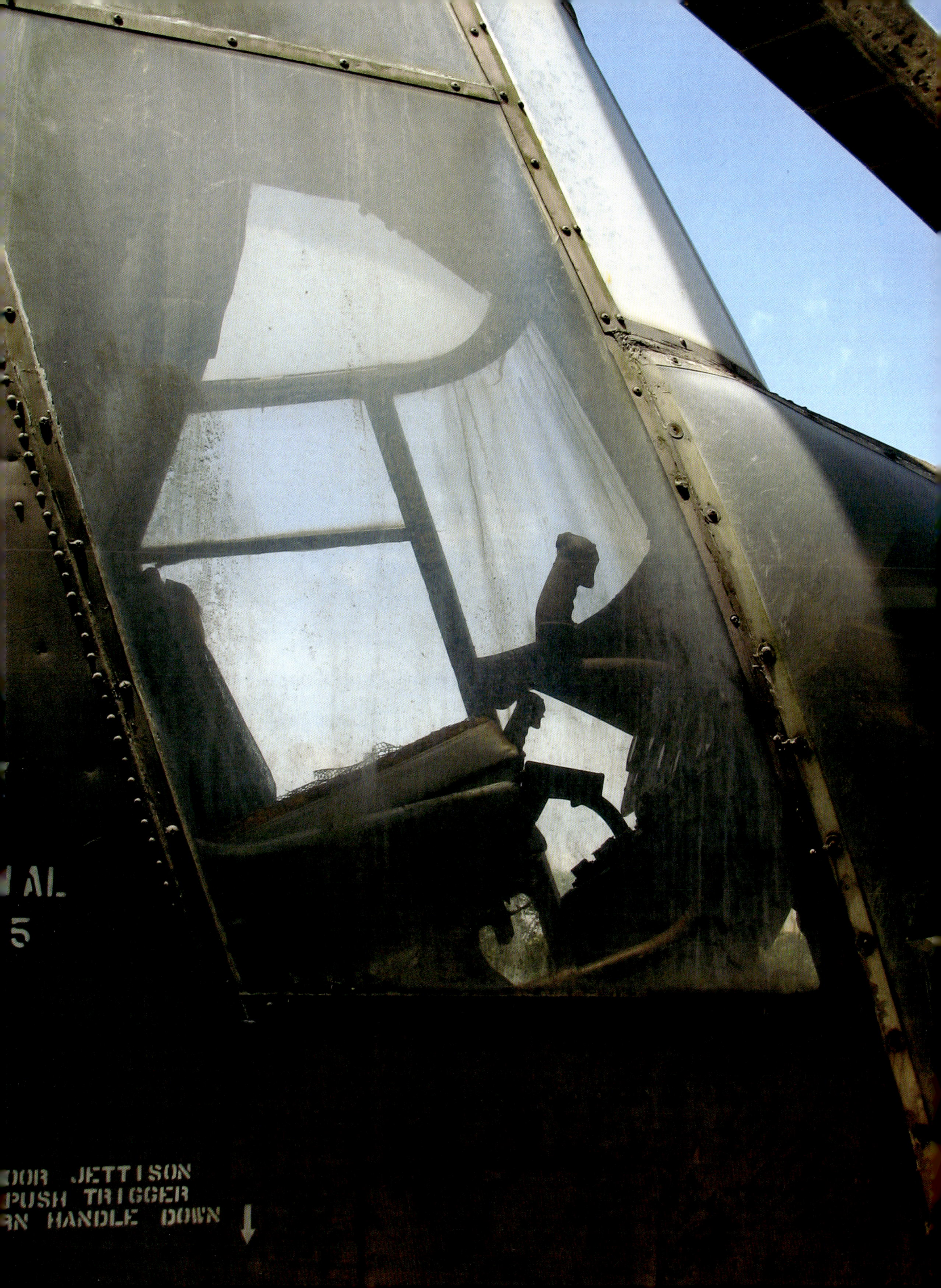
AL
5
OOR JETTISON
PUSH TRIGGER
RN HANDLE DOWN

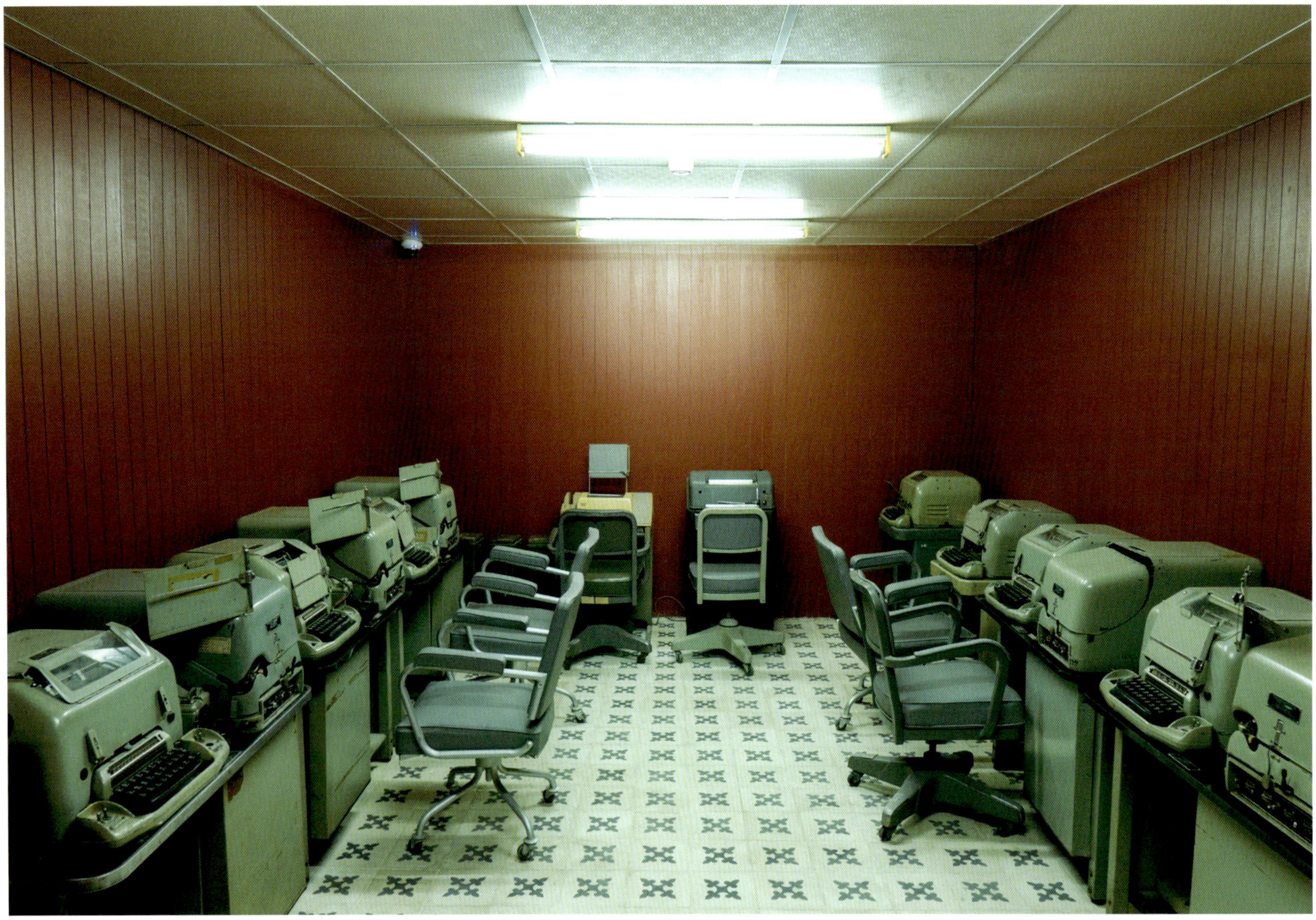

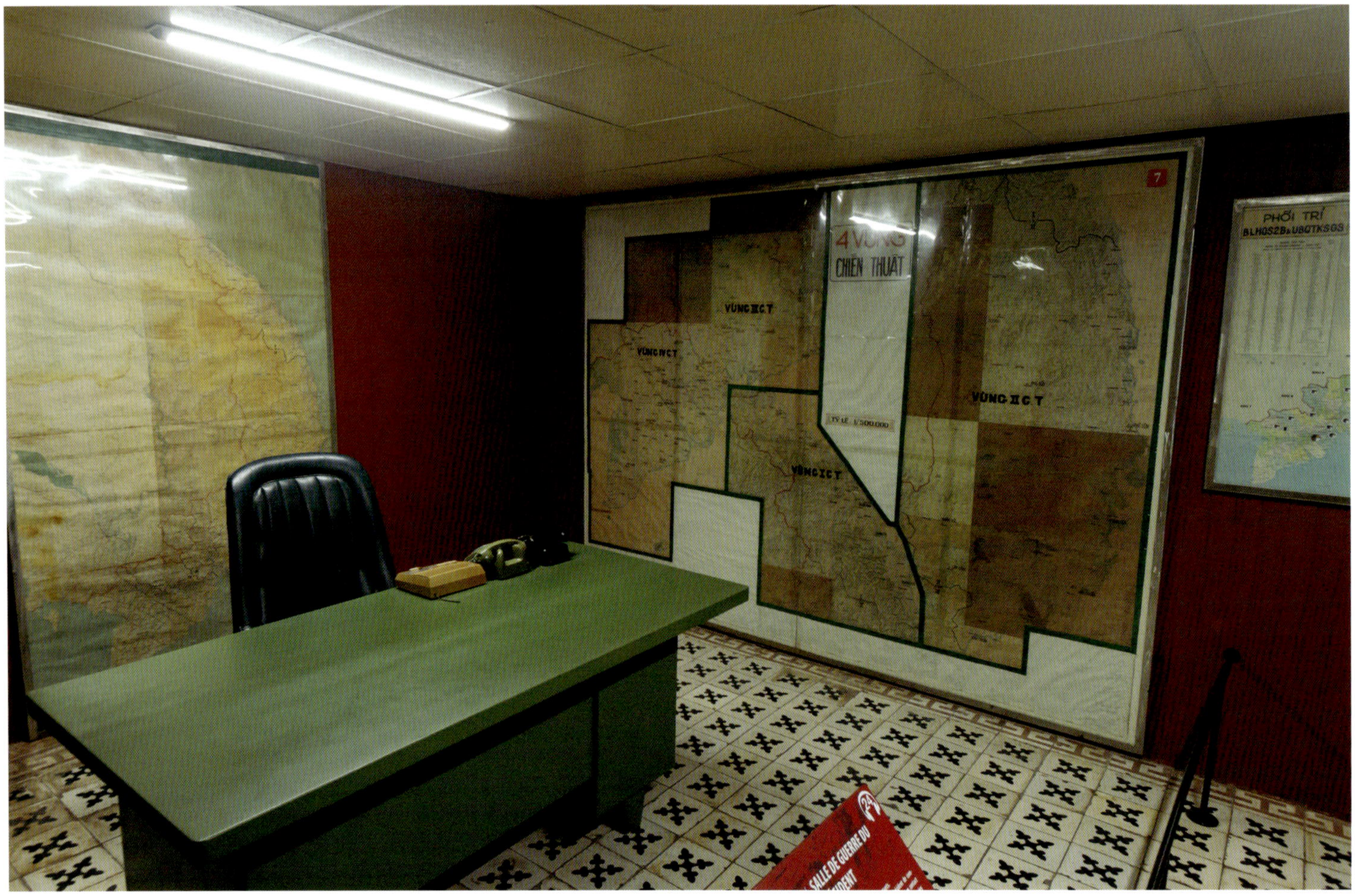
7
CHIÊN THUẬT
PHỐI TRÍ

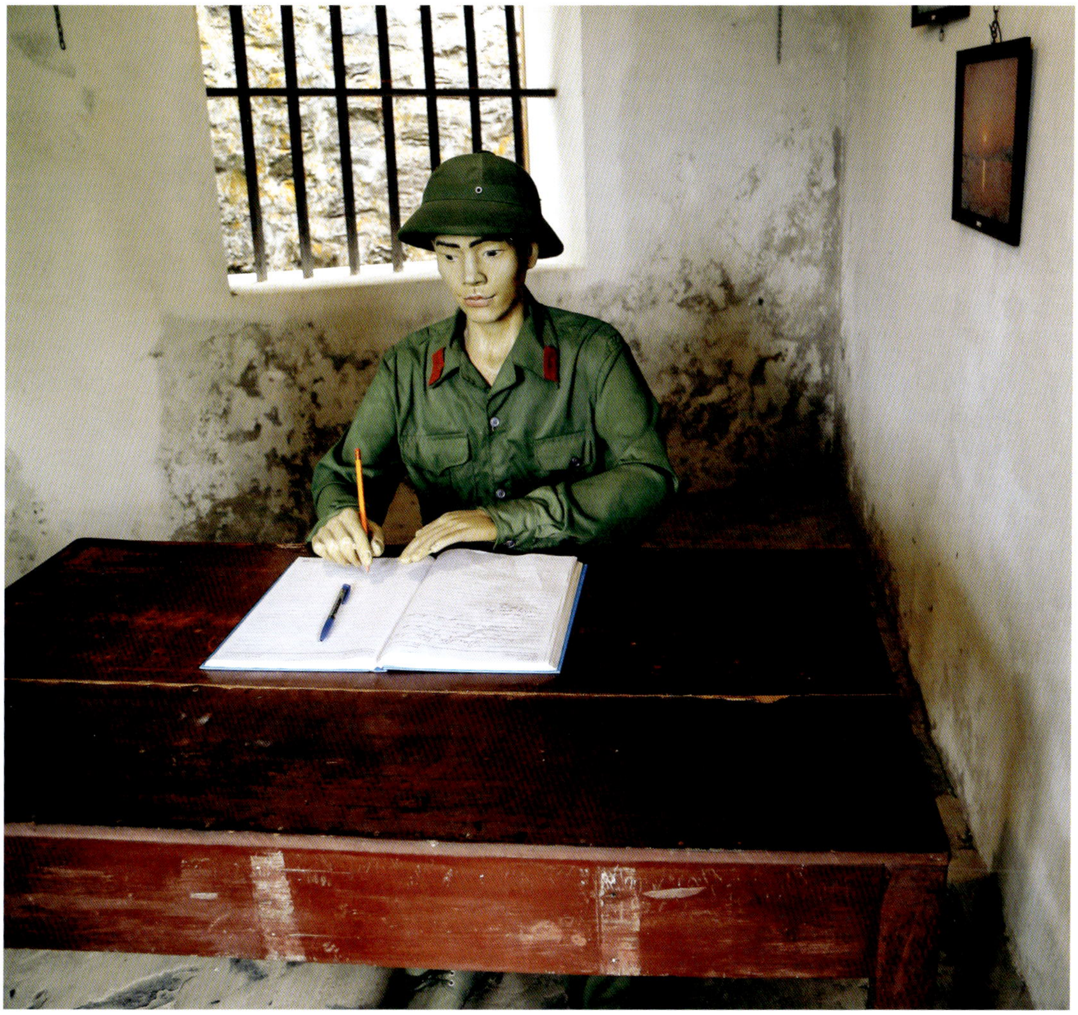

OPPOSITE TOP:
Communications Room, Independence Palace Bunker, Ho Chi Minh City, Vietnam
Teletype machines line the walls of this former communications hub in what was South Vietnam's Presidential Palace.

OPPOSITE BOTTOM:
Map Room, Independence Palace Bunker, Ho Chi Minh City, Vietnam
From 1967 until its gates were breached by a North Vietnamese tank in April 1975, the South Vietnamese war effort was directed from this bunker under the palace.

LEFT AND BELOW:
Cannon Fort, Cat Ba Island, Vietnam
The Japanese built a series of tunnels on Cat Ba island during World War II to help protect the approaches to Haiphong harbour. After the war the French installed several guns on the site, which the North Vietnamese later utilized.

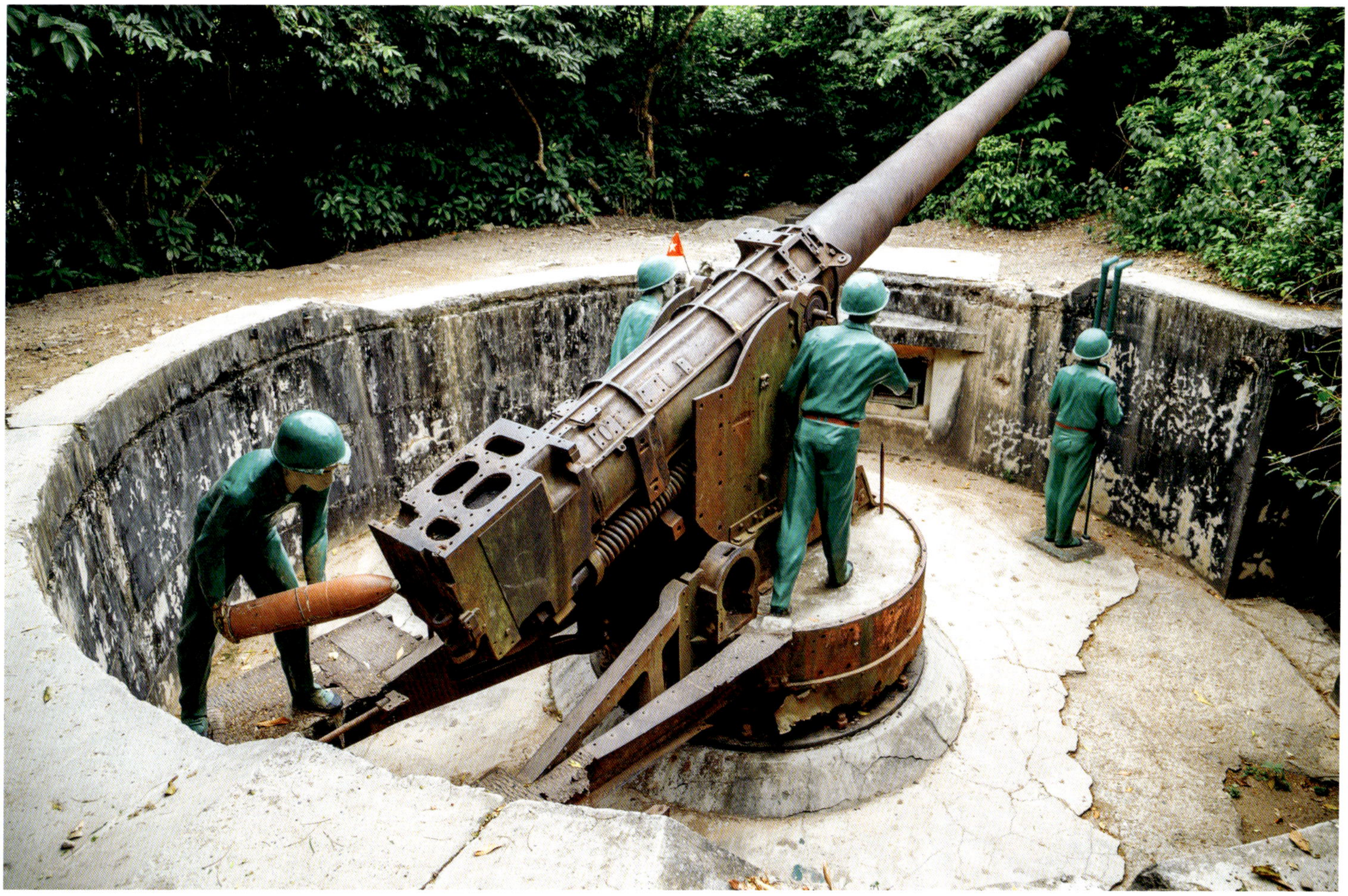

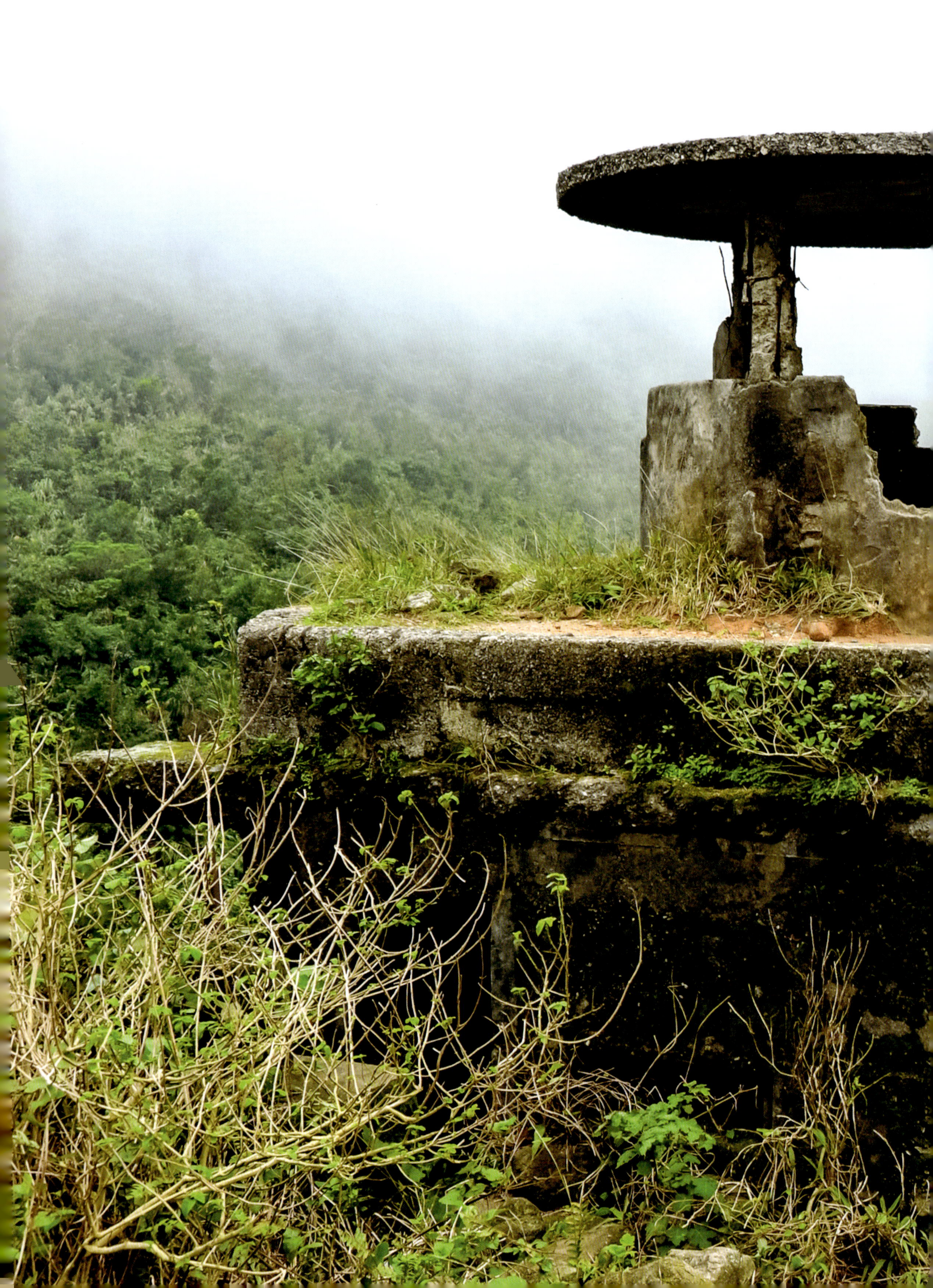

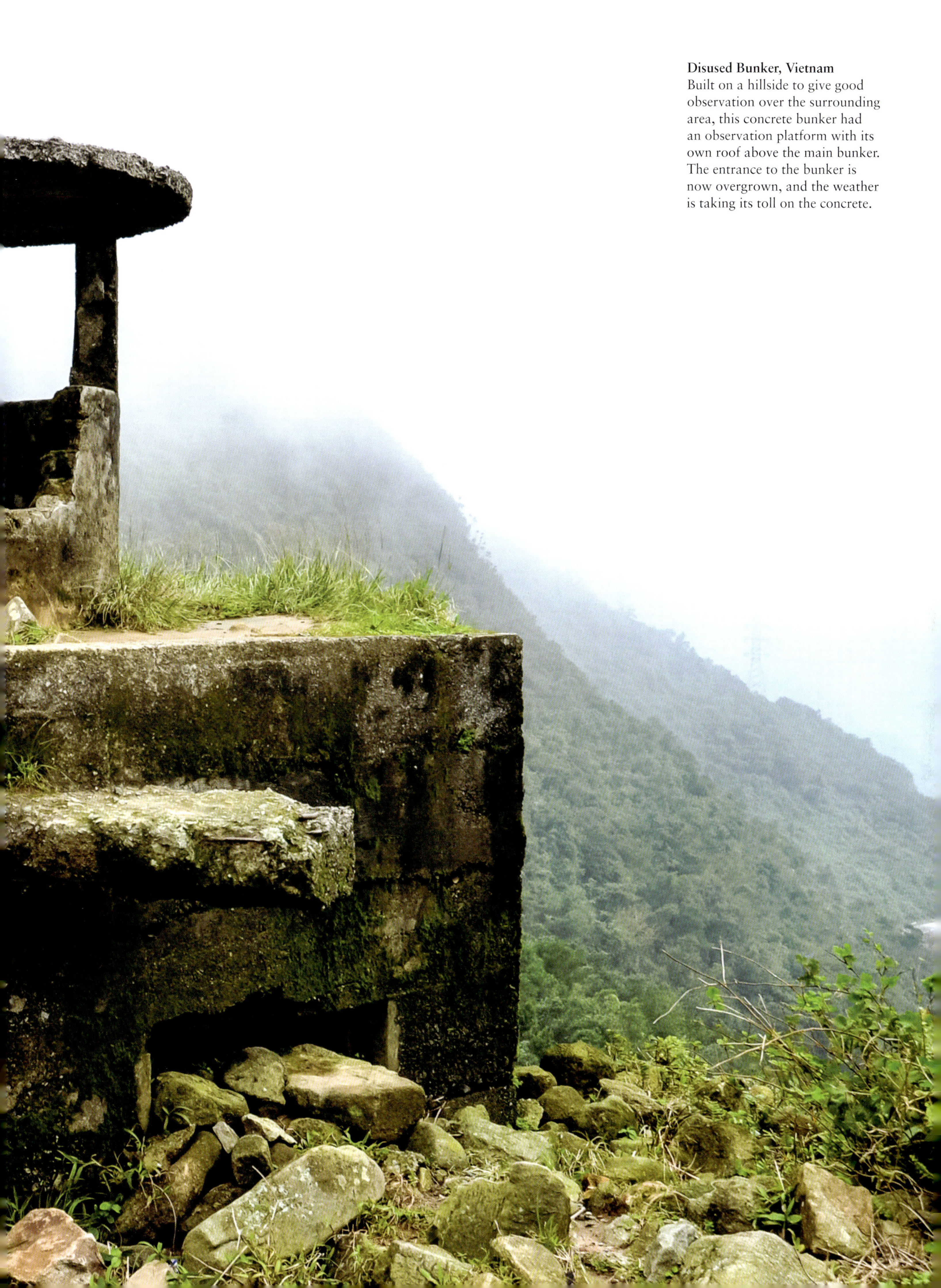

Disused Bunker, Vietnam
Built on a hillside to give good observation over the surrounding area, this concrete bunker had an observation platform with its own roof above the main bunker. The entrance to the bunker is now overgrown, and the weather is taking its toll on the concrete.

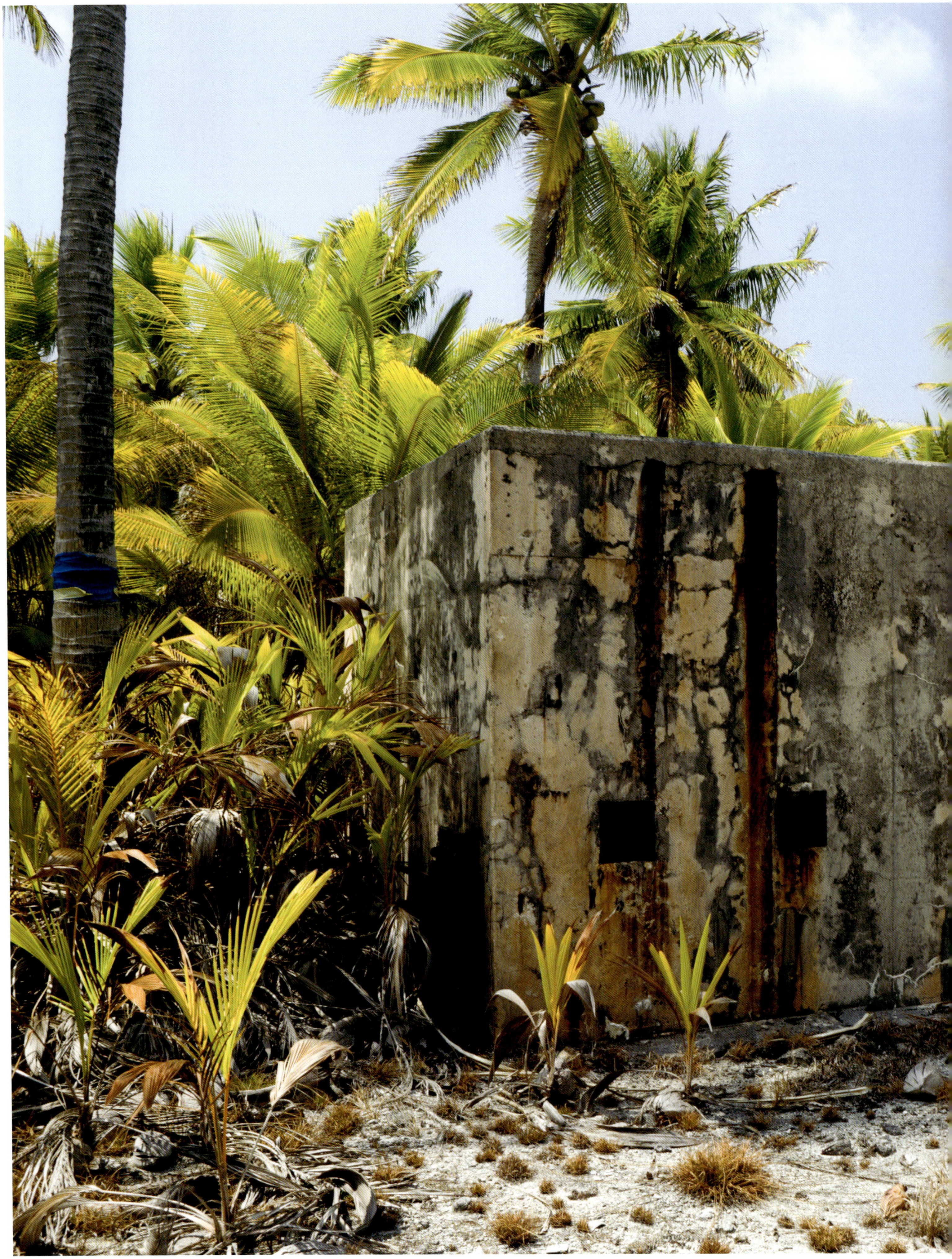

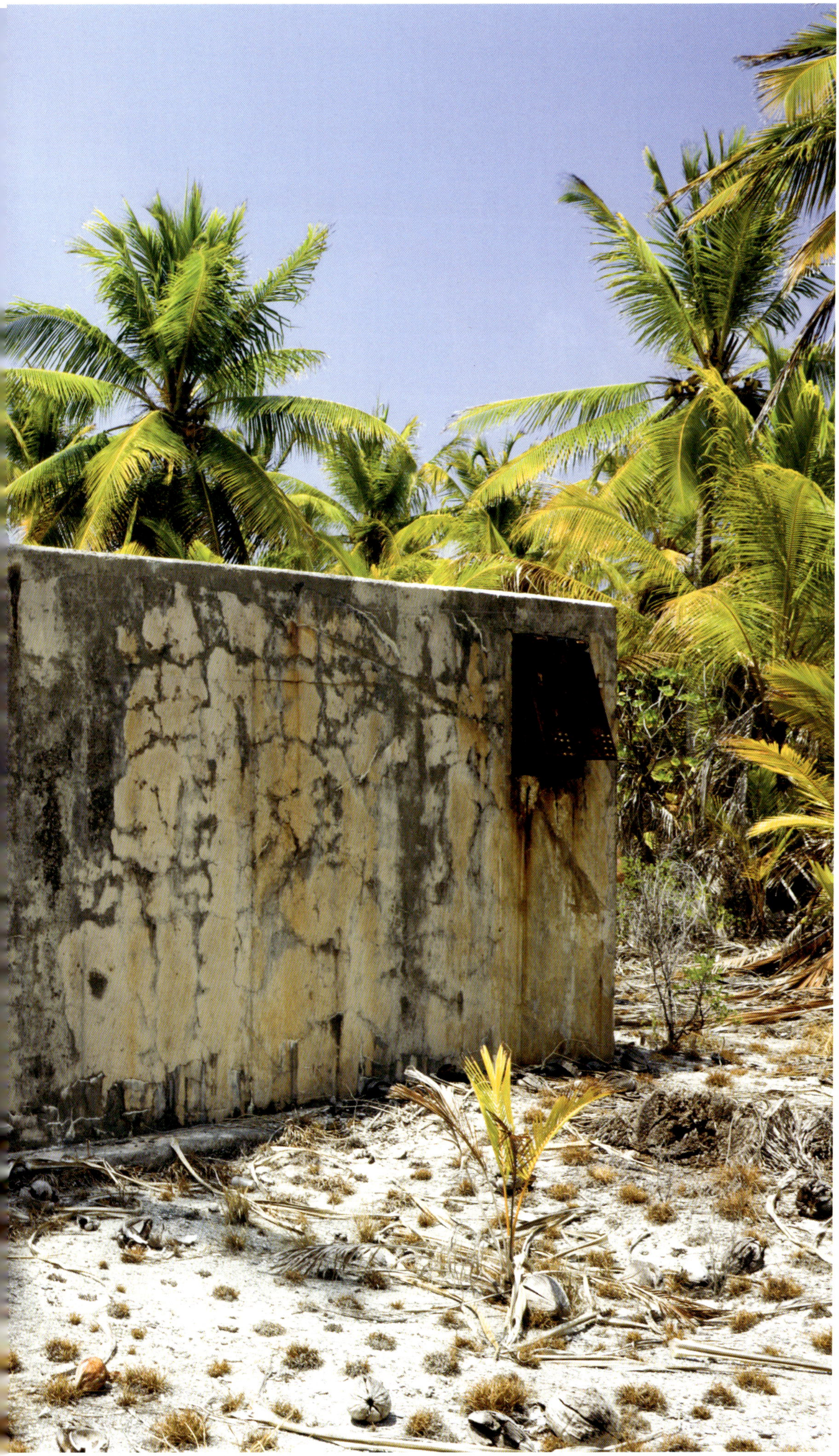

Former Observation Bunker, Bikini Atoll, Marshall Islands, Micronesia

Bikini Atoll was used as the site of a series of nuclear tests after World War II. Two atom bombs were dropped to test the effects of a nuclear blast on warships in 1946, while 21 hydrogen bombs were detonated between 1954 and 1958. Shown here is one of the observation bunkers built to monitor the explosions and their after effects.

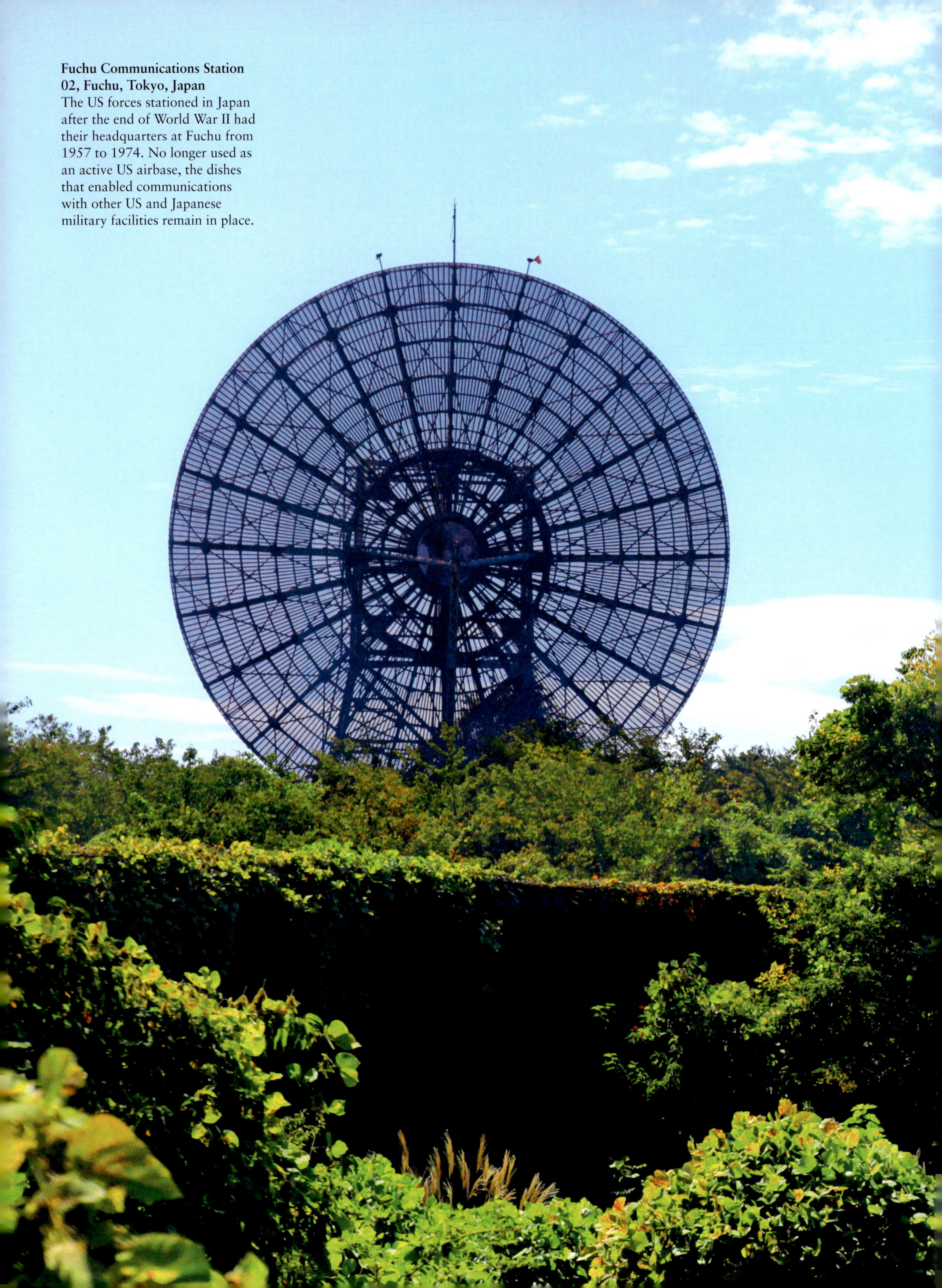

Fuchu Communications Station 02, Fuchu, Tokyo, Japan
The US forces stationed in Japan after the end of World War II had their headquarters at Fuchu from 1957 to 1974. No longer used as an active US airbase, the dishes that enabled communications with other US and Japanese military facilities remain in place.

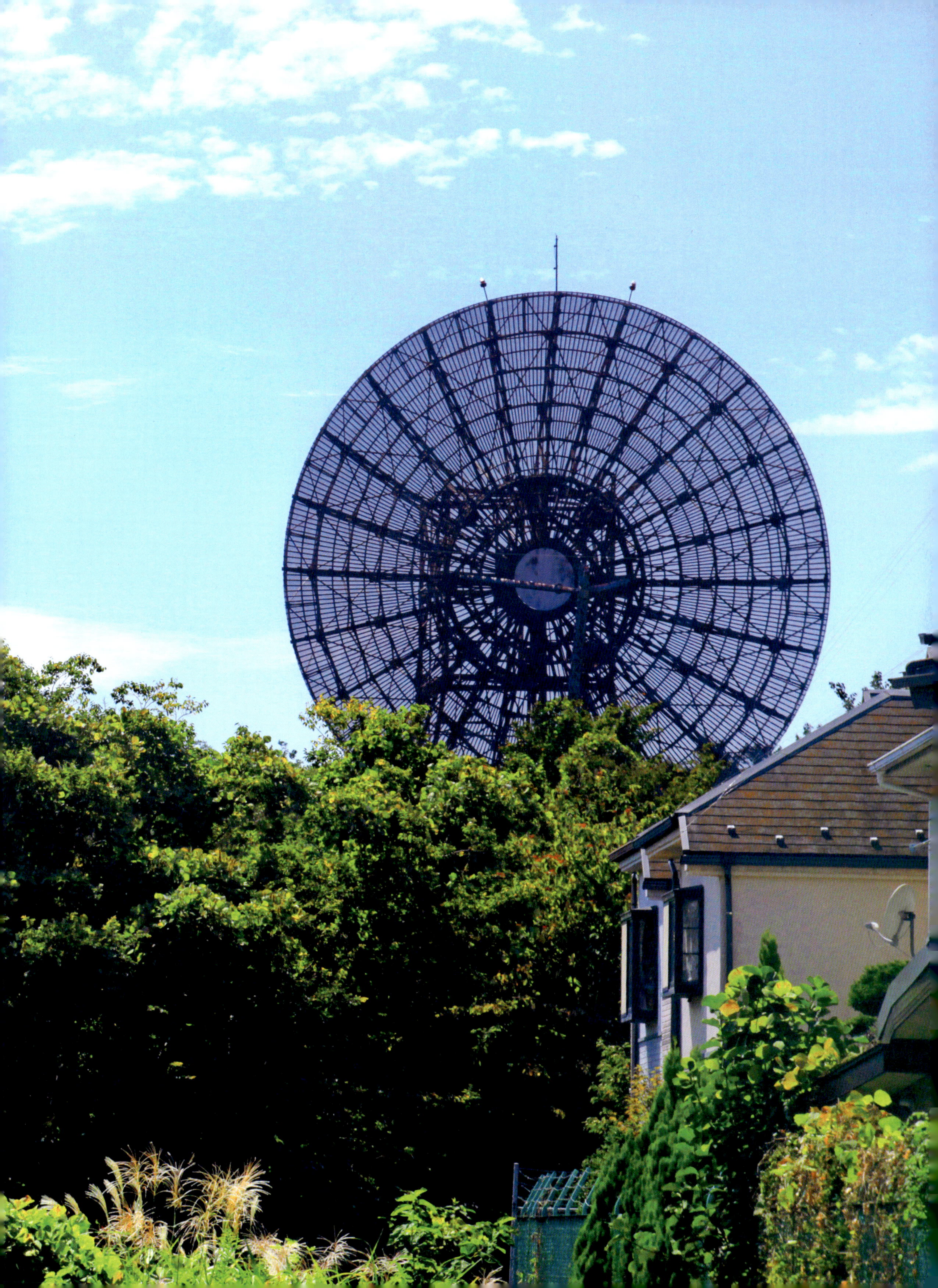

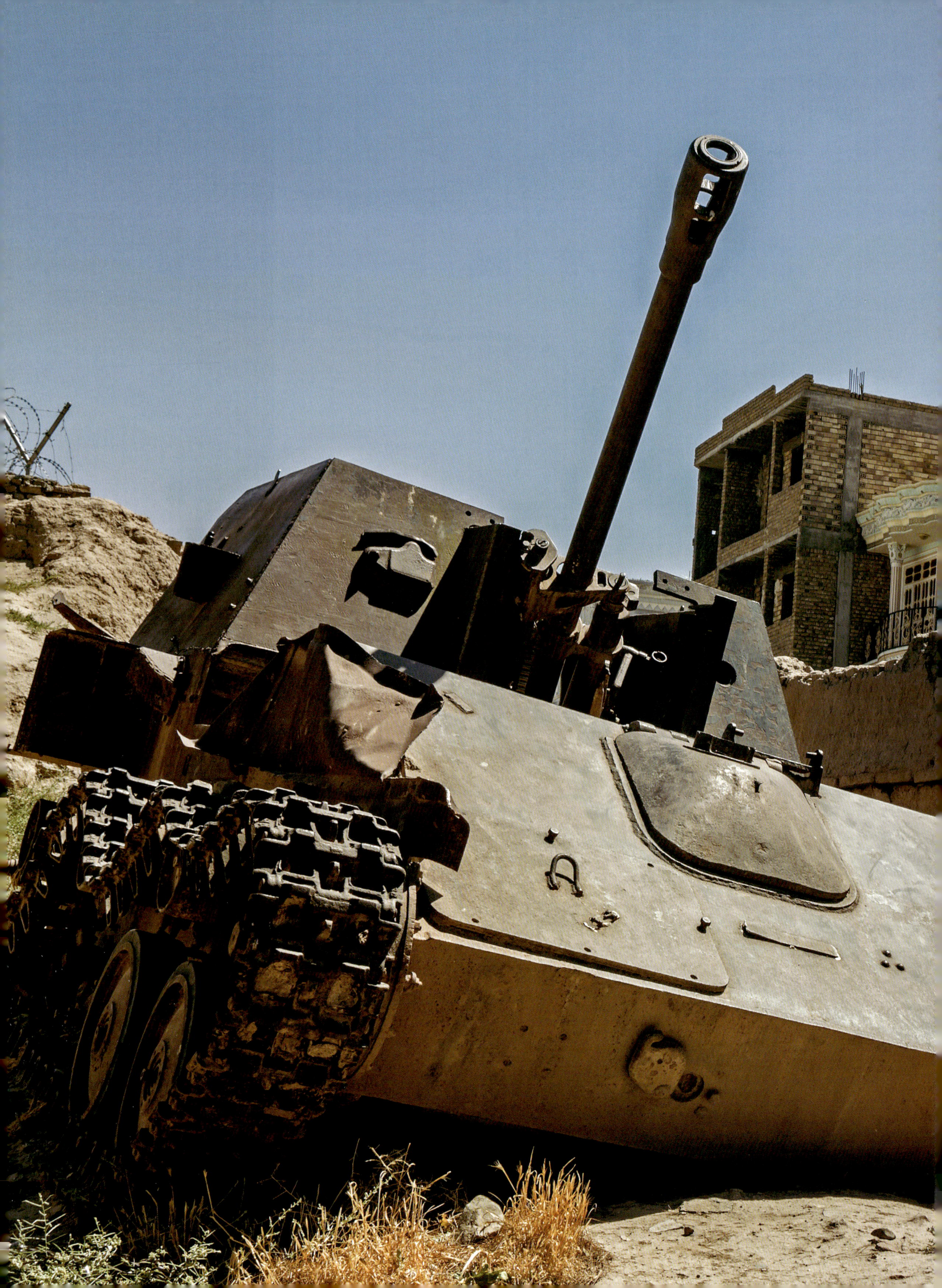

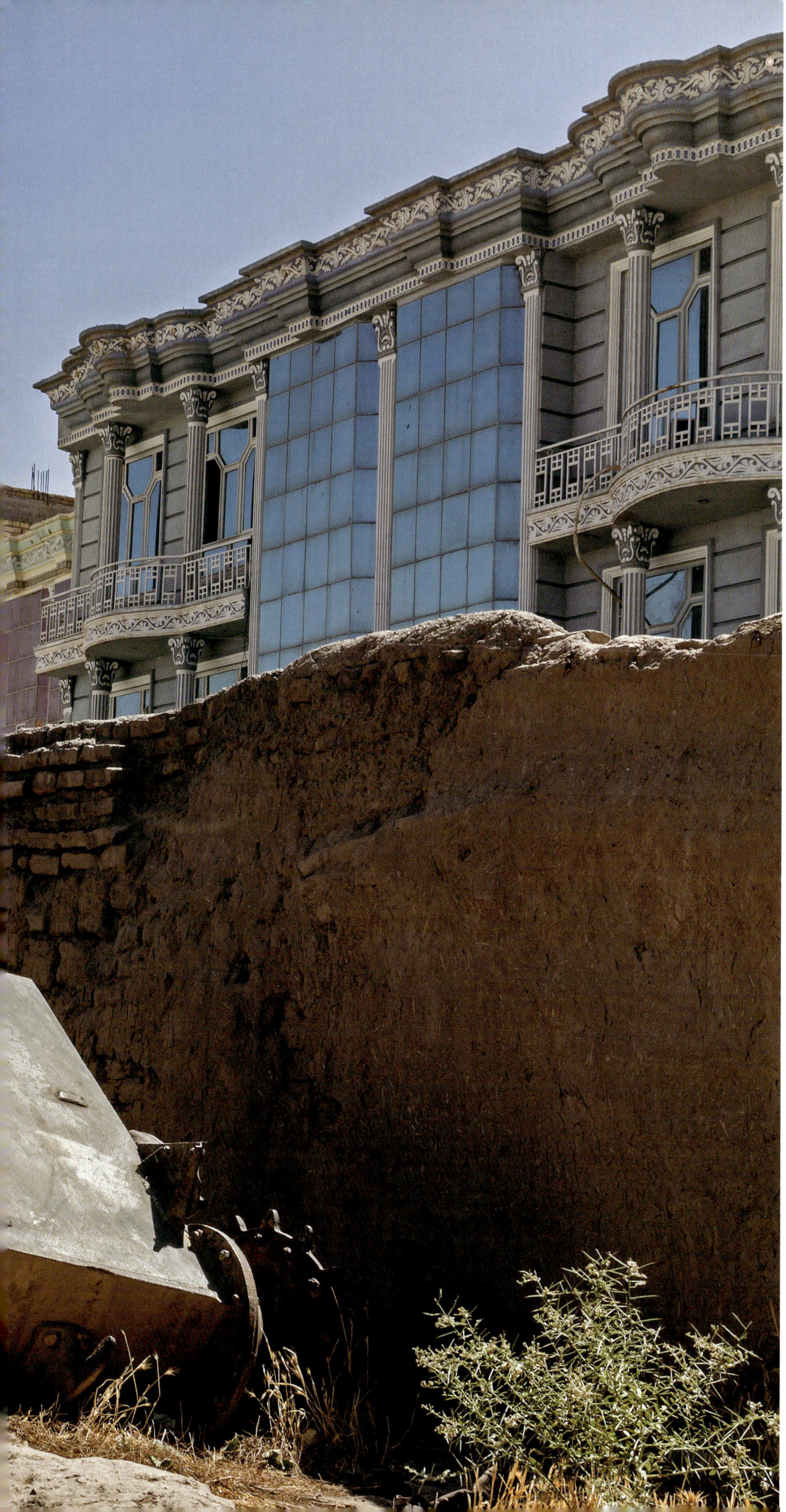

LEFT:

Soviet-made SU-76 Self Propelled Gun, Afghanistan

Decades of conflict in Afghanistan since the 1970s has left much of the country littered with destroyed weapons. Shown here is a knocked-out self-propelled gun with one track missing, suggesting it may have hit a mine. The SU-76 was first used by the Soviets against German forces in World War II.

OVERLEAF:

An-12 Wreck, Bagram Air Base, Afghanistan

A heavily-autographed An-12 cargo aircraft wreck rests on a tracked chassis, with what appears to be the remains of a (wheel-less) BTR-70 wheeled infantry vehicle in the foreground. Although built with American help in the 1950s, Bagram became a key military base in Afghanistan for Soviet forces after their invasion.

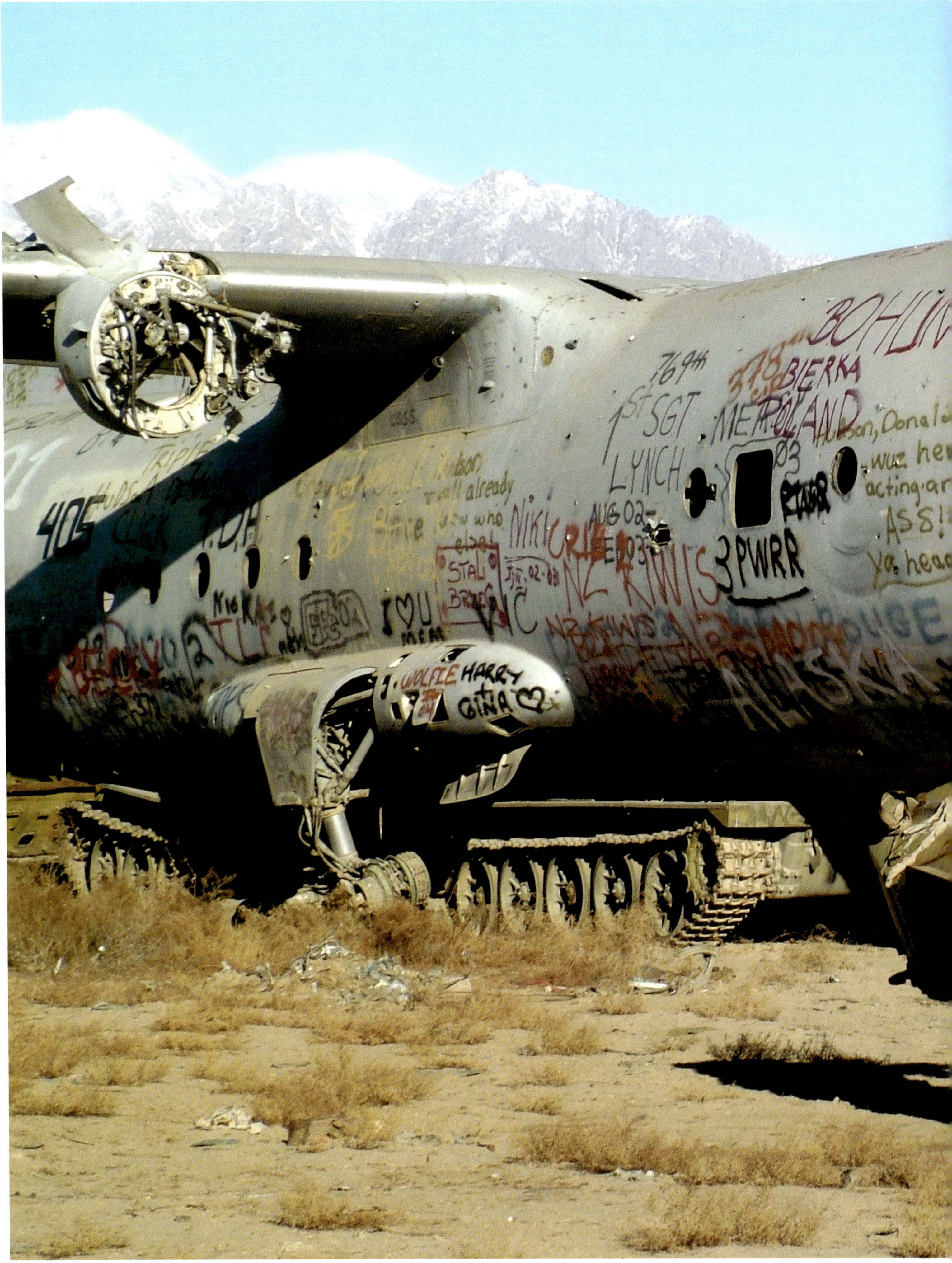
405
769th
1st SGT
LYNCH
BIERKA
AUG 02-
3 PWRR
NZ KIWIS
WOLFIE
HARRY
GINA

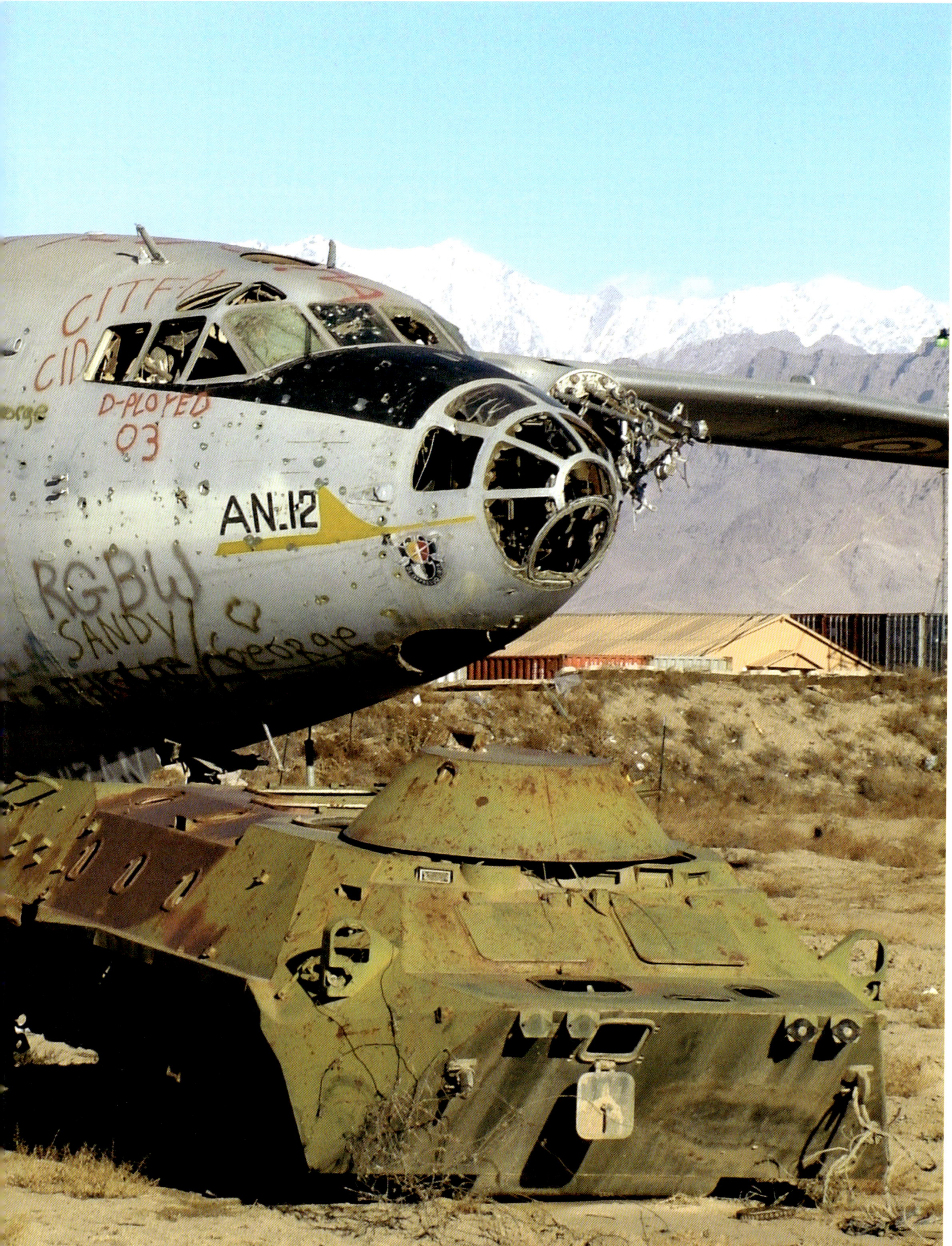
AN-12
D-PLOYED
03
RG-BW
SANDY

TOP RIGHT:

Mil Mi-8 'Hip' Carcass, Afghanistan

Once the US began supplying the Afghan rebels fighting Soviet forces with Stinger anti-aircraft missiles, large numbers of Soviet and Afghan government helicopters began to be shot down. This particular aircraft, however, shows no signs of an impact. It was likely abandoned after being damaged by artillery, or stripped for parts.

BOTTOM RIGHT:

Afghan Air Force MiG-21 Wreck, Bagram, Afghanistan

Little is left of this MiG-21, a Soviet-era fighter jet that first flew in 1956 and was subsequently widely exported throughout the world. Over 150 MiG-21s were sent to Afghanistan from 1973 until the Soviet withdrawal. This example has Afghan Air Force markings.

OPPOSITE:

D-1 152mm Howitzer, Afghanistan

Built by the Soviets in Sverdlovsk during the 1940s, the D-1 howitzer was capable of firing a 152mm shell 12.4km (7.7 miles). As with much Soviet equipment, it was sold or distributed to many Warsaw Pact and Soviet client states during the Cold War.

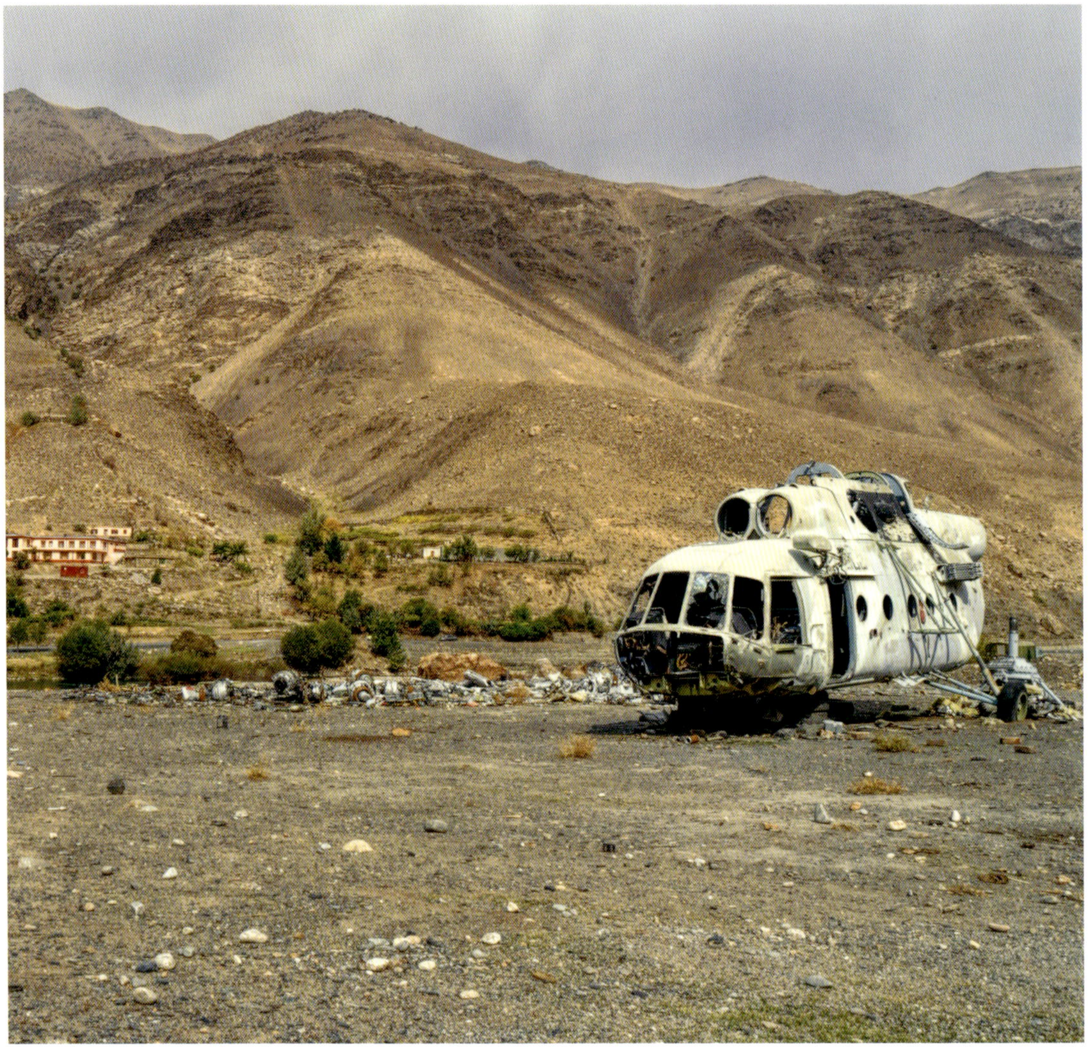

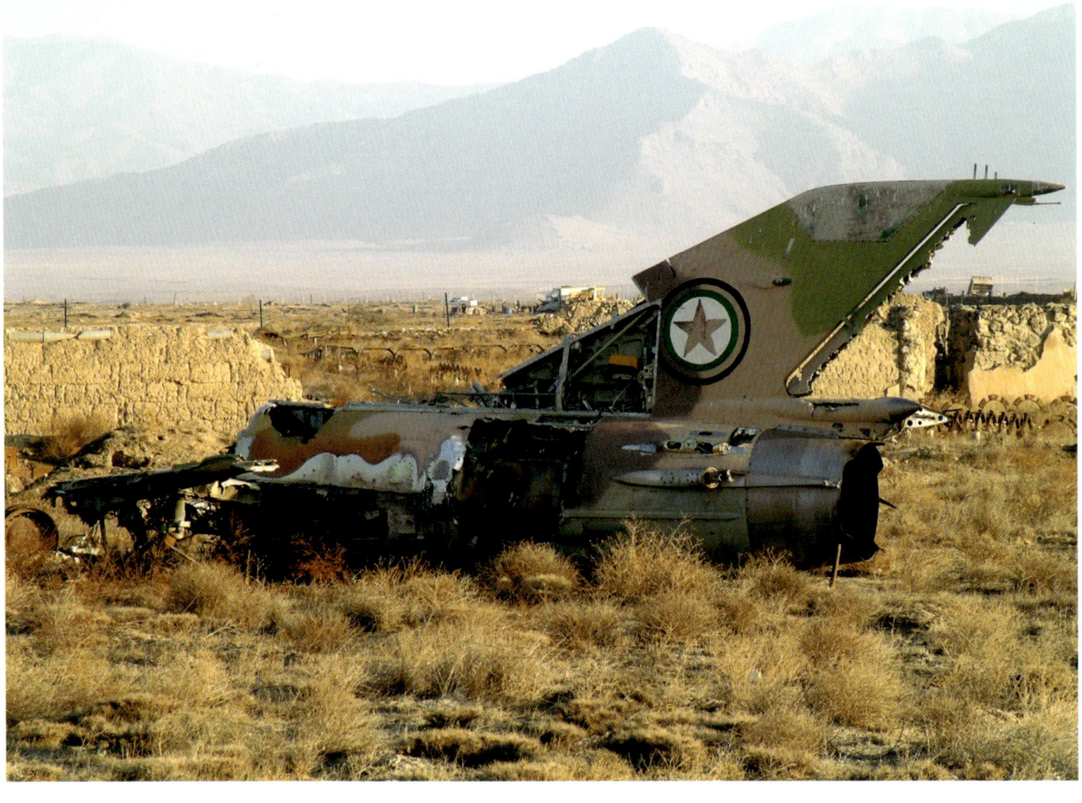

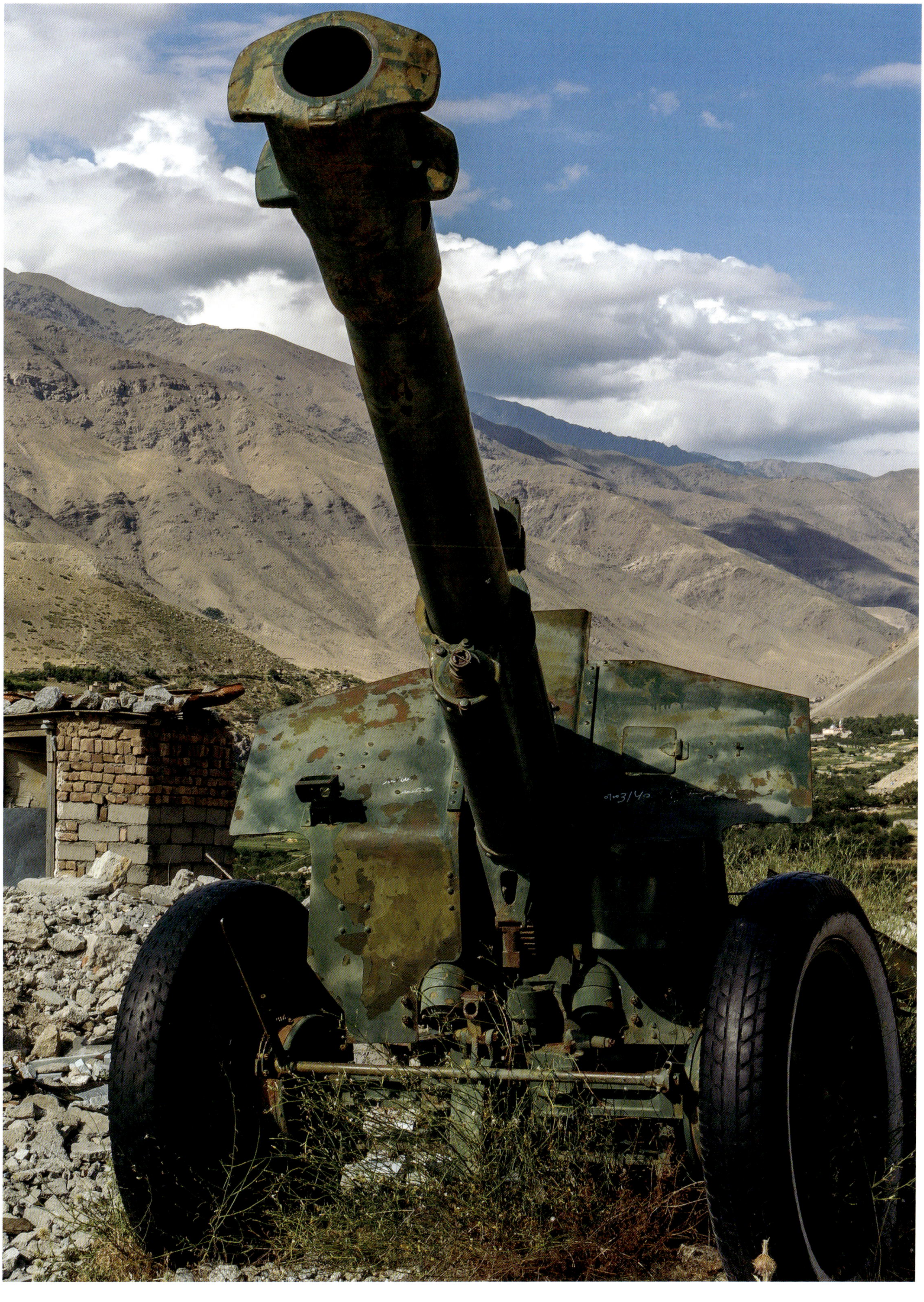

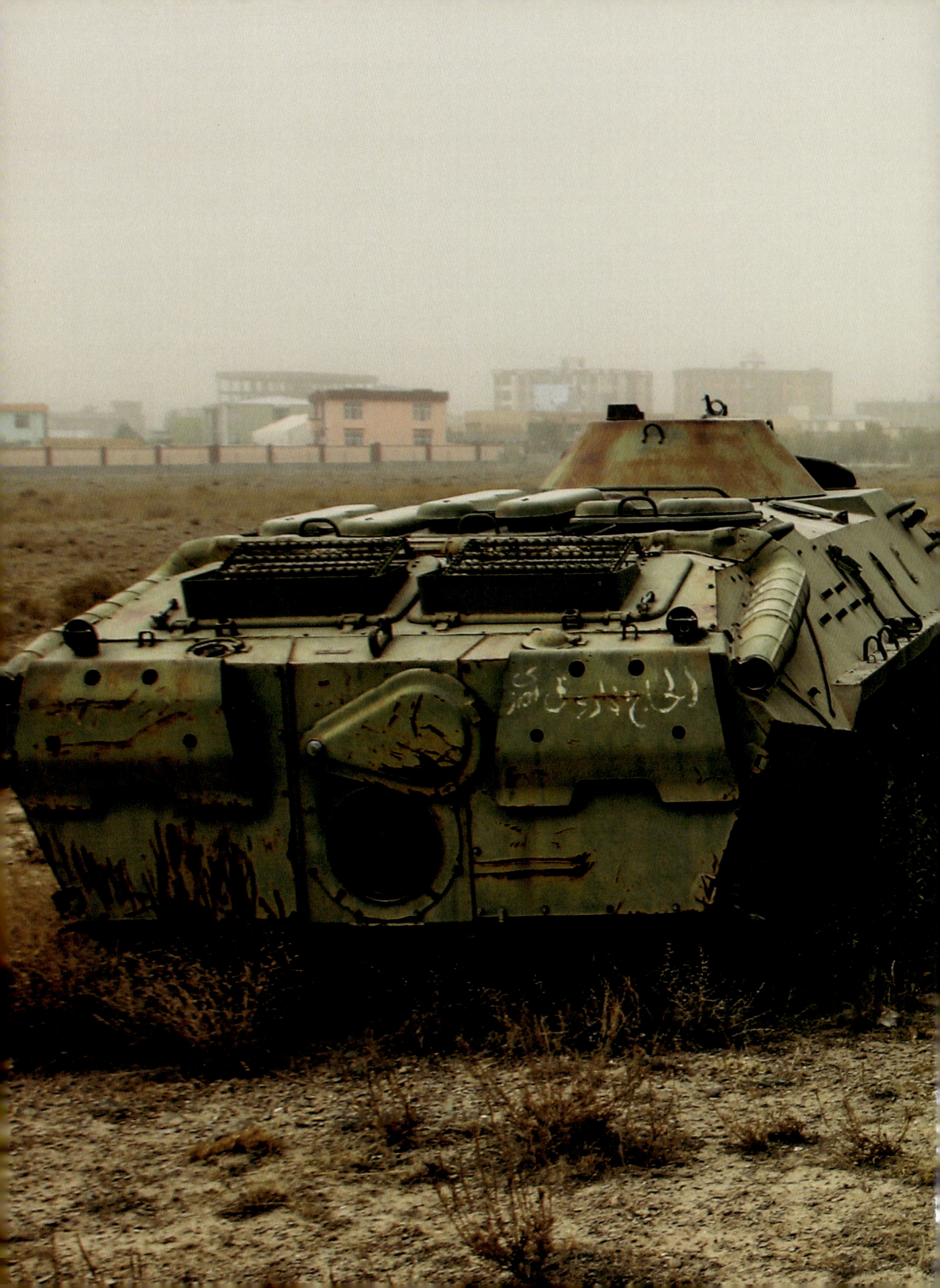

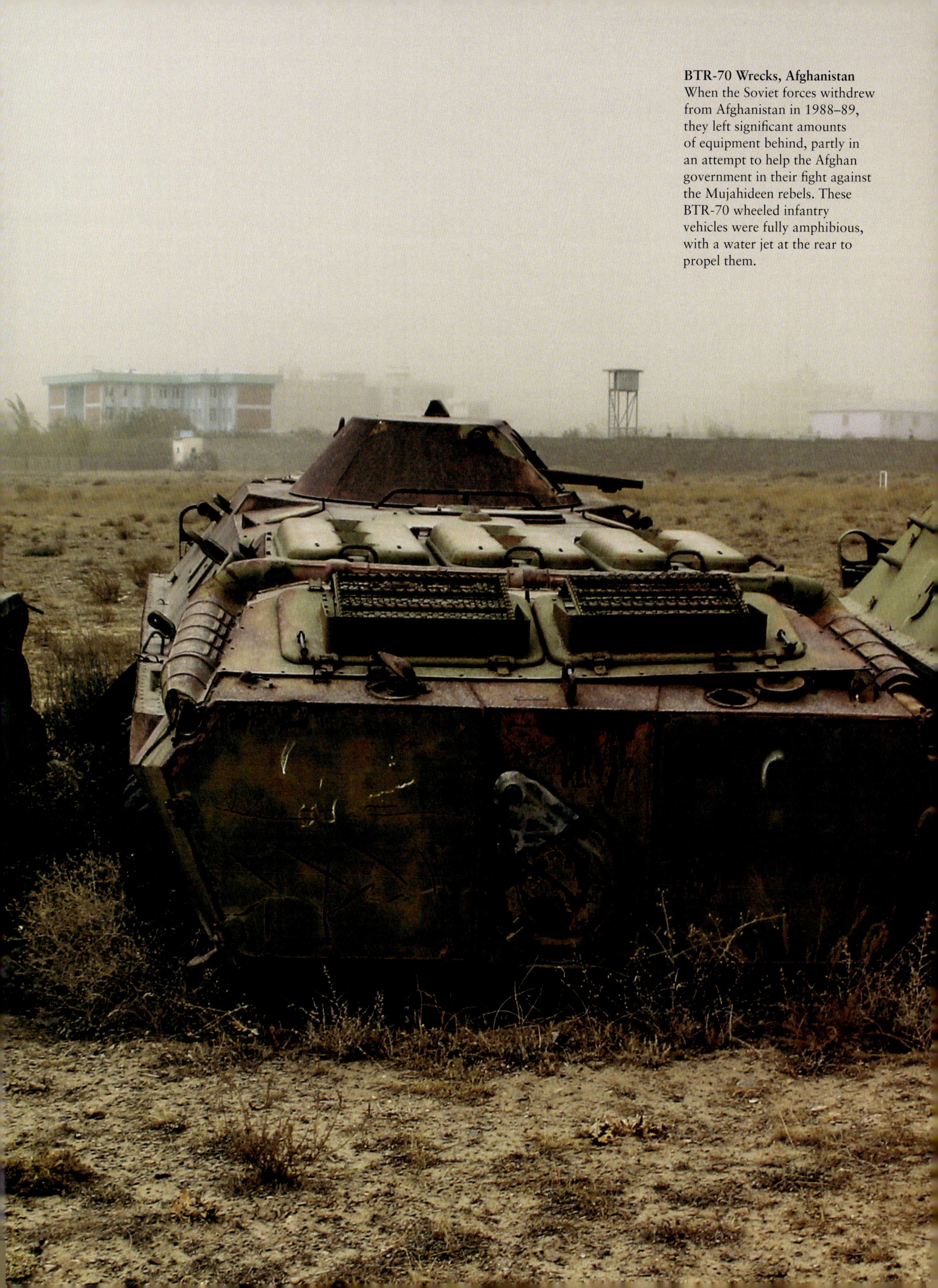

BTR-70 Wrecks, Afghanistan
When the Soviet forces withdrew from Afghanistan in 1988–89, they left significant amounts of equipment behind, partly in an attempt to help the Afghan government in their fight against the Mujahideen rebels. These BTR-70 wheeled infantry vehicles were fully amphibious, with a water jet at the rear to propel them.

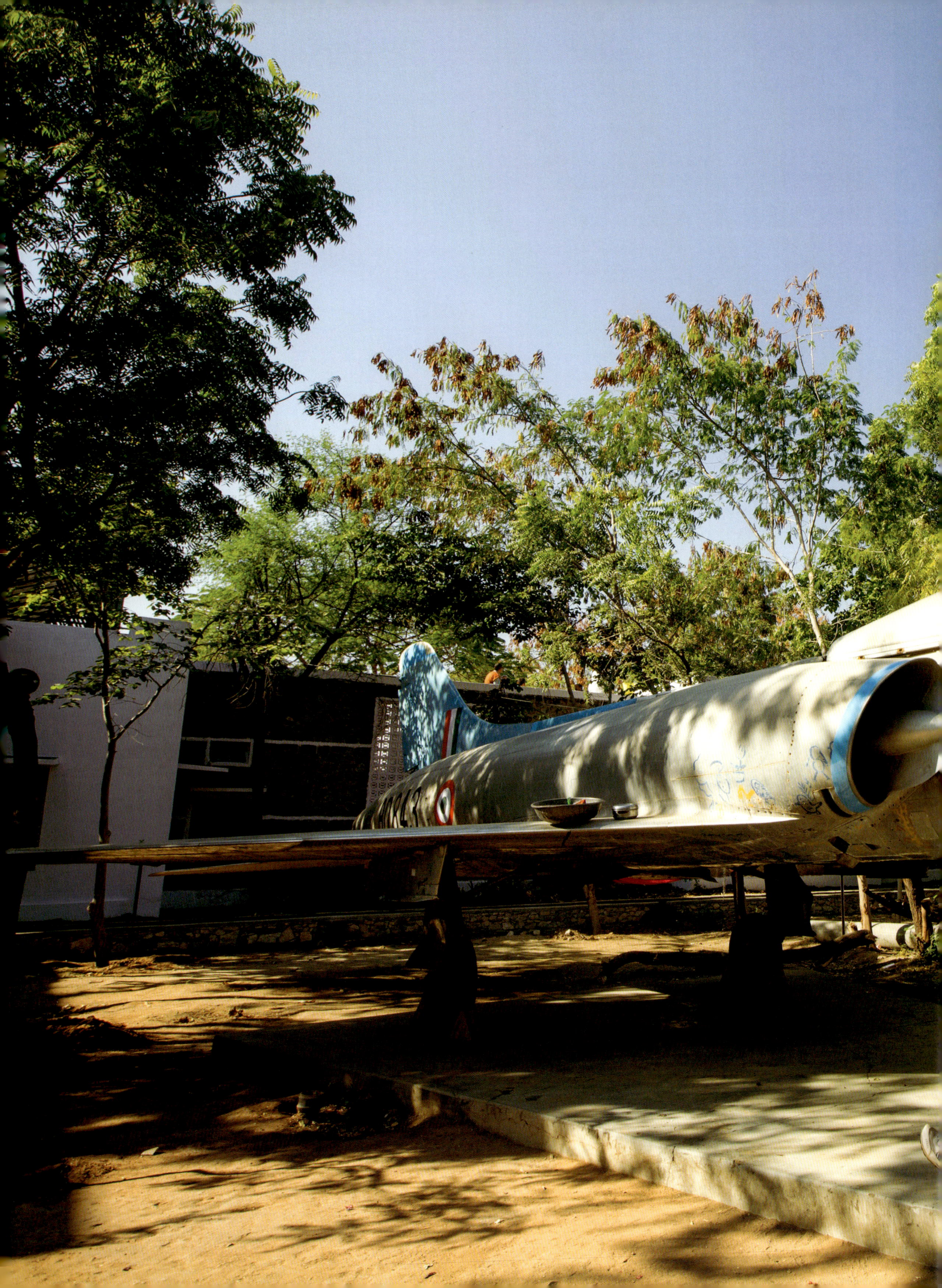

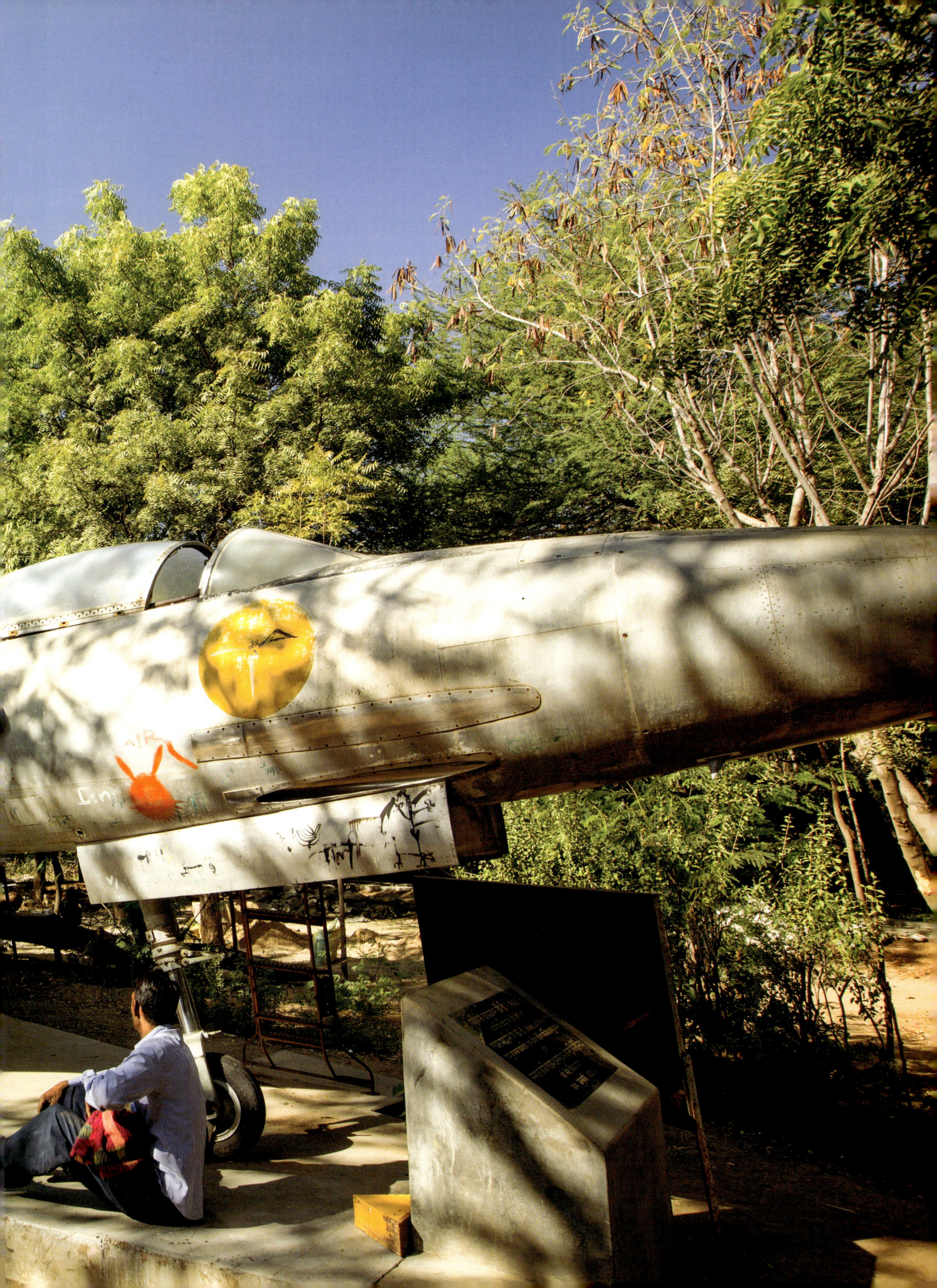

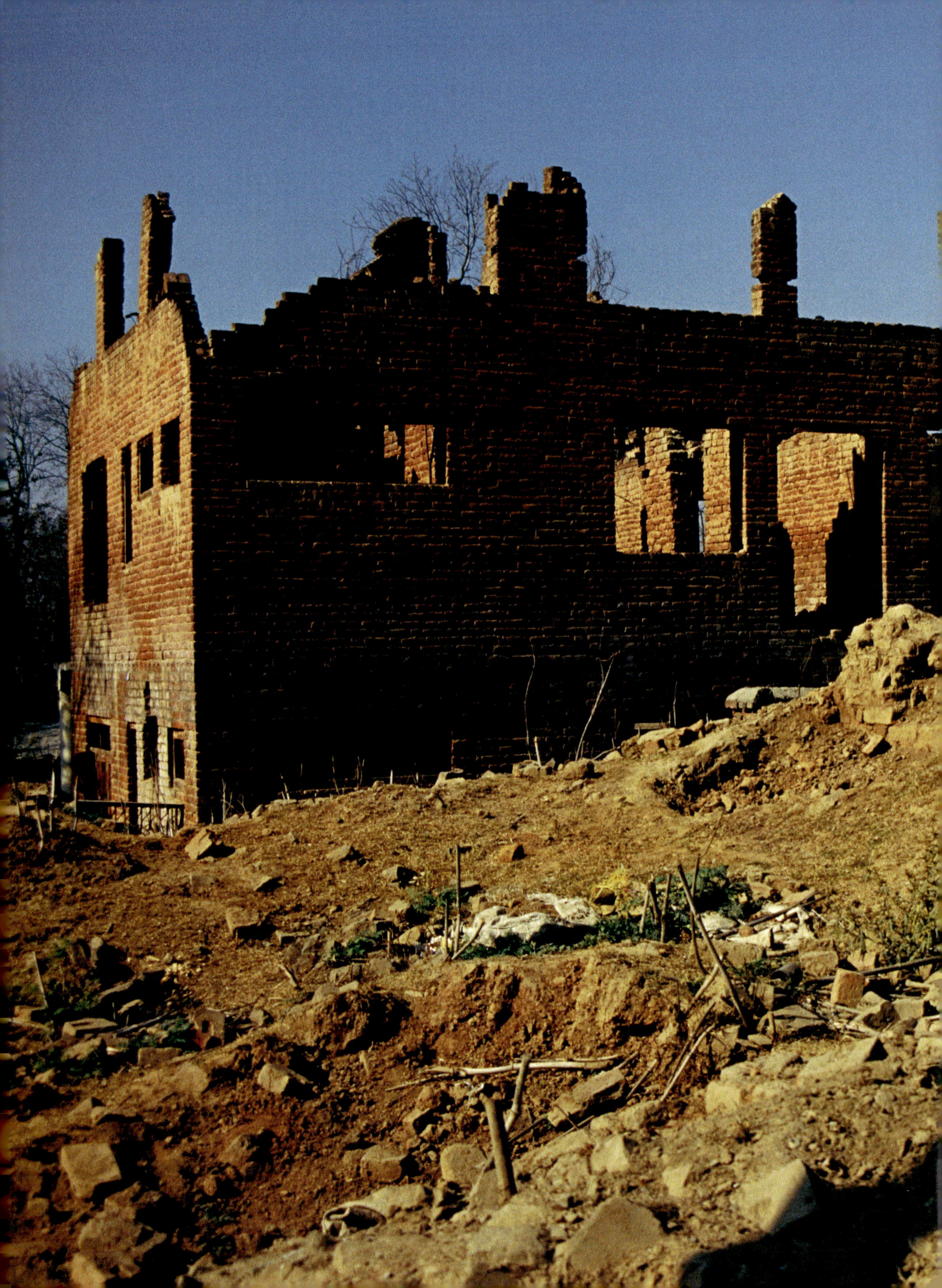

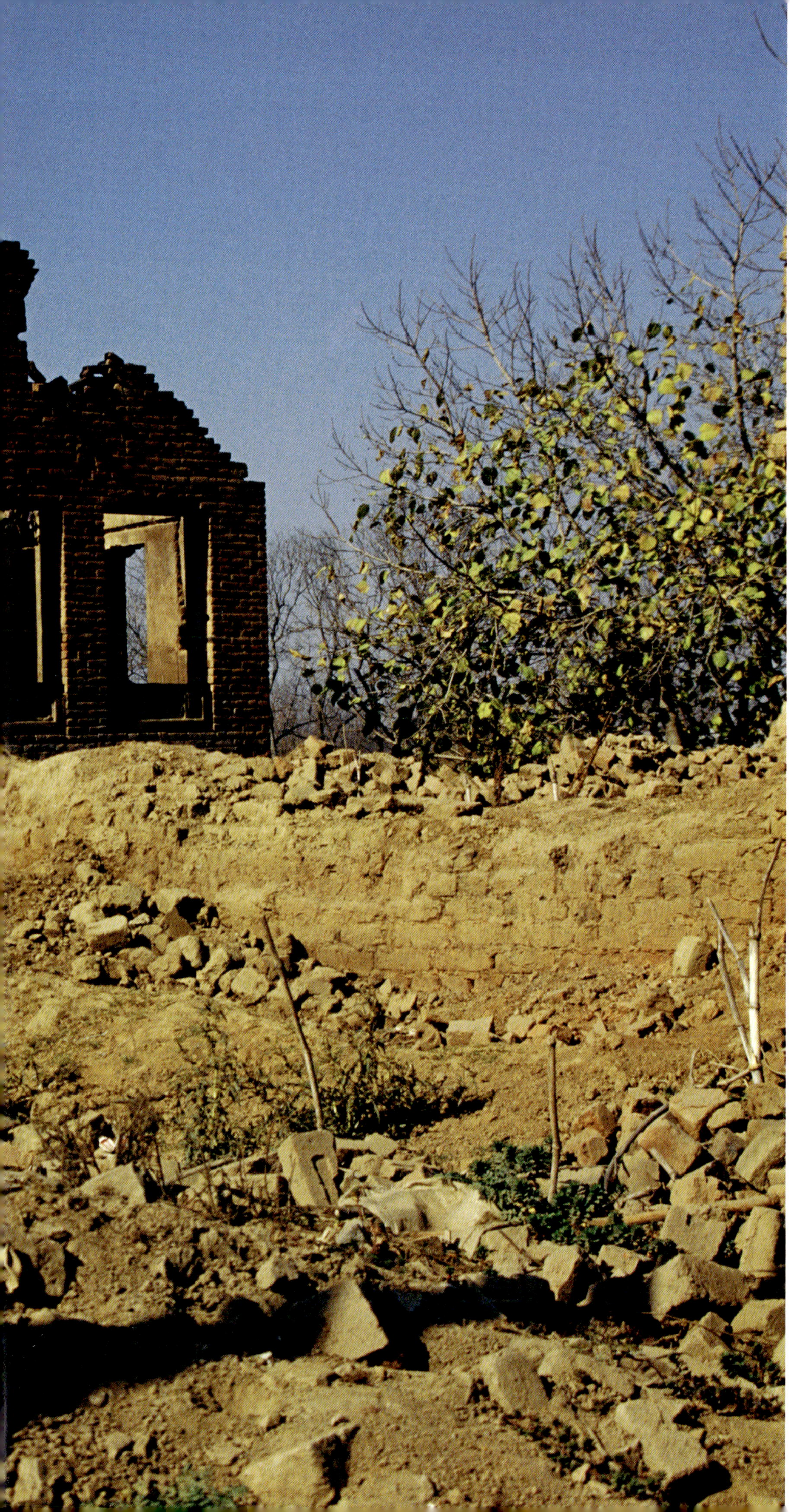

PREVIOUS PAGES:
Hal HF-24 Marut, Tilonia, Rajasthan, India
The HF-24 fighter-bomber was used by the Indian Air Force during the 1971 Indo-Pakistani War. The first Indian-developed jet aircraft, it was designed by Kurt Tank, the designer of some of the most successful German aircraft in World War II. Hampered by an underpowered engine, it was withdrawn from service in the 1980s.

LEFT:
Damaged House, Kashmir, India
Since the partition of India in 1947, there has been repeated conflict between India and Pakistan in the Kashmir region, with war breaking out in 1947, 1965, 1971 and 1999. As a result the area is littered with the remnants of houses, the original occupants forced to flee their homes to escape the violence.

T-34/85 Medium Tank, Socotra Island, Yemen

With an excellent field of fire covering any attempted landing, this T-34/85 has been 'dug in' to lower the tank's silhouette and provide better protection to its crew. An island strategically placed between Africa and the Arabian peninsula, Socotra was used as a base by the Soviet Navy between 1971 and 1985.

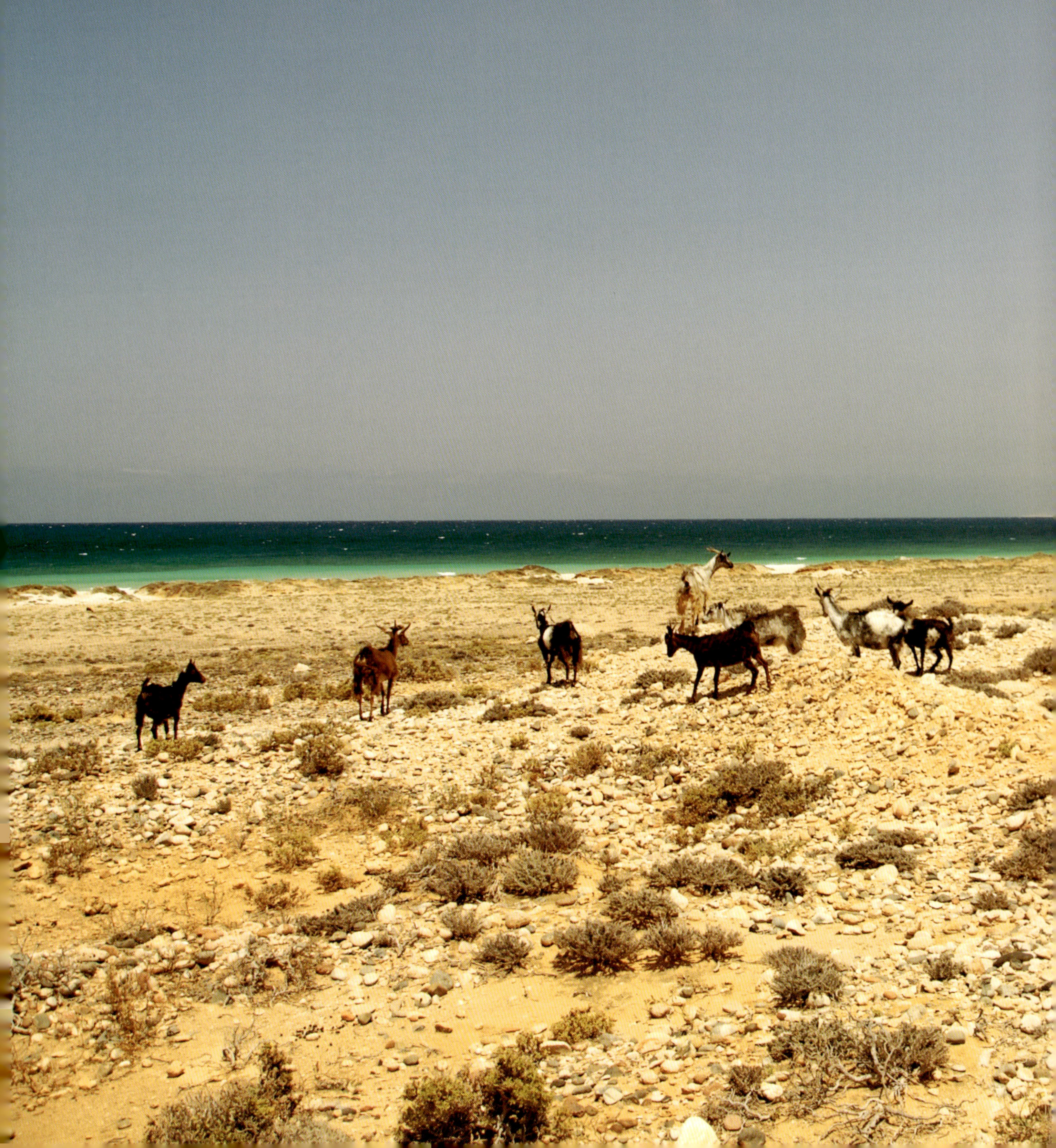

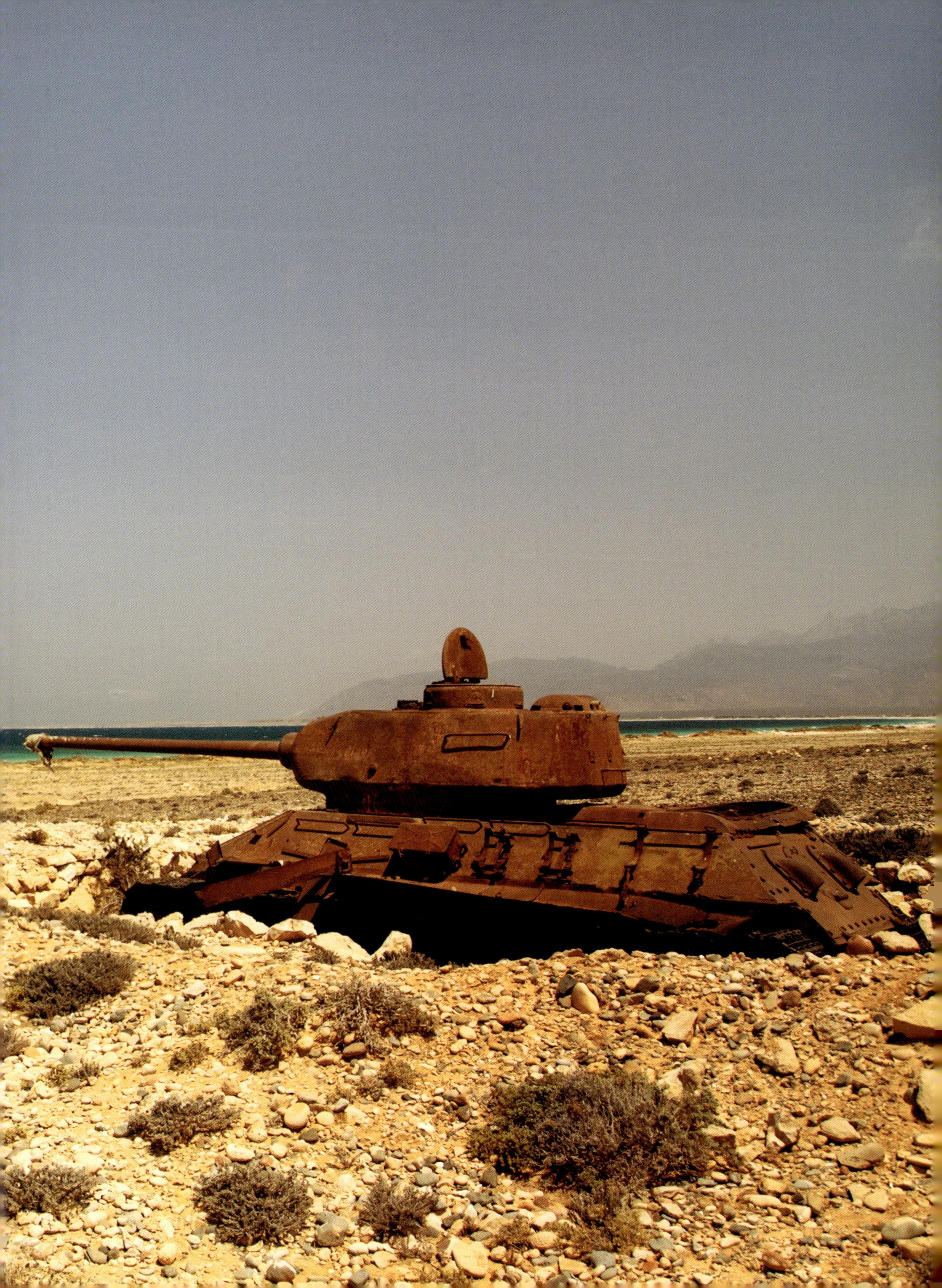

RIGHT:
Israeli Centurion Tank, Golan Heights, Israel
The Golan Heights dominate northern Israel and southern Syria. During the Yom Kippur war of October 1973, fierce tank battles took place to control the heights between the Israeli and Syrian forces. The Israeli Defence Force used Western equipment, the Syrians Soviet arms and material. The clash was seen as a proxy for what might happen if war broke out in Germany.

BELOW:
Syrian D-30 122mm howitzer, Golan Heights, Israel
This Soviet-made howitzer is not in an emplacement but has been abandoned whilst configured for towing.

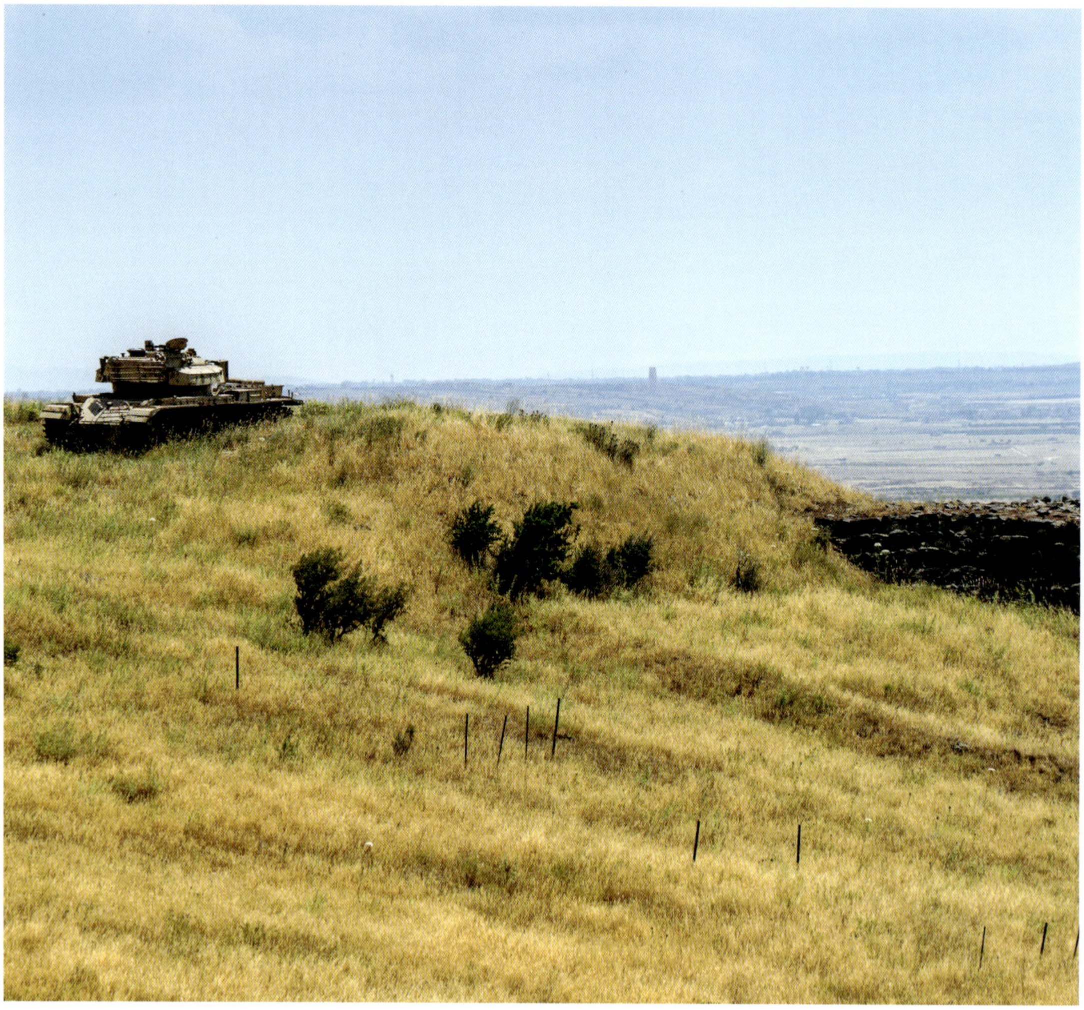

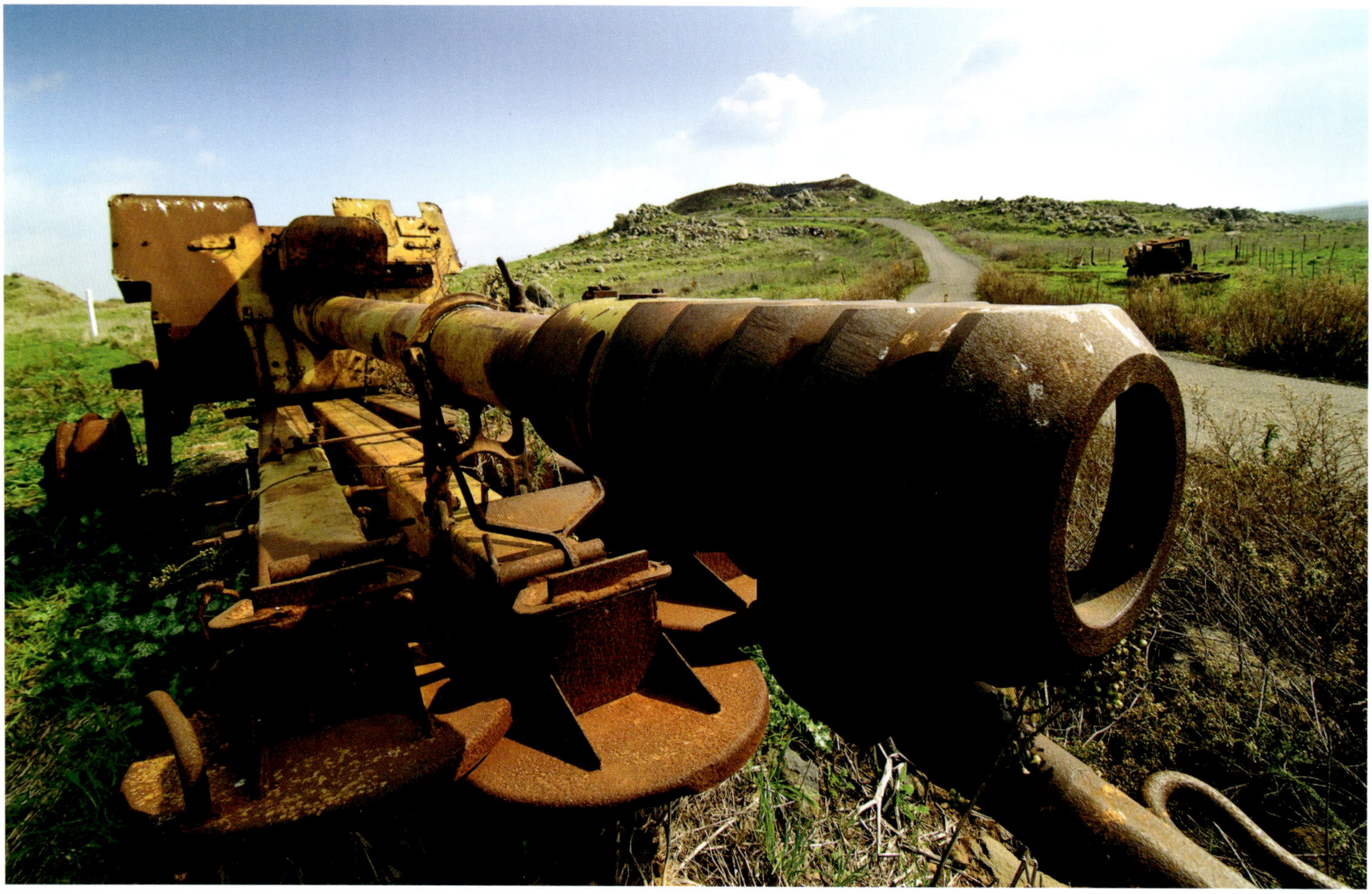

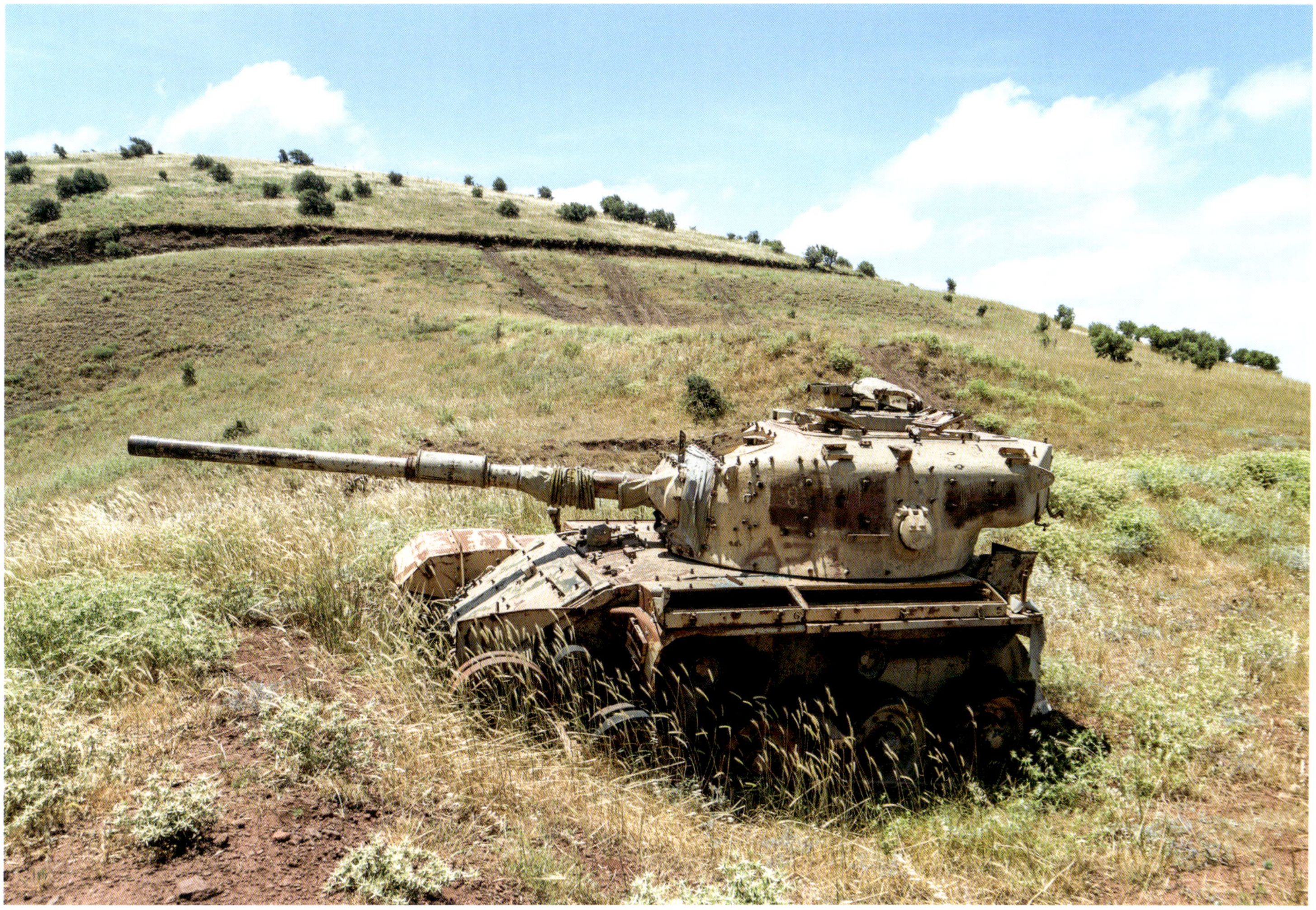

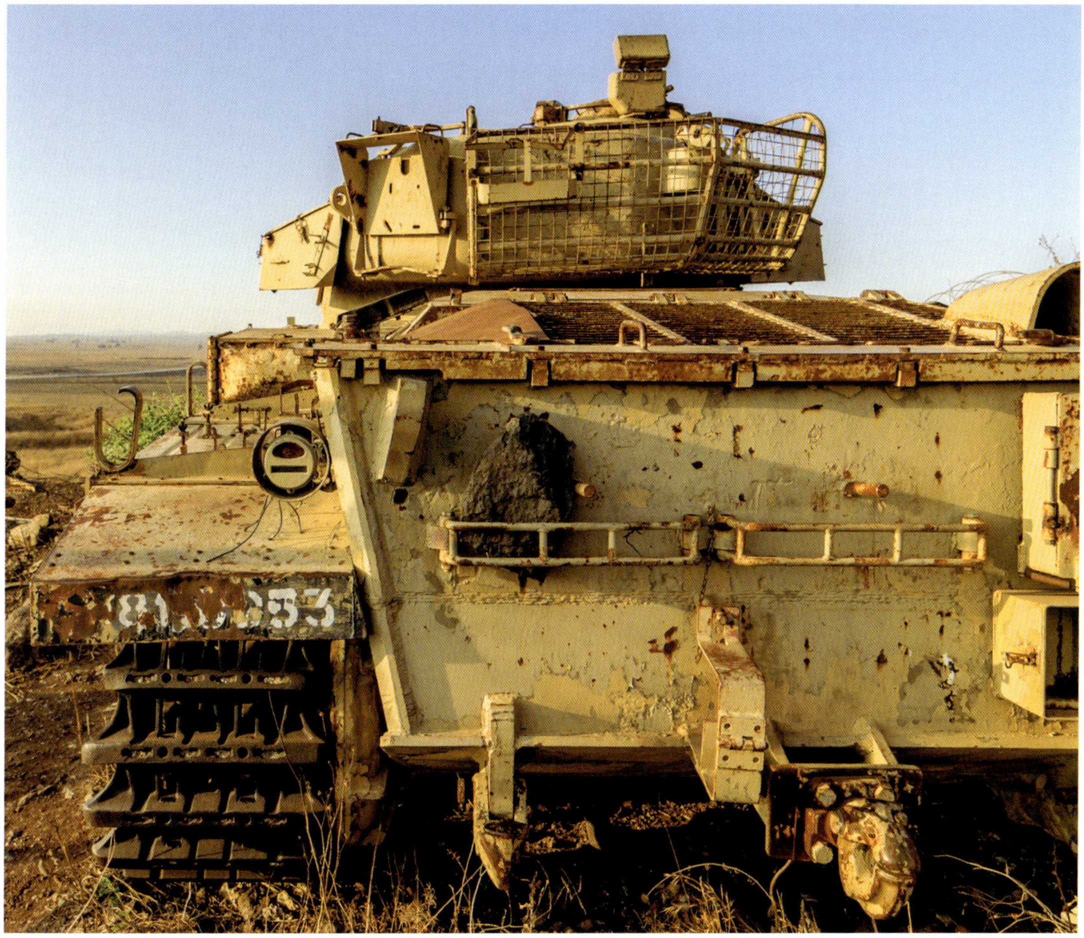

ABOVE:
Israeli Centurion Tank Wreck, Golan Heights, Israel
Behind an earth bank to give added protection to its armour, this Centurion has seen better days, as it is missing its entire engine, at least one track and various other stowage bins. The Centurion originally entered British service at the end of World War II, but the Israeli models had been upgraded substantially with a 105mm (4.13in) gun.

LEFT:
Israeli Centurion Tank, Tel Saki, Golan Heights, Israel
Israeli tank commanders were taught to fight with their heads out of their open hatches, which led to higher casualty rates but a much better level of awareness of what was happening on the battlefield.

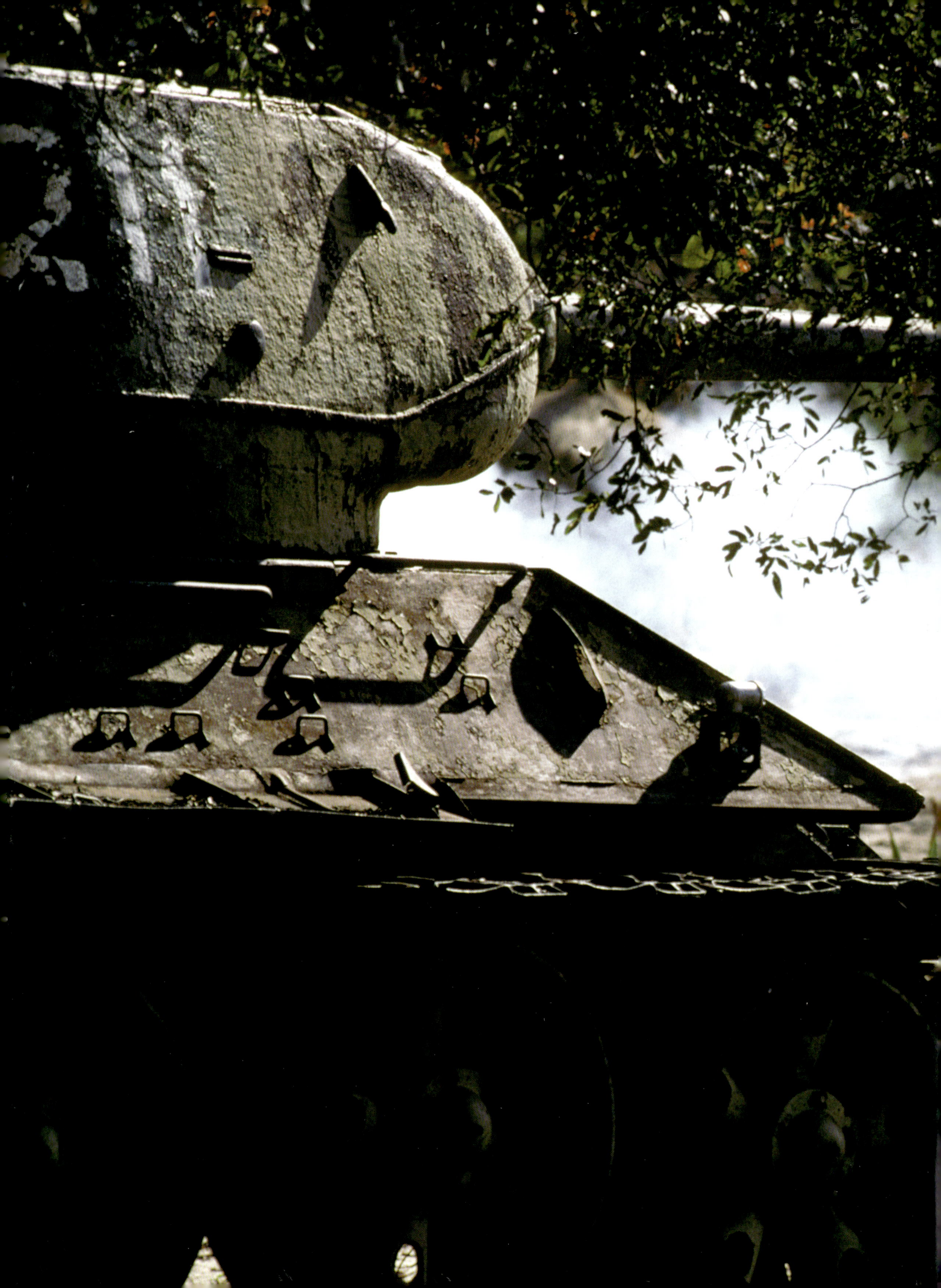

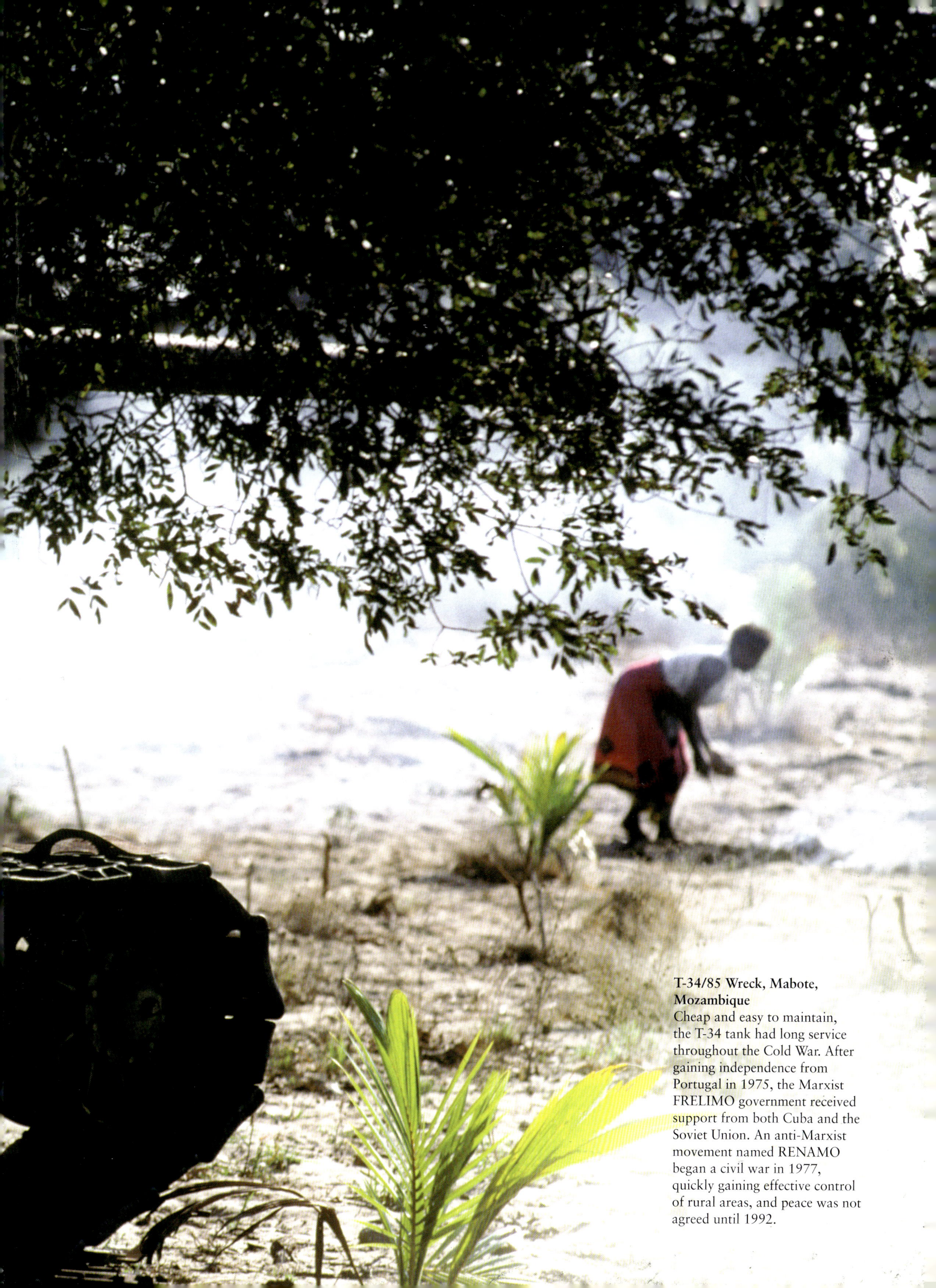

T-34/85 Wreck, Mabote, Mozambique
Cheap and easy to maintain, the T-34 tank had long service throughout the Cold War. After gaining independence from Portugal in 1975, the Marxist FRELIMO government received support from both Cuba and the Soviet Union. An anti-Marxist movement named RENAMO began a civil war in 1977, quickly gaining effective control of rural areas, and peace was not agreed until 1992.

Acknowledgements

PICTURE CREDITS

Alamy: 6, (ITAR-TASS), 7 (Andrew Gray), 26/27 (Corentin LE GALL), 32/33 (Raul Touzon), 34/35 (Yuri Smityuk), 44–45 all (Corentin LE GALL), 48 (age fotostock), 51 top (Maciej Bledowski), 54/55 (Ian Bottle), 56 (Martin Lindsay), 59 top (kpzfoto), 62/63 (Dimitrina Lavchieva), 64/65 (Haidyn Bushnell), 66/67 (Andrew Lloyd), 68/69 (Martin K), 72/73 (Petr Bonek), 76–77 all (VPC Photo), 90/91 (Kpzfoto), 92 (EyeEm), 98 bottom–99 all (Eden Breitz), 100–103 all (imageBROKER), 104/105 (Germany Images David Crossland), 106/107 (dpa picture alliance archive), 110 (Panther Media), 111 bottom (Wojciech Strozyk), 112/113 (Ger Beekes), 122 & 124/125 (Robert Stainforth), 126 (Rick Strange), 127 (Clynt Garnham Environmental), 128–131 all (Stuart Aylmer), 133 (George Impey), 134/135 (Media Drum World), 136 top (brenton west), 136 bottom (The National Trust), 137 (Kevin Bonnett), 138/139 (Mark Lees), 141 top & bottom (military cold war), 142/143 (Guy Corbishley), 144/145 (SCheckley), 146/147 (AJB), 150/151 (incamerastock), 152–157 all (Cosmo Condina), 156 top (001 Images), 156 bottom (Military Collection), 158/159 (Erin Paul Donovan), 160 & 161 (Randy Duchaine), 166/167 (Zuma Press Inc), 168 & 169 (Ian Patrick), 172 top (001 Images), 172 bottom (military collection), 174/175 (RichardBakerUSA), 177 top (Rick Pisio\RWP), 177 bottom (Jon Lord), 178/179 (Media Drum World), 182/183 (Zuma Pics Inc), 184/185 (Danita Deilmont), 186 top (Tom Salyer), 187 top (Frances Howorth), 188/189 (dpa picture alliance), 200/201 (Octavian Barabas), 202/203 (Reinhard Dirscherl), 204/205 (Hideo Kurihara), 206/207 (Torsten Pursche), 208/209 (Petr Titmuss), 210 top (Michal Knitl), 210 bottom (Peter Titmuss), 211 (Torsten Pursche), 214/215 (Ashley Cooper Pics), 216/217 (Dinodia Photos), 222/223 (AfriPics.com)

Dreamstime: 8 (Anna Arinova), 10/11 (Roman Nedoshkovskiy), 12/13 (Vladimir Zapletin), 14 top & bottom (Ivan Boryschak), 15–17 all (Vladimir Zapletin), 18/19 (Amadeustx), 20/21 (Olgavolodina), 22/23 (Vladimir Zapletin), 24 top (Nordroden), 24 bottom (Vladimir Zapletin), 25 top (Bravissimos), 25 bottom (Anton Garin), 28/29 (Blissss), 36 & 38/39 (Sergeyussr), 40/41 (Vladimir Zapletin), 46/47 (Llvllagic), 49 (Fotokon), 50 bottom (Roberts Vicups), 51 bottom–53 all (Maciej Bledowski), 57 (Fabrizio Mariani), 58, 59 bottom–61, 70/71 & 74/75 all (Fotokon), 82/83 (Vladimir Zapletin), 84/85 (Oleksеii Hlembotskyi), 86/87 (Ua2mosfet), 88/89 (Fotokon), 94/95 (Andrea La Corte), 96/97 (Stefan90), 98 top (George Tsartsianidis), 114–119 all (Thomas Males), 120/121 (Annachizhova), 140 (Burnstuff2003), 160 (Andrei Gabriel Syanescu), 162–163 (Andrey Nyrkov), 164/165 (Joe Sohm), 173 (crackerclips), 176 (Andrei Gabriel Stanescu), 180/181 (Joe Sohm), 186 bottom (Daniel Montesi), 190 bottom (marc-andre Le Tourneux), 191 (Steve Allen), 192 (Trentiness), 194/195 (Robert Paul Van Beets), 196 top & bottom (Pierre Aden), 197 (Claude Berthelot), 198 top (Piccaya), 198 bottom (Stuart456), 199 top & bottom (Sanga Park), 212/213 (Trentinness), 218/219 (Ganna Stryzhekin), 220 top (Dmitriy Feldman), 220 bottom (Vadil), 221 top (Dmitriy Feldman), 221 bottom (Nancy Anderson)

Getty: 30/31 (Scott Peterson), 42/43 (John Van Hasselt-Corbis), 50 top (Doug McKinlay), 78/79 & 80–81 all (Matej Divizna), 108/109 (Andreas Rentz), 111 top (Andreas Rentsz), 132 top (Heritage Images), 132 bottom (Richard Baker), 148 (Hans Hansen), 170–171 (Bloomberg), 187

Rex by Shutterstock: 164–167 all (Jim Lo Scalzo)

Shutterstock: 190 top (Johnny Giese)

SELECT BIBLIOGRAPHY

McCamley, Nick. *Cold War Secret Nuclear Bunkers: The Passive Defence of the Western World during the Cold War.* Pen & Sword Military: Barnsley, 2013.

Miller, David. *The Cold War: A Military History*. Pimlico: London, 2001.

Ozorak, Paul. *Underground Structures of the Cold War: The World Below*. Pen & Sword Military: Barnsley, 2012.